Tourism, Power and Space

Edited by
Andrew Church and Tim Coles

LONDON AND NEW YORK

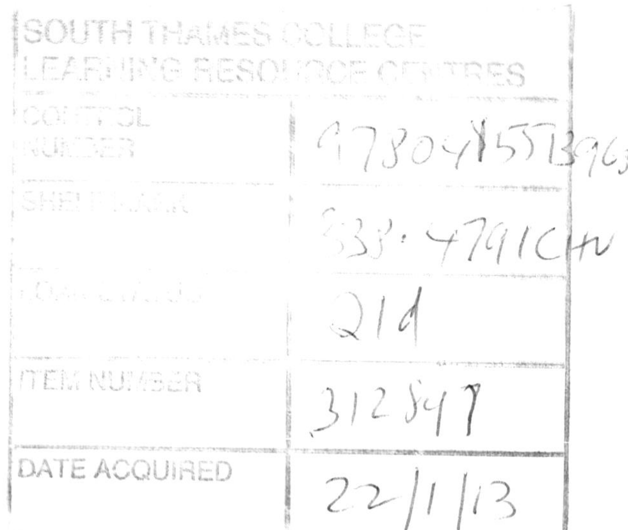

First published 2007
by Routledge
2 Park Square, Milton Park, Abingdon, Oxon OX14 4RN

Simultaneously published in the USA and Canada
by Routledge
711 Third Avenue, New York, NY 10017

Routledge is an imprint of the Taylor & Francis Group, an informa business

First issued in paperback 2011

Typeset in Times New Roman
by Keystroke, 28 High St, Tettenhall, Wolverhampton

British Library Cataloguing in Publication Data
A catalogue record for this book is available from the British Library

Library of Congress Cataloging in Publication Data
Tourism, power, and space / edited by Andrew Church and Tim Coles.
p. cm. – (Routledge contemporary geographies of leisure, tourism and mobility)
Includes bibliographical references and index.
1. Tourism–Political aspects. 2. Tourism–Social aspects. I. Church, Andrew.
II. Coles, Tim. III. Series: Routledge studies in contemporary geographies of leisure, tourism, and mobility.
G155.A1T592434 2007
338.4′791–dc22 2006012907

ISBN13: 978–0–415–32952–1 (hbk)
ISBN13: 978-0-415-51396-8 (pbk)
ISBN13: 978–0–203–39209–6 (ebk)

Tourism, Power and Space

Power is one of the most important and contested concepts in the social sciences but it has been routinely and conveniently overlooked in critical discussions of tourism. Although power is an elusive, challenging and contested concept, it offers major rewards for those prepared to embrace its potential. Key thinking and major approaches to unravelling the complexities of power are outlined in this collection and their relevance to current and future tourism studies is discussed.

Tourism, Power and Space deepens our understanding of the social world by connecting debates about tourism with discourses of power. Perspectives from leading power theorists (Parsons, Lukes, Giddens, Foucault, Clegg, Morriss) are used to inform critical readings of tourism but the contributions also reveal tourism as a major cultural, social and economic construct that is vital to the ways in which we consider notions of power in the contemporary human condition. The first part of the book explores exciting recent progress in tourism studies on the negotiation and experience of power through embodiment and performance in tourism. In the second part the connections between power, property and resources are surveyed. The pivotal nature of technologies of power and state institutions in mediating power relations is identified. The final part considers issues of power and governance. Particular attention is paid to the concept of empowerment and the manner in which tourism is manipulated to engineer ideological aspirations and to construct new social realities.

Tourism scholars have grappled with issues of power in the past, albeit all too frequently in indirect or uncertain ways. Advancing community involvement in sustainable tourism requires an understanding of contextual power relations that shape tourism developments. As this stimulating collection of essays show, critical considerations of power in tourism have a central role to play in social science analysis in years to come in the areas of governance and mobilities.

Andrew Church is Professor of Human Geography at the University of Brighton.

Tim Coles is University Business Research Fellow and Senior Lecturer in Management in the School of Business and Economics at the University of Exeter.

Contemporary Geographies of Leisure, Tourism and Mobility

Series Editor: C. Michael Hall

Professor at the Department of Tourism, University of Otago, New Zealand

The aim of this series is to explore and communicate the intersections and relationships between leisure, tourism and human mobility within the social sciences.

It will incorporate both traditional and new perspectives on leisure and tourism from contemporary geography, e.g. notions of identity, representation and culture, while also providing for perspectives from cognate areas such as anthropology, cultural studies, gastronomy and food studies, marketing, policy studies and political economy, regional and urban planning, and sociology, within the development of an integrated field of leisure and tourism studies.

Also, increasingly, tourism and leisure are regarded as steps in a continuum of human mobility. Inclusion of mobility in the series offers the prospect to examine the relationship between tourism and migration, the sojourner, educational travel, and second home and retirement travel phenomena.

The series comprises two strands:

Contemporary Geographies of Leisure, Tourism and Mobility aims to address the needs of students and academics, and the titles will be published in hardback and paperback. Titles include:

The Moralisation of Tourism
Sun, sand . . . and saving the world?
Jim Butcher

The Ethics of Tourism Development
Mick Smith and Rosaleen Duffy

Tourism in the Caribbean
Trends, development, prospects
Edited by David Timothy Duval

Qualitative Research in Tourism
Ontologies, epistemologies and methodologies
Edited by Jenny Phillimore and Lisa Goodson

The Media and the Tourist Imagination
Converging cultures
Edited by David Crouch, Rhona Jackson and Felix Thompson

Tourism and Global Environmental Change
Ecological, social, economic and political interrelationships
Edited by Stefan Gössling and C. Michael Hall

Routledge Studies in Contemporary Geographies of Leisure, Tourism and Mobility is a forum for innovative new research intended for research students and academics, and the titles will be available in hardback only. Titles include:

1. Living with Tourism
Negotiating identities in a Turkish village
Hazel Tucker

2. Tourism, Diaspora and Space
Tim Coles and Dallen J. Timothy

3. Tourism and Postcolonialism
Contested discourses, identities and representations
C. Michael Hall and Hazel Tucker

4. Tourism, Religion and Spiritual Journeys
Dallen J. Timothy and Daniel H. Olsen

5. China's Outbound Tourism
Wolfgang Georg Arlt

6. Tourism, Ethnic Diversity and the City
Jan Rath

7. Tourism, Power and Space
Andrew Church and Tim Coles

Contents

Figures

Tables

Contributors

Carl I. Cater is Lecturer in Tourism and Coordinator of Postgraduate Programs in the Department of Tourism, Leisure, Sports and Hotel Management at Griffith Business School, Queensland, Australia.

Andrew Church is Professor of Human Geography in the School of the Environment at the University of Brighton, UK. He was formerly Honorary Chair of the Geography of Leisure and Tourism Research Group of the Royal Geographical Society (with Institute of British Geographers).

Tim Coles is University Business Research Fellow and Senior Lecturer in Management in the School of Business and Economics at the University of Exeter, UK, where he was the co-founder of the Centre for Tourism Studies. He was formerly Honorary Secretary of the Geography of Leisure and Tourism Research Group of the Royal Geographical Society (with Institute of British Geographers).

David Crouch is Professor of Cultural Geography, Tourism and Leisure at the University of Derby. He is Visiting Professor in Geography and Tourism at the University of Kalmar and of Cultural Geography at the University of Karlstad, Sweden.

Alison M. Gill is a Professor at Simon Fraser University in Vancouver, British Columbia, where she holds a joint appointment in the Department of Geography and the School of Resource and Environmental Management. She is currently Associate Dean in the Faculty of Arts and Social Sciences.

C. Michael Hall is Professor of Marketing at the College of Business and Economics, University of Canterbury, New Zealand. He is also Docent in the Department of Geography, University of Oulu, Finland and a Visiting Professor at the School of Service Management, Lund University, Sweden.

Alan A. Lew is Professor and Chair of the Department of Geography, Planning and Recreation at Northern Arizona University, USA.

Neil Ravenscroft is Professor of Cultural Policy in the Chelsea School at the University of Brighton. He is managing editor of *Leisure Studies*, the journal of the Leisure Studies Association.

Nicolai Scherle is Senior Lecturer in the Department of Cultural Geography at the Catholic University of Eichstätt-Ingolstadt (Germany). His monograph on the presentation of cultural aspects in German-language travel guides was awarded a research prize by the International Tourism Fair (ITB) in Berlin in 2000.

Gareth Shaw is Professor of Retail and Tourism Management in the School of Business and Economics and co-founder of the Centre for Tourism Studies at the University of Exeter, UK.

Dallen J. Timothy is Associate Professor in the Department of Recreation Management and Tourism at Arizona State University, USA.

Caroline Winter is Lecturer in the School of Sport, Tourism and Hospitality Management at La Trobe University, Victoria, Australia.

Preface

The purpose of this book is to stimulate a greater connectivity between power theory and tourism analysis. In the early stages of the book's development we were struck by the willingness by which our colleagues within tourism and the social sciences were prepared to discuss power. Conceptually, the term 'power' assumed a kind of mythical status of its own. Practically each time power was named in a presentation, a paper or a report, it was taken for granted what it actually meant. On occasions it was used as a simple explanatory variable, a property in simple vectored relationships between individuals, groups or societies. Frequently, power was conceptualised as existing on its own as practically a commodity or a capacity or a currency which could be traded or fought over. In other instances, it was used in more subtle ways largely inspired by Foucault's canon of 'power–knowledge' following the fashion of the so-called 'cultural turn'. Power, it seemed, was all around us and we could not ignore or dismiss its routine presence in mediating everyday lives and social relations.

Increasingly, and to our disappointment, we noticed that conceptual clarification or theoretical discussion were routinely absent in presentations, papers and commentaries. A single 'brand' of power theory was employed and the possibilities presented by alternative approaches didn't warrant further treatment or critical discussion. Such selectivity may have been in the interests of time or even to beat the constraints imposed by an editor's preferred word count. Whatever the reason, the presumption was that the audience members or readers would have a basic awareness of the concepts and theoretical bodies to which the presenters or writers were referring. Power was, it seemed, a concept which all social scientists, especially those involved in the study of tourism, were assumed to have encountered in their training or their readings.

In fact, power may be one of the most important concepts in the social sciences but it is also one of the most routinely under-theorised and ambiguously conceptualised on a day-to-day basis. Power should appear in the conceptual armoury of students and scholars. That said, the specifics of these ideas are applied or put to the test all too infrequently, especially within tourism studies. Perhaps this is one of the reasons why tourism studies is often conveniently bemoaned as either non- or, more charitably, atheoretical? We do not wish to rehearse this argument again and, as we show in the opening chapter, the theoretical momentum in tourism studies has built up in recent years. Rather, we wish to focus on the exciting potentials, possibilities and imperatives of the multiple relationships between tourism and power. Political acts involving a complex combination of economic, cultural and social exchanges are at the heart of tourism whether these are in the form of host–guest encounters, business-to-business relationships or consumer–business transactions. Tourists are not automatons who always conform in strict and predictable ways to positivist forecasts; nor are tourism producers, governors or members of the host community. They encounter, they manipulate, they valorise these exchanges, and power cannot be artificially isolated from these lived experiences.

In our view, a more detailed treatment of power is vital to a fuller understanding of tourism. However, the relationship does not have to be a linear one just from power to tourism. Tourism as part of a wider mobility is inescapably a central feature in the current human condition. Contemporary power theory and concepts are usefully informed and shaped by a more extensive empirical treatment of tourism. Tourism, and for that matter leisure, are significant elements in our time–space budgets, how we undertake our business and spend our free time, and how we mediate our identities in our individual worlds. Tourism, travel and mobility are so deeply embedded in our lives that our theorisations of power cannot be dislocated from human movement.

In these respects, there are encouraging signs which this collection serves to project. Power does feature widely in the tourism research agenda, albeit too often in indirect or uncertain ways. Syntheses of this work, especially on empowerment, serve to benchmark progress and they are suggestive that particular research agendas in tourism will only be progressed in the future by greater engagement with power theory. This is evidenced further in those cases where different strands and approaches to power theory have been adopted. Here we identify the possibilities of power theory with respect to performances and practices, discourses, tactics, property and resource mobilisation, and empowerment and governance. Diverse theoretical starting points are used, but the thoughts of Foucault and Lukes are well represented. The collection represents a first concerted attempt to connect power and tourism and, as the introduction makes clear, there are significant possibilities for those prepared to engage with and foray beyond the ideas under the spotlight here.

Andrew Church
Brighton, UK

Tim Coles
Exeter, UK
March 2006

Acknowledgements

Inevitably with a book of this size and scope, production has been a major task. We would like to thank Andrew Mould (Senior Editor at Routledge) and C. Michael Hall (Series Editor) for their interest and enthusiasm in this project from a very early stage. Zoe Kruze at Routledge supported us through the frustrations of the editorial process and showed a zeal for project management way beyond the call of duty. We should like to thank the support staff at the University of Exeter for their excellent contribution to the manuscript. In the Department of Geography, Cathy Aggett and Helen Pisarska fought their way through the text. Helen Jones and Sue Rouillard turned our doodles into the excellent maps, diagrams and figures in this volume. In the Department of Management, Carole Marshall provided excellent support as the manuscript assumed a life of its own! At the University of Brighton, Nikki Simmonds in the School of the Environment also provided calm and highly effective assistance with editing the numerous drafts. Julian Gallagher also helped track down the more elusive articles in different libraries and journals. Many kind people have shared thoughts, ideas and wisdom which have help shaped the final form of this book. In particular, we would like to recognise comments and observations of Cara Aitchison, Kath Browne, Peter Burns, Sean Carter, David Crouch, David Duval, Becky Elmhirst, Rebekka Goodman, Michael Hall, Kevin Hannam, Rupert Holzapfel, Jillian Litster, Nigel Morgan, Lesley Murray, Simon Penney, Darren Smith, Marcus Stephenson, Alan Tomlinson, Lizzie Ward and Allan Williams. Some of you may not have realised it at the time, but all of you have contributed to this project in one way or another. Of course, the usual caveats apply.

Carl I. Cater would like to thank Destination Queenstown for permission to reproduce Figure 3.1 and Dart River Safaris for permission to reproduce Figure 3.4. Caroline Winter would like to thank the Great Southern Railway for granting permission to reproduce the image in Figure 5.1. Alison M. Gill would like to thank the Social Sciences and Humanities Research Council of Canada who have provided several research grants over the past decade to support her work in Whistler. She is grateful to the graduate students who have worked with her on these projects, notably Ann Hawkins, Julia Marcoux, Angela Xu, Sean Moore and Neil Chura, and to her colleague Peter Williams, with whom she has collaborated on several occasions. She would like to acknowledge the excellent collaboration from the planning department at the Resort Municipality of Whistler and to record her gratitude to the many residents of Whistler who provided her with information. Andrew Church and Neil Ravenscroft would like to thank all those who took part in their studies of leisure and tourism on Britain's inland rivers and the differing government bodies that funded the research.

1 Tourism, politics and the forgotten entanglements of power

Tim Coles and Andrew Church

Tourism and the predicaments of power

Kenya, 28 November 2002: an Arkia airlines charter jet was narrowly missed by two surface-to-air missiles as it started its ascent from Mombassa airport. On board were 261 passengers, the majority of whom were Israeli citizens on their way home after their vacations. Just as it was attacked, 15 people died in a bomb attack on the Paradise Hotel on the Indian Ocean coast. Nine Kenyans and three Israelis, two of whom were children, were killed along with the three suicide bombers. Eighty people were injured, many badly (BBC 2002). Al-Qaeda operatives in Kenya claimed responsibility in the aftermath of the attack (CNN 2002).

The Kenya atrocity is a distant memory now but at the time it acted as a chilling reminder of what had happened in New York and Washington in the previous year and in Bali the month before. Shortly afterwards, Steve Bell, a political cartoonist, published in *The Guardian* newspaper an image of George W. Bush standing at the Presidential lectern. Behind Bell's signature portrayal of Bush flies what appears to be a B52 bomber and an airliner which a missile has just missed. For a bemused-looking Bush, apparently 'only an all-out war on turrism can bring to an end this war on turrism!' (Bell 2002). The joke hinges on Bush's Texan drawl, his particular pronunciation of 'tourism', its closeness to his verbalisation of 'terrorism', and a seeming inability on the part of the US authorities to differentiate between the two. Published relatively early into the President's 'Global War on Terrorism', the cartoon is all the more poignant for its early connection of the two concepts. Foreshadowing the findings of the 9/11 Commission (2004), Bell was one of the first commentators to connect the practice of tourism and acts of terrorism. Tourists are not only the subjects of acts of terrorism as recent attacks in Bali and Sharm el Sheik are a reminder; tourism has become a central component in the mediation of terrorism. The 9/11 attacks were facilitated by temporary mobilities as the attackers travelled around the world training and preparing themselves, garnering knowledge and collecting intelligence, meeting contacts and probing the weaknesses in the security apparatus of the United States. Masquerading as business travellers, on 9/11 the four sets of hijackers set off on routine commuter flights from Boston and New York with such devastating effects. Other forms of transitory migrations have been, and indeed remain, a medium used by Al Qaeda for prosecuting its conflict (09/11 Commission 2004).

Bell's image is all the more striking because it articulates issues about the multiple and complex connectivities between tourism and power. It points to the sometimes hidden but also often highly visible presence of power relations in the production, governance and consumption of tourism, as well as the importance of tourism in political, cultural and social practices that empower individuals and organisations. Tourism has assumed a central position in the power politics of the unfolding world order as new alliances have been forged on both sides in the 'War on Terror'. The vulnerability of the seemingly powerful is exposed by a cartoon which simultaneously points to the apparent capacity of a minority group to impose its agenda. Unfolding power relations are further evident in the desire to reassert authority and domination. From a position of appearing powerless in the face of attacks, the American government seeks, for a range of specified and unspecified motives, to empower itself by drawing on a range of technologies and measures to prevent attacks and enhance Homeland Security (09/11 Commission 2004). New border controls and immigration tools have indeed been progressively introduced to regulate visitor flows. Images and objects of revised security arrangements have the potential to build greater consumer confidence (Hall *et al.* 2004), but narratives that portray these responses as draconian may also alienate overseas visitors and frustrate the rebuilding of the tourist economy in the United States (Cochrane 2005). Thus, while the authority of government is brought to bear on the tourist sector, as is so often the case in the contemporary world, the subtleties and complexities of the interactions between power and tourism remain elusive and for the most part hidden away behind political rhetoric.

Connecting tourism and power

The purpose of this book is to strengthen the connection between tourism research and conceptualisations and theorisations of power. As we will show, linkages exist but they are very unevenly developed. We wish to place constructs of power more firmly at the centre of the agenda of critical tourism research. We would contend that tourism studies should be explicitly engaged with power, practically to be rewired more extensively into discourses and conceptualisations of power. Our choice of words is quite deliberate. Far from being total strangers, tourism and power are often mentioned in the same breath. Issues of power, empowerment and disempowerment permeate many aspects of tourism research. Sometimes these incursions are explicit and direct, at other times they are indirect and latent. Whatever the mode of infiltration, the intricate connections and feedbacks between constructs of tourism and power have been recognised. Tourism clearly plays a role organising and governing social life. Our challenge is to progress beyond often infrequent, partial and even plainly opportune treatments of power in tourism. Power is not a convenient conceptual 'port of call', a loosely defined notion that handily serves to explain ambiguous asymmetries among different stakeholders in the development process, or which helps to describe vaguely the unequal allocation of resources. Power has for some time been one of the major concepts in the social sciences (Clegg 1989: xviii), and as Prus (1999: 3) puts it 'few terms in the social sciences have engendered as much mystique (fascination, curiosity, fear) as "power"'. Yet tourism analysis has become only selectively linked with established and emergent discourses of power, usually those influenced by post-modern and post-structural social theory. Within this collection, we aim to map some of the diverse intersections between tourism and power. We explore how power manifests itself, how it is expressed, and the multiple ways in which it is articulated, circulated and (deliberately) not even used in tourism. In so doing, we intend to illustrate the potential and potency of the full range of theorisations of power towards developing deeper understanding in critical issues of tourism.

That said, the intention is not to present an exclusively uni-directional portrayal of the relationship between power and tourism. By this, we mean that we do not solely consider discourses of tourism to be informed and shaped by debates on power. Rather, the relationship between discourses of tourism and power is a more fluid and reflexive one. Scholarly analysis of tourism has much to contribute to the understanding of power in contemporary societies. As Britton (1991: 458) recognised, tourism has become a 'major internationalised component of Western capitalist economies . . . one of quintessential features of mass consumer culture and modern life'. The study of tourism offers social scientists a greater insight into the nature of modern-day life. In this respect, it too provides an ideal empirical setting in which to appraise the value of current thinking on power. One of the most remarkable features of accounts in tourism that do engage directly with power discourses is the degree to which constructs of power are accepted practically as given. Writings of Weber, Lukes or Foucault may be tactically deployed but they are largely just reported (cf. Reed 1997; Hollinshead 1999; Cheong and Miller 2000; Kayat 2002; Sofield 2003). There is often no comment from tourism scholars on the epistemological, ontological or methodological implications of conceptualisations of power although the writings of Urry (2002, 2003), Aitchison (2003) and Franklin (2004) are important instances where this challenge has been accepted. As we shall demonstrate below, major positions on power are far from unproblematic and they have been the subject of compelling critiques. Many of the prevailing ideas on power are a function of the time and context of their emergence; they may well be of enduring relevance yet their appropriateness to contemporary conditions is routinely left uncontested (cf. Thomas and Thomas 2005). For example, the nature of American local government which so influenced the writings of C. Wright Mills (1959), Robert Dahl (1961a, 1961b), Steven Lukes (1974) or other power theorists in the 1960s and 1970s is not the same as that encountered by Judd and Simpson (2003) or Laslo (2003) nearly thirty years later in their respective inspections of the politics of the development process in urban tourism.

Power in tourism

As a basic construct, power has featured frequently and repeatedly in tourism discourses over the years. As far back as the mid-1970s, Doxey's (1976) oft-cited work, as well as Bjorklund and Philbrick's (1975), focused on the development of standardised views of resident reactions in the face of tourism (Shaw and Williams

2004: 178), and hence the relative power of residents to devise effective strategies to deal with tourism. Other work has explored the nature of host–guest encounters through the lens of social exchange theory (Ap 1992; Kayat 2002) whereby asymmetries of power in the social relations of tourism are manifest in the underlying assumption that residents 'behave in a way that maximises the rewards and minimises the costs they experience' (Madrigal 1993: 338 in Shaw and Williams 2004: 178). Although heavily critiqued for a variety of reasons (cf. Butler 2006), Butler's (1980) tourist area life-cycle model postulated transitions in the power relations of tourism between local people and external actors as the development process unfolds. Power asymmetries are also evident in de Kadt's (1979) early exploration of the social and cultural effects of tourism in developing countries. Drawing on wider thinking on development at the time, de Kadt (1979: xii) stressed the importance of not focusing solely on growth for growth's sake but also the potential for growth to address wider social inequalities within developing countries. As part of this aspiration, the 'development community is searching for means that will enable the poor to provide for their basic needs through more productive work, more widely available social services, and increased participation in political decisionmaking' (de Kadt 1979: xii). Tourism may, he argued, contribute to these wider policy aspirations but lamentably 'a pro- or anti-tourism stance might be taken up without real evidence to support it' (de Kadt 1979: xiii).

A wide variety of disciplinary positions has been evident among those with an interest in the relationship between tourism and power. From a sociological grounding, Morgan and Pritchard's (1998: 7) examination of marketing, promotion and branding explores the way in which 'tourism processes *manifest power* as they mirror and reinforce the distribution of power in society' (italics original). Basch (2004) has explored power relations in the customer service encounter between the tourist and the accommodation provider from a perspective of social psychology. Thurlow and Jaworski (2003) have used applied linguistics to demonstrate the power of language in inflight magazines in mediating what they term a 'globalization of nationality' and the promotion of 'global lifestyles'. Richter (1983) has exposed the tripartite relationship between tourism, power and international relations, while Timothy's (1997) work identifies boundaries and the practice of border crossings by tourists as an expression of the current status quo in geopolitics. From a political science perspective, Judd and Simpson (2003) note how public–private partnerships involved in urban tourism projects function as independent centres of power outside traditional local government structures. Mayors forge alliances with such groups and bypass democratic processes leading to considerable potential for reconstructing the local state.

Accounts of the *de jure* practice of power in tourism have accompanied discussions of the de facto operation of power. For instance, Arino (2002) outlines how the state and its opponents exercised power in the introduction of the ecotax in the Balearic Islands. Historians have adopted longer-term perspectives on the connectivities between tourism and power in some cases to legitimate the role of the state (cf. Baranowski and Furlough 2000; Koshar 2002). Cocks (2000) has drawn our attention to the role of local chambers of commerce in the early promotion of urban tourism in America. Festivalisation at the turn of the previous century was accompanied by the propensity to empower local bourgeois elites further by restricting access to political power and cultural capital. In more extreme circumstances, Baranowski (2000) describes the role of tourism in the National Socialist agenda for Germany in the 1930s. Travel was used as a form of political coercion by the ruling elites and tourism became a medium through which to articulate dominant ideologies (cf. Keitz 1991). Increased travel opportunities for individuals were presented as a benefit of the consumerism induced by the Nazi regime and a phenomenon to be directly equated with the programme *Kraft durch Freude* (KdF – Strength through Joy) (cf. Semmens 2005).

Interest in the spatialities of power has been notable in writings as varied as those on political economy and tourism (Britton 1991), the cultural and performative geographies of tourism (Crouch 1999), identities and power relations (Aitchison 2003) and tourism, public sector policy and planning (Elliot 1997; Hall 1994, 2000). In the context of developing countries Bianchi (2002) exposes how power relations in tourism are central to the way in which the social practices of global patterns of production and consumption are constituted (cf. Britton 1991; Mowforth and Munt 1998). Inspections of global commodity chains in the tourism sector reinforce this perspective (Clancy 1998; Mosedale 2005). Tourism is viewed as a commodity fashioned in an articulated system of producers, suppliers and intermediaries. The allocation and distribution of benefits (often money and knowledge) inevitably result in winners, loosers and rivalries among the individuals and/or groups involved. Ioannides (1998) has mapped the gatekeepers, the principal nodes in the chain, as power brokers in the consumption of tourism (see also Klemm and Martín-Quirós 1999; Bastakis *et al.* 2003), while Crase and Jackson (2000) have explored the idea of information asymmetry as a form of power in the economics of market operation.

The body is a key source of social difference and power. The importance of performative and embodied perspectives to the analysis of tourism has been stressed (Aitchison 2000; Crouch 2000 and Chapter 2 in this volume; Franklin 2003; Cater, Chapter 3 in this volume). Feminist writings and studies concerned with sexualities have played a significant role in opening up power as an issue for tourism research, especially the interactions between power and identities. These have indicated how tourism and mobility both reflect and contribute to power relations linked to age, class, disability, gender, race and sexuality. A number of studies have revealed the role of tourism in maintaining patriarchal structures (Swain 1995). Indeed, Aitchison (2003: 83) has observed that,

> both poststructural feminism and postcolonial feminism have placed emphasis on the textual, discursive and performative construction of the Other in the reinscription of gender-power relations. Together these post-positivist perspectives have laid bare tourism's inherent paradox: although associated with a globalised melting pot where postmodern deconstruction and reconstruction have induced the breakdown of previous boundaries . . . the global tourism industry simultaneously serves to inscribe the Otherness of culture and particularly, the Otherness of women and black people.

Feminist writing has challenged power relations within the tourism academy. Aitchison (2001) has identified the gendered features of academic tourism research publications. Pritchard (2004) argues that the 'malestream' of research and academic posts in the tourism field, along with the dominance of 'masculinist' epistemological and ontological perspectives, has resulted in the marginalisation of feminist research on tourism. Furthermore, Pritchard (2004) claims that the tourism academy has been remiss in its treatment of sexuality. While reviews in the 1990s identified a general lack of interest in the topic (Markwell 1996; Pritchard *et al.* 1998; Veijola and Jokinen 1994), there is now an emergent literature situated in a number of disciplines delivering insights into the interactions between power, tourism and sexualities (Browne 2007). The injustices, in the form of the discriminations and exclusions that gay men and lesbian tourists experience, have been increasingly documented. In some cases, such as gay cruises to the Bahamas, this involves governments using state powers to deliberately exclude non-heterosexual visitors (Puar 2002a). Empirical studies have also highlighted the role of gay tourism marketing and discourses in the propagation of Western-centric post-colonial discourses (Alexander 1998). Puar (2002b) argues that an embodied and performative understanding of tourism, sexuality and power will be most fruitfully developed through an increased engagement with post-colonial and queer theories. Travel and tourism can empower some individuals through opportunities for fulfilment, embodied performance, transgression and escape. Gay and lesbian tourist destinations can enable practices, performances and identity building not possible in 'home' locations. Simultaneously, gay and lesbian tourists may also be subjected to constraints that arise not only from the capitalist commodification of tourism spaces but also from a series of disciplining hetero-normative gazes that can limit the opportunity to develop sexual citizenship and identity (Brown 1999; Browne 2006; Cantu 2002; Johnston 2005; Pritchard *et al.* 2000; Puar 2002b). In tourism studies, as in other disciplines (Hubbard 2002), research into sexuality has tended to focus on the 'other' and considered only homosexualities whilst ignoring heterosexualities in the production of tourism spaces. Moreover, in nominally homosexual leisure and tourism spaces excluding processes function further where the homopatriarchy of gay male practices can exclude lesbians (Pritchard *et al.* 2002) and other males with marginalised sexual identities (Binnie 2004).

The paradox of tourism and power

Despite this growing engagement with power there remains an important paradox in that power and power relations are frequently invoked as pivotal features in the production of tourism, the negotiation of tourist experiences, and the administration and governance of tourism; however, they are routinely under-conceptualised in tourism discourses. With notable exceptions among some of the studies mentioned above, established discourses of power by major theoreticians have by and large failed to feature prominently in contemporary studies of tourism. In many instances, tourism commentators are prepared to deploy the explanatory virtues of power (often as a capacity or authority) but seldom do they progress beyond elementary, and hence apparently uncontentious, conceptual simplifications (cf. Haugaard 2002; Morriss 2002). More worrying still, power is being taken for granted; that is, as an implicit or implied feature, lurking as it does in the background of many studies, invoked at convenient moments in the narrative but not the subject of rigorous identification, evaluation or analysis. Rather, it serves as an enforcement measure to add immediate and seemingly extra conceptual substance to the argument. Few tourism scholars recognise the rich intellectual genealogies of power discourses, the highly contested and nuanced approaches to understanding power, and the significance (as well as potential) of debates on key theorists' ideas. Instead, power is often perceived as an obvious and self-justifying concern. It is conceptualised in a vague and generalised manner, and there exist in this approach several real dangers (see Morriss 2002). As a single salutary example, the apparently mutually implicated nature of power and resources is routinely invoked, such that power is equated simply as access to and control over resources including images and representations (cf. Kayat 2002; Murphy and Murphy 2004: 350; Dessewffy 2002; Fyall and Garrod 2005: 145; Henderson 2003; Hunter 2001; Trist 1999; Williams 2002). This may well characterise the attributes of the powerful and manifest changes in the 'balance' of power between identifiable groups; however, it may grossly simplify the full complexities and potentialities of the practices and performances of power leading to potentially misleading conclusions. In a stark warning, Allen (2003) observes that the relationship between power and resources is not always an obvious or simple one. It is crucial to distinguish between the exercise of power and the control of resources because the two do not always go hand in hand in a causal manner as, for instance, power may not be utilised. Perceptions of power and its significance may also differ markedly among stakeholders, as may their strategies and tactics for employing it (Buchanan and Badham 1999).

Nowhere perhaps is this perceptible paucity of theoretical and conceptual engagement with power discourses more emphatically exposed than in studies of so-called 'sustainable tourism'; that is in, arguably, the most high-profile topic within cross-disciplinary studies of tourism. Interest in more responsible and inclusive modes of tourism development and management has been accompanied by a commitment to comprehensive approaches to development which are flexible and dynamic, integrative and inclusive, and oriented towards the community and the goals of all stakeholders (Simpson 2001). In turn, this has raised questions about how to build enduring and viable partnerships of, and collaborations among, diverse stakeholders which will enhance the effectiveness, efficiency, harmony and equity of tourism development (Teo 2002; Timothy 1998; Selin and Chavez 1995; Selin 1999; de Araujo and Bramwell 1999; Bramwell and Lane 2000; Bramwell and Sharman 1999; Burns 1999, 2004).

Ultimately, as both Scheyvens (1999) and Sofield (2003) have demonstrated, tourism may lead to the empowerment of local communities in multiple (i.e. economic, social, psychological and political) ways (see also Timothy, Chapter 9 in this volume). While more equitable, fair and locally empowering forms of tourism production, governance and consumption remain the aspiration, inevitably they require interaction among human beings; in other words, they are political processes and they are the subject of power relations among constituencies. Contestation, consensus and dissonance among competing participatory interests are inevitable features of development in this manner (Simmons 1994; Fallon 2001). Almost by definition, such issues necessitate an interest in how power is exercised, by whom, in what manner of political arrangement and to what end. Furthermore, as Ryan (2002) points out, if there is to be equity related to tourism development, there must be an element of power sharing. This requires individuals (or groups) to take responsibility for (i.e. power over) the delivery of equity. Jamal and Getz (1995: 190) have introduced the idea that adjustments in the relative access to, and exercise of, power may be required in community-based tourism planning to achieve a more equitable, fairer set of outcomes (cf. Ashley 1998; Forstner 2002). However, Thomas and Thomas (2005) argue that the redistributive aspects of power relations in stakeholder coalitions are notably absent from such discussions. Similarly, Bramwell and Lane's (2000: 8-9) wide-ranging review emphasises that the nature of power, its dispersal among stakeholders, and its ability to contribute to, or frustrate, the operation and outcomes of collaborations is only generally conceptualised and by largely instrumental means.

Regrettably, such generality leads to limited conclusions. As Reed (1997: 657) has observed, these may be a function of initially imprecise and limiting premises (see Hall, Chapter 11 in this volume). For instance, she contends that 'while power relations are included with collaborative theory, it is frequently assumed that collaboration can overcome power imbalances by involving all stakeholders in a process that meets their needs'. Her work draws our attention to the disappointing situation whereby different theoretical approaches to power (beyond partial, simplified and implied Weberian readings) are all too infrequently invoked in accounts of empowerment when there is clear evidence of their potential validity as interpretative frameworks (Sofield 2003). Rather, as Timothy (Chapter 9 in this volume) demonstrates, research towards unravelling how power features in empowerment through tourism appears to be driven by a largely inductive approach. A rich collection of case studies of the relationship between tourism and empowerment now provides a strong empirical basis from which to deepen the understanding of empowerment by contemplating the mutual implications of power theory and tourism in the contemporary world.

Thus, only recently, according to Hannam (2002: 229), has 'research into tourism development . . . begun to focus more explicitly upon the concept of power' as opposed to more general articulations in which power features as a much vaguer and even implied notion. To accompany interest in power as a concept, Hannam identifies a relative shift away from the political and economic as dominant arenas of power towards investigation of the social and cultural relations of power. This shift has been accompanied by the introduction into tourism studies of the ideas of Michel Foucault and other post-structural social theorists (Urry 2002; Veijola and Jokinen 1994; Wearing 1995; Morgan and Pritchard 1998; Aitchison 1999; Cheong and Miller 2000; Franklin and Crang 2001; Edensor 2000, 2001; Franklin 2004; Winter, Chapter 5 in this volume). Hollinshead (1994, 1999) has, in particular, been an enthusiastic advocate of Foucault's thinking. For him, Foucault's contribution might,

> not always be a completely fresh approach to matters of discursive power and disjunctive effect in tourism, [but] the very depth, range and ubiquity of its investigative assault on matters of dominance, subjugation and normalisation could conceivably be of multiplicative value and differentiative potency in and across tourism studies.
>
> (Hollinshead 1999: 10)

In order to understand the advances resulting from this increased engagement with theories of power, the next section sets out the key conceptual debates and theoretical traditions within the study of power, and considers the insights provided by tourism researchers who have drawn on the different theoretical perspectives.

Conceptualising power

The question 'what is power?' is immensely problematic. It is a conundrum that many scholars of power have struggled to solve, and it is one that we cannot hope to answer fully or definitively here. There appears to be almost as many definitions of power as scholars writing on the subject. As other power theorists have pointed out (Haugaard 2002; Morriss 2002; Lukes 2005), it is an almost impossible (and largely unrewarding) task to attempt to review fully the extensive bodies of writings on power (see Mann 1986, 1993; Hindess 1996; Prus 1999). An historiographical overview does, however, point to four broad features that are vital to the further introduction and effective development of power discourses in tourism. The first is the plurality of approaches to understanding power, and a second allied feature is the essential contestablility of power as a concept. The third feature concerns disagreements over the language used to discuss power. And fourth, the relevance of, but highly overlooked nature of, debate over the use of the power discourse and why concepts of power are analytically valuable.

The plurality of approaches is indicated by the many established and contrasting theoretical statements on the nature of power from, among others, Max Weber, Hannah Arendt, C. Wright Mills, Talcott Parsons, Steven Lukes, Anthony Giddens, Barry Barnes, Stewart Clegg, Michel Foucault and Pierre Bourdieu to list but a few (see Clegg 1989; Hindess 1996; Prus 1999; Haugaard 2002; Allen 2003). From these widely varying and contested theorisations, Haugaard (2002: 4) attempts to identify 'certain generalized perceptions of power'. These include what he terms, 'power over' and 'power to' (associated with the analytic tradition); 'conflictual power' and 'consensual power' (social theory of the modern variety); and 'power as constitutive of reality' (postmodern social theory). These general perceptions of power are explicitly and implicitly presented in discussions of tourism. For example, a variety of studies of the political economy of tourism in developing countries have identified the power of interest groups 'over' local and regional governments (Elliott 1997; Mowforth and Munt 1998). By contrast, studies from a performative perspective have highlighted the empowering, 'power to' aspects of tourism activities (Coleman and Crang 2002; Crouch 2004). Fallon (2001) argues that the uneven patterns of tourism development in Lombok (Indonesia) reflect local differences in power relations between developers, tour operators and local communities, with some locations developing consensual relations between these interest groups whereas in other locations conflict has occurred. A variety of the studies from a Foucauldian perspective consider the constitutive and productive nature of power (e.g. Morgan and Pritchard 1999).

Haugaard (2003: 89) proposes a seven-fold typology of theorisations on the production of power. Power is created from social order (Parsons, Luhmann, Barnes, Haugaard, Clegg, Giddens); bias (Bachrach and Baratz); systems of thought (Foucault); 'false consciousness' (Lukes); power/knolwedge, obligatory passage points (Foucault, Clegg); discipline (Foucault); and coercion (Weber, Dahl, Bachrach and Baratz, Mann, Poggi). There are, therefore, notable and sometimes extensive differences in substance among contested interpretations of power. Begg (2000: 14–15) easily identifies 19 definitions of power from different sources without claiming to be exhaustive (Table 1.1). All grand statements of this type are problematic. They contain limitations to one degree or another based on their intellectual framing. Each position reflects its author's biases and sympathies. Lukes has, for example, critiqued his original (1974) thesis and decried its now widely cited definition of power as a 'mistake' in his subsequent (2005: 12) writings. Such differences are far from trivial. As he observes:

> Disagreements matter because how much power you see in the social world and where you locate it depends on how you conceive of it, and these disagreements are in part moral and political, and inescapably so.
>
> (Lukes 2005: 12)

Each major theorist's work has potential application for the production of insights into power relations in tourism. Reminiscent of Orwell's *Animal Farm*, it seems that to date some theorists have had more potential than others. As argued below, the ideas of both Foucault and Lukes have been more commonly used to unravel critical issues in tourism. However, the application of their thinking does not necessarily imply an exclusivity or ease of potential application, nor does it suggest that they are any the less problematic than other conceptualisations (Clegg 1989; Stewart 2001). In the case of Foucault, the emergence of his thoughts on power in tourism partly reflects trends within the social sciences.

Not only do distinctive theoretical and conceptual positions have particular methodological and empirical consequences, but they have also been accompanied by the emergence of distinctive vocabularies of power. Herein lies the third feature of the discourse relating to power: namely, that a more precise use of language is crucial to the application of power discourses to research problems in tourism. Sloppy usage has the potential to induce misleading conclusions. Although there may be some slippage or overlap, to conflate terms and their distinctive meanings may provoke misunderstandings over the diverse ways in which power achieves its effects (Allen 2003: 30).

Table 1.1 A selection of definitions of power

Author(s)	*Year*	*Definition of power*
Russell	1938	The production of intended effects.
Weber	1947	Power is the probability that one actor within a social relationship will be in a position to carry out his own will, despite resistance and regardless of the basis on which this probability rests.
Bierstedt	1950	Power is latent force . . . Power itself is the prior capacity which makes the application of force possible.
Parsons	1956	Power we may define as the realistic capacity of a system-unit to actualise its interests within the context of system-interaction and in this sense exert influence on processes in the system.
Dahl	1957	A has the power over B to the extent that he can get B to do something that he would not otherwise do.
Blau	1964	Power is the ability of persons as groups to impose their will on others despite resistance through deterrence whether in the form of withholding regularly supplied rewards or in the form of punishment inasmuch as the former, as well as the latter, constitutes in effect negative sanction.
Kaplan	1964	The ability of one person or group of persons to influence the behaviour of others, that is, to change the probabilities that others will respond in certain ways to specified stimuli.
Lukes	1974	A exercises power over B when A affects B in a manner contrary to B's interests.
Foucault	1982	A way in which certain actions may structure the field of other possible actions.
Carter	1992	There is a power relation when an individual or a group of individuals can ensure that another or others do not do something, want or do not want something, believe or do not believe something, irrespective of the latter's intertests.

Sources: adapted from Bacharach and Lawler (1980: 16–17) and Begg (2000: 14–15).

To achieve precision is no easy task and especially not in a book with contributions from several authors with different theoretical positions! We are not so much advocating here a strict terminological policing, but rather we are trying to promote a greater awareness of the dangers of the casual use of language in discussions surrounding power in tourism. Lukes (2005: 61) comments that the word 'power' is polysemic – like the words 'social' and 'cultural' – and its precise meaning depends on the context of application. Morriss (2002) points out that there is an extensive anglophone literature on power but many of the constructs of power stem from the discussion of terms originally from other languages. As such, views of power may be coloured by the subjectivities of translation (see also Poggi 2001; Clegg 1989; Lukes 2005). Stewart (2001: 6) has noticed a propensity among research workers to use words interchangeably when they are not in fact synonymous. 'Power' is all too often conflated with 'domination' such that the concepts of 'power over' and 'power to' (see below) are implicated logically and empirically, and the pursuit of power becomes the search for strategic success through resource mobilisation.

The last and by no means least important feature in the theorisations of power pertains to the use of power discourse. Simply put, why do we have concepts of power and why are they useful? As Morriss (2002) reminds us, there has been a willingness to discuss what power is or should be. All too often, though, there has been a failure to recognise that we use the concepts of power for several different purposes. Morriss (2002) identifies three contexts in which we talk of power and in which concepts of power function (see also Hall, Chapter 11 in this volume). The first is the practical context, the second moral and the third evaluative. By the practical context, Morriss (2002: 37) is referring to the desire to evaluate the extent of an agent's powers; in other words, what can be brought about, or what can't. Behaviour and its practicalities will be conditioned by the assessment of power. Estimation is not limited to the agent's own powers. Comparisons are made with competitors' powers which provide benchmarks in the process of self-evaluation, as well as clues as to what they may do for the agent, or what they may require of the agent. With the moral context, the pivotal idea is the ascription of responsibility (Morris 2002). Power is used to bring about certain outcomes, and to say that somebody or something is powerful is to assign responsibility for particular outcomes, or 'the powerful are those whom we judge or hold to be responsible for significant outcomes' (Lukes 2005: 66). As Morriss (2002: 39) notes, the connection between power and responsibility is essentially negative because responsibility can be denied by demonstrating a lack of power.

The third, evaluative context refers to the use of the concept of power to appraise social systems. It refers to the distribution of power within society and the degree to which the population's interests and expectations are met (Morriss 2002: 40). Of the three contexts, this is held to be the most complicated because the range of interests and things citizens value is so great. Evaluation depends on the identification of the extent to which citizens have the power to satisfy their own needs, and to which they are subject to the power of others. Thus, evaluation of social systems depends on the resolution of freedom to act against the limitations of domination. This context is important because it uncovers significant contradictions. Morriss (2002) identifies the hypothetical instance of a group of otherwise powerless individuals who voluntarily co-operate with one another to achieve their goals. They come to assume greater control over their own lives and their power is increased by collective action. Simultaneously, each individual has power over, but is subject to the power of, others. As Lukes (2005: 68) points out, issues of powerlessness and domination are not necessarily so separate. Among the powerful are those who are able to contribute to a reduction in others' powerlessness.

The contents of the book are designed to reflect the diversity of theorisations of power and the concluding chapter revisits the three contexts identified by Morriss (2002) and the challenges that arise for tourism research. What follows considers the plurality of power theories in more detail and identifies which conceptualisations of power have been most influential in tourism studies.

Genealogies and bloodlines of power discourse

A number of commentators have argued that two conceptions of power have dominated Western political thought since the time of Thomas Hobbes and John Locke (Hindess 1996: 1). In the first, power is viewed as a simple quantitative phenomenon. In effect, it is treated as a tangible entity which is open to empirical observation and measurement. Power is conceptualised as a capacity; it has a currency, a valency which determines the extent of any resulting action. The great allure of this conceptualisation for social scientists is that it appears to offer an easy means of identifying who has power and who is powerless (Hindess 1996: 27). In the second, more complex, conceptualisation, power involves not only a capacity to act, but simultaneously a right to act. Within this position, capacity and right function alongside one another because they rely on the consent of those over whom power is being exercised.

Dowding (1996: 4) suggests that power discourse may be reduced to two apparently even simpler prevailing concepts: 'power to' and 'power over'. Power is a capacity but, as he points out, in practical terms there seems little sense in saying that actor A 'has power' in an abstract, isolated (purely dispositional) sense. Rather, power is used in some basic way ('power to') to obtain a particular outcome. Power is never really power in general but it has specific application. 'Power over' implies 'power to' but it has specific connotations in so far as A has 'power over' B to secure an outcome from B. Dowding (1996: 2–3) further describes 'power to' as 'outcome power', or 'the ability of an actor to bring about or help bring about outcomes'. 'Power over' is given the coda 'social power' because it implies a social relation between at least two actors. Social power is, therefore, 'the ability of an actor deliberately to change the incentive structure of another actor or actors to bring about or help bring about outcomes' (Dowding 1996: 3). Allen (2003: 5) suggests that 'power over' is essentially the outcome of instrumental ties between actors, whereas 'power to' is manifested in associational ties. The former refers to a form of leverage in order to induce outcomes and is based on will and the potential for conflict, whereas the latter refers to a collectivity of action to facilitate a common aim through mutual action, and in so doing raises the prospects for or of mutual empowerment.

For Stewart (2001: 31) one of the problems of theorisations of power has been a concentration on power over rather than power to resulting in a discourse that sees power and domination 'as integrally related concepts'. The view of social power as causal power, however, is endorsed by Scott (1991: 4–5) and he notes (after Dahl 1968) the importance of differentiating between those exercising and holding power (see also Latour below). Since power is a disposition, the anticipation of its use may mean that power can have significant social consequences even where there is no explicit and overt intervention on the part of the principal (Scott 1991: 5). Scott (1991: 6) notes that this differentiation is evident in two streams of power research: his so-called 'mainstream tradition' that focuses on the episodic exercise of power by one actor over another; and the 'second stream' which concentrates on the capacity to do something.

A number of writers have drawn on these different streams in attempts to produce synthesised theorisations of power. In one of the more high-profile accounts of power, Clegg (1989) attempted to encapsulate the fluidity, diversity and 'outflanking' nature of power. His conceptualisation was based on the observation of 'three circuits of power': episodic, dispositional and facilitative. The last is linked to the systemic features of capitalism and involves domination. Episodic power is based on agency and actors seeking to achieve their goals which Clegg (1989: 208) referred to as the 'normal power of social science' considered in the writing of Dahl and others. Dispositional power provides the context to episodic power and concerns the 'rules of the game' which are based on a variety of changing social and political practices. Clegg (1989) drew on conceptualisations of translation and obligatory passage points developed by Callon and Latour to describe the processes which come to fix the 'rules of the game':

> dispositional and facilitative power respectively, constitute the field of force in which episodic agency conceptions of power are articulated. Fixing these fields of force is achieved through enrolling agencies' obligatory passage points. Power involves not only securing outcomes, which is achieved in the episodic circuit of power, but also securing and reproducing the 'substantively rational' conditions within which the strategies espoused in the circuit of episodic power make contextual sense.
>
> (Clegg 1989: 210)

Haugaard (2002) notes the value of Clegg's circuits but argues that, like Lukes's (1977) earlier work, the circuits tend to present structural inequalities as unaffected by agency when in fact they are continually contested and evolving. Clegg's (1989) circuits of power do, however, suggest that, although seemingly straightforward, even the concepts of 'power to' and 'power over' mask several further complexities. In the case of 'power over', the resulting relationship between A and B is asymmetric in A's favour. However, it does not suggest that the relationship between A and B resolves only in a single direction. The result is a function of the resolution of both A's ability to dominate B and B's ability to resist A. Max Weber (1978: 53) recognised this in his extensive treatise on the nature of society and operation of organisations at the beginning of the last century. He viewed power (*Macht*) as, 'the probability that one actor within a social relationship will be in a position to carry out his own will despite resistance, regardless of the basis on which the probability rests'.

According to Galbraith (1983: 20), Weber's views on power represent the 'common perception' of power: namely, that 'someone or some group is imposing its will and purpose or purposes on others, including those who are reluctant or adverse'. As Galbraith (1983) goes on to note, it is precisely because power has such an apparently 'common sense' meaning that so often it is used in all aspects of life (not just in the social sciences) with little regard for the need for definition or its complexities. Weber's comments are, though, instructive because they point to the different modes by which one may secure compliance over another. Weber

(1978: 53) further distinguished power from domination (*Herrschaft*), or 'the probability that a command with a specific given content will be obeyed by a given group of persons'. Power was also differentiated from discipline, or 'the probability that by virtue of habituation a command will receive prompt and automatic obedience in stereotyped forms, on the part of a given group of persons'.

A number of writers have endeavoured to build the lexicon of power by clarifying terms and establishing their meanings with greater accuracy and certainty. Bachrach and Baratz (1970) presented one of the earliest attempts to widen the vocabulary of power (Table 1.2). This was no easy task because, as Lukes (1974: 17) maintains, there was some confusion in their conceptual map. With this misunderstanding eliminated, Lukes (1974) argues that their original typology of 'power', in fact, identified five forms of control, or ways in which power manifests itself, each with subtle but highly important differences (cf. Lukes 1974: 17–18). Allen (2003) has since attempted to inject a further degree of precision. He makes the very particular point that authority and domination should not be confused. Domination is one among several 'modalities' of power in addition to authority, coercion, manipulation or seduction. Based on his readings of Weber, domination is 'a more tigthly orchestrated means of influencing the conduct of others'. If constraint and imposition are central to exercise of domination, then in fact '*close discipline, continuous control* and *supervision* represent the organizational means by which domination may be achieved' (Allen 2003: 28, italics original). He adds further lexical clarity by the identification of seduction as offering the prospect that an action is in fact optional and that the subject is not in fact compelled to participate. It introduces the possibility of choices. As a result, it involves a 'renunciation of total domination, not its propagation; it is a modest form of power which is intended to act upon those who have ability to opt out' (Allen 2003: 31).

Galbraith (1983) mentions the threat of physical punishment or the possibility of pecuniary reward as an organisational means of power. This is somewhat simplistic and Hannah Arendt has extensively explored the relationship between power and violence. Violence may not be regarded as a modality of power. She does note that 'power and violence, though they are distinct phenomena, usually appear together. Wherever they are combined, power, we have found, is the primary and predominant factor' (Arendt 1970: 52). She maintains, though, clearly 'power and violence are opposites; where one rules absolutely, the other is absent. Violence appears where power is in jeopardy but left to its own course it ends in power's disappearance . . . Violence can destroy power; it is utterly incapable of creating it' (Arendt 1970: 56).

Arendt's work was rooted in her experience of politics from the 1940s to 1970s and, in particular, she explored the distinction between, what Haugaard (2002: 132) elegantly describes as, the 'politics which prevent human flourishing' and 'a form of virtue politics which contributes to human freedom and emanicipation'. Power is vital to this latter form of politics but violence is characteristic of the former (Haugaard 2002). Violence may be used as a coercive measure but actually it signifies the diminution or erosion of power, not its reinforcement, because, for those in power, the referral to violence signifies the dissipation of their capacity to

Table 1.2 A typology of power and related concepts based on Lukes's (1974) reading of Bachrach and Baratz (1970)

Concept	*Meaning*
Power	All forms of successful control by A over B – that is, of A securing B's compliance. It embraces coercion, influence, authority, force and manipulation.
Coercion	Exists where A secures B's compliance by the threat of deprivation where there is a conflict over values or course of action between A and B.
Influence	Exists where A, without resorting to either a tacit or overt threat of severe deprivation, causes B to change B's course of action.
Authority	B complies because he recognises that A's command is reasonable in terms of his own values. This is either because A's command has content which is legitimate and reasonable, or because it has been arrived at through a legitimate and reasonable procedure.
Force	A's objectives are achieved in the face of B's non-compliance by stripping B of the choice between compliance and non-compliance.
Manipulation	Is an 'aspect' or sub-concept of force (and distinct from coercion, power, influence and authority) since here compliance is forthcoming in the absence of recognition on the complier's part either of the source or the exact nature of demand upon the complier.

Sources: abridged from Lukes (1974: 17–18) based on Bachrach and Baratz (1970: 24, 28, 30, 34, 37).

carry out their will over others. The socialisation of power which an analysis of violence implies is evident in Arendt's (1970: 44) view that,

> Power corresponds to the human ability not just to act but to act in concert. Power is never the property of an individual; it belongs to a group and remains in existence only so long as the group keeps together. When we say of somebody that he [*sic*] is 'in power' we actually refer to his being empowered by a certain number of people to act in their name. The moment the group, from which the power originated to begin with . . . disappears, 'his power' also vanishes.

For Arendt (1970: 44–46), power is different from strength, force, authority and violence. It is also a collective capacity and it is based on consensus. Power is, as Allen (2003: 53) succintly puts it, rooted in mutual action. In this respect, there are some immediate similarities with Talcott Parson's views. Parsons studied in Heidelberg in the 1920s where he was influenced by the writings of Simmel and, more so, Weber (Levine 2000) whose work he later translated (see Weber 1978). Parsons (1963, 1967) is credited with developing a more positive, utilitarian view of power as a capacity to achieve collective goals (Clegg 1989) through collective action via institutionalised political leadership based on binding obligations and sanctions (Allen 2003: 53). Power is tied to authority and consensus, but it is distanciated from conflicts of interest, coercion and force (Lukes 1974). This simple clarity is not, though, perhaps immediately apparent from Parsons's (1967: 308) unwieldy definition of power as a

> generalized capacity to secure the performance of binding obligations by units in a system of collective organization when the obligations are legitimized with reference to their bearing on collective goals and where in the case of recalcitrance there is a presumption of enforcement by negative situational sanctions – whatever the actual agency of that enforcement.

For Haugaard (2002), one of the defining features of Parsons's earlier (1963) contribution was that he challenged the dominant view at the time that power was a zero-sum game (cf. Giddens's 1968 critique). Weberian views of power contain a tacit assumption that there is a given and fixed 'quantity' (Parsons 1963: 233) of power. The gains made by the winners were therefore apparently at the direct expense of the losers. However, the question arises as to why power has to involve a zero-sum game? Power does not just exist. Even as a disposition it has to be created and its creation is a function of the relativities between agents; that is, there is a social production of power. If power is not constant, then the possibility exists that the gains made by the powerful need not be at the expense of the less powerful. In this respect, the existence of powerful individuals or organisations need not necessarily be invidious (Haugaard 2002: 67), but power could contribute positively to the general accomplishment of order and civility (Clegg 1989: 131).

Parsons (1963, 1967) advanced his views on power by drawing an analogy with money as a generalised symbolic medium. Money circulates within an economy but the value of money is based on consensus. A similar system of belief establishes the legitimacy of power. Symbolic legitimacy enables the holders of power (and money) to call forth obligations from others (Clegg 1989: 130). According to Haugaard (2002: 68), consensual power as self-reinforcing and self-perpetuating is based on the belief in legitimacy as a key element in the conceptualisation of power more generally. Furthermore, as social systems become more complex, consensual power becomes more vital to multiple goal attainment. Over time, the social system in which power functions becomes more effective in its operation and in the attainment of goals. As a result, members of the social system become more willing to comply with those in authority. Coercion may be involved in the process of power formation but, as the system begins to operate ever more effectively, it becomes divorced from power. Just how effectively the system operates to deliver the attainment of goals is essentially a determinant of the production and hence amount of power in the system (Haugaard 2002).

Bruno Latour (1986) revisited the argument that power is a relational effect, and in so doing he suggested that there are limits to viewing power purely as disposition. For him, power cannot be realised without connectivities and networks among actors who transact with one another to one degree or another. He describes a paradox such that,

> when you simply *have* power – *in potentia* – nothing happens and you are powerless; when you *exert* power – in actu – *others* are performing the action

> and not you . . . Power is not something you may possess and hoard. Either you have it in practice and *you* do not have it – others have – or you simply have it in theory and you do not have it.
>
> (Latour 1986: 264–265, emphasis original)

Latour (1986: 265) invoked the example of Amin Gemayel, the former president of the Lebanon. Elected head of state in 1982 (cf. Moubayed 2001), he had power over the country, but because very few people acted when he ordered things, he was powerless in practice.

There are counters to this view that power is meaningless without performance or practice. For instance, Lukes (2005) notes that it is possible to be powerful by the non-exercise of power. He draws attention to the almost adversarial nature of many conceptualisations of power. One party may indeed exert power over another by acting against its interests. Instead, he notes, the (non-)exercise of power in the interests of others 'may, but also may not, be among the most effective and sometimes the most insidious forms of power' (Lukes 2005: 110).

Latour's emphasis on the relational and the practices of power draws our attention to tensions between the application of constructs of power to understanding real-world and the philosophical assumptions of power. Morriss (2002) maintains strongly that power is a disposition and that, while it may be a tempting contingency to connect power with its manifestation, it is false just to reduce potentialities to actualities. As he puts it, 'episodic concepts report happenings or events, whilst dispositionals refer to relatively enduring capacities of objects' (Morriss 2002: 14). He identifies two fallacies in discourses of power – exercise fallacy and vehicle fallacy – and in so doing stresses that it is the disposition that we should ultimately seek to assess, not just the events or vehicles of power.

Exercise fallacy refers to 'the claim that the power to do something is nothing more than the doing of it' (Morriss 2002: 15); simply put, the exercise fallacy reduces the identification and assessment of power to the observation of its exercise, not the observation or measurement of the disposition, power itself. Dispositions can, after all, remain forever unmanifested. Exercise fallacy is evident in the work of those interested in decision-making (Lukes 2005: 109), but such behaviourists (see below) tend to avoid the so-called 'vehicle fallacy'. This refers to the association of power with a thing (its vehicle), a tangible entity; that is, power is identified with the resource/s that give rise to it (Morriss 2002: 19). As Allen (2003: 5) argues, resources are the technologies through which power is exercised and sustained. Resources are an obvious starting point from which to study power, but there is a subtle distinction in operation: the appraisal of a catalogue of resources provides only useful evidence in the assessment of power (Morriss 2002: 19); assessment of resources is *not* the assessment of power. As Morriss (2002: 18–19) neatly puts it, 'wealth is not political power . . . whilst some people use their wealth to collect politicians, others can only collect paintings'.

Towards a radical view of power: Lukes and tourism

Weber's thinking formed a major starting point from which American political theorists and sociologists in the post-war years started to think about power and of how to study it empirically (Lukes 1974: 1). This body of work concentrated on mapping the location and relative capacity of power as part of a wider project on the nature of democracy in the United States. The central concerns were who ran the community and who made the big decisions. The ensuing debates over power were also initially heavily influenced by neo-Marxist considerations of the role of the state in capitalist societies (Haugaard 2002).

Questions of powerlessness and domination were evident in two of the earliest contributions in this genre. According to Lukes (1974, 2005), C. Wright Mills's (1956) *The Power Elite* and Floyd Hunter's (1953) *Community Power Structure* drew attention to the power vested in elites in American society. For Mills (1959: 3–4), the power elite is, 'composed of men [*sic*] whose positions enable them to transcend the ordinary environments of ordinary men and women; they are in positions to make decisions having major consequences'. They are in 'command of the major hierarchies and organizations of modern society' and they are 'not solitary rulers' (Mills 1956: 4). Hunter's book explored the leadership patterns in a city of a half a million people, a 'regional city'. Hunter concluded how 'the men [sic] of real power controlled the expenditure for both the public and private agencies devoted to health and welfare programs in the community' (Hunter 1953, in Lukes 2005: 3–4).

Such an approach was attacked by Dahl (1958) as the 'ruling elite model' (Lukes 1974, 2005), with an undue willingness to concede that the local community had lost its social solidarity, and as behaviourist with an accent on decision-making (Dahl 1961a, 1961b). For Lukes (1974, 2005: 5), Dahl's ideas assumed a central position in the emergence of 'pluralist' views of power. Unlike the elitists, the pluralists maintained that power was distributed more widely through society and not vested exclusively in a single overall ruling power elite. Pluralist views noted that power relations were played out over multiple and often interlocking issues (Lukes 2005: 5). They also rejected the view that power structures were stable over time (Bachrach and Baratz 1962: 947). Since different actors assumed different positions of power relative to the issues, there could be no singular ruling group. The idea of a single, stable group was undermined by oscilliations in the importance of issues. As Lukes (2005: 5) notes, as straightforward as these ideas may nowadays seem, at the time they precipitated complex – and it might be added, enduring – methodological dilemmas regarding how to operationalise power in social sciences research. In particular, they raised the tantalising issues of how power is defined (by different groups), how should it be investigated (by which techniques?), and how and where is it distributed (i.e. how plural or democratic is it?).

Counter-criticisms suggested that the 'pluralists' had limited methodological horizons of their own. Bachrach and Baratz (1962: 948) noted that pluralists were not only interested in the sources of power, but also in one specific nature of its exercise. Their accent was on decision-making and their method was the analysis of concrete (i.e. observable) decisions, or 'a choice among alternative modes of action' (Bachrach and Baratz 1970: 39; Lukes 1974: 18). In Bachrach and Baratz's view however, the pluralists missed the 'Second Face of Power'; that is, the dimension of non-decision-making or,

> the extent to which and the manner in which the *status quo* oriented persons and groups influence those community values and those political institutions . . . which tend to limit the scope of actual decision-making to 'safe' issues.
> (Bachrach and Baratz 1962: 952)

Simply put, what is kept off the agenda is as much an expression of power as what is included. The ability to limit decision-making to reasonably uncontroversial, uncontentious subjects is an expression of power. Non-decisions offer the

> means by which demands for change in the existing allocation of benefits and privileges in the community can be suffocated before they are even voiced; or kept covert; or killed before they gain access to the relevant decision-making arena; or, failing all these things, maimed or destroyed in the decision-implementing stage of the policy process.
> (Bachrach and Baratz 1970: 44)

Power was not solely evident in tangible decisions (Lukes 2005: 6) but 'the extent that a person or group – consciously or unconsciously – creates or reinforces barriers to the public airing of policy conflicts, that person or group has power' (Bachrach and Baratz 1970: 8). In his seminal text *Power: A Radical View*, Steven Lukes (1974: 11–20) describes these as one- and two-dimensional views of power respectively (Table 1.3). The two-dimensional view is notable, according to Lukes (1974: 17), because Bachrach and Baratz's (1970) work introduced the 'mobilization of bias' into discussions of power and the 'Second Face of Power' is, hence, an expression of a prevailing set of subjectivities (i.e. values, beliefs, procedures, etc.) that function to the advantage of one group (in power) at the expense of another (the powerless). This raised the possibility for the first time that power is mutually implicated with ideology because non-decisions were decisions that result 'in suppression or thwarting of a latent or manifest challenge to the values or interests of the decision-maker' (Bachrach and Baratz 1970: 44).

As helpful as it may be, though, in advancing the power agenda, Lukes (1974: 21) argued that the 'Second Face' was still limited by virtue of its focus on observable conflict. It may be important to identify potential issues which non-decision-making had been prevented from becoming actual (Lukes 1974: 19) but, methodologically, the emphasis was still on observable conflict – whether overt or covert – as the articulation of power. This two-dimensional view discounted the possibility that conflict may be latent; it may not be necessary at all to have conflict in order to observe the operation of power (Lukes 1974: 23), and non-decisions presuppose the existence of grievances which are denied their airing in the political process (Lukes 1974: 24). Instead, Lukes (1974: 23) argued that 'the most effective

Table 1.3 The three dimensional views of power as originally outlined by Steven Lukes in *Power: A Radical View*

View of power	*Nature of view*	*Focus on*
One-dimensional	Behaviouralist, pluralist	• Behaviour • Decision-making • (Key) Issues • Observable (i.e. overt) conflict • (Subjective) Interests, seen as policy preferences revealed by political participation.
Two-dimensional	(Qualified) Critique of behavioural focus	• Decision-making and non-decision-making • Issues and potential issues • Observable (overt or covert) conflict • (Subjective) Interests, seen as policy preferences or grievances.
Three-dimensional	Critique of behavioural focus	• Decision-making and control over the political agenda (not necessarily through decisions) • Issues and potential issues • Observable (overt or covert) conflict and latent conflict • Subjective and real interests.

Source: Lukes (1974: 13, 25).

and insidious use of power is to prevent such conflict from arising in the first place'. Power is exercised not just to restrict what enters the agenda for political discussion. Rather, power can be used to prevent people from having grievances,

> by shaping their perceptions, cognitions and preferences in such a way that they accept their role in the existing order of things, either because they can see or imagine no alternative to it, or because they see it as natural and unchangeable, or because they value it as divinely ordained and beneficial.
>
> (Lukes 1974: 24)

Instead, Lukes (1974: 21ff.) postulated a three-dimensional view of power which 'allows for a consideration of the many ways in which *potential issues* are kept out of politics whether through the operation of social forces and institutional practices or through individuals' decisions' (Lukes 1974: 24, emphasis original). In the three-dimensional view inaction, unconscious and collective (not individual, as in previous theorisation) operation may feature (Lukes 1974: 50) and there need not necessarily be observable conflict (Table 1.3). In contrast, there may be 'a *latent conflict*, which consists in a contradiction between the interests of those exercising power and the *real interests* of those they exclude'. On this basis, Lukes (2005: 30, 37) concluded that 'A exercises power over B when A affects B in a manner contrary to *B*'s interests'. Latent power operates ideologically to shape people's thoughts and wishes so that (otherwise apparent) differences of interest are obviated (Vogler 1998: 699). Such orchestration of their needs in this manner is practically contrary to their 'real' interests, and hence the freedom to choose may yield different sets of choices of action.

As compelling as this radical approach may at first sight seem, *Power: A Radical View* has been the subject of criticism. As Lukes (2005: 12) has recently recalled, his original thesis concentrated solely on power as domination (power over), and domination is, as he puts it, 'only one species' or, as Allen (2003) might term it, 'modality' of power. It also reduced power to an analysis of binary relations between two actors when actors' interests and hence power relations concern multiple issues (Lukes 2005: 12–13). He has also noted that it is a good example of Morriss's (2002) 'exercise fallacy' because it concentrates on the application rather than the dispositionality of power (Lukes 2005: 109). Benton (1981: 180) has argued that the thesis has more fundamental shortcomings; the concept of 'interests' should be abandoned from a position of realist epistemology because it proves unworkable. What are the real interests of the powerless, how are these determined, and how are these to be uncovered by research workers? Under Lukes, power implies what Haugaard (2002: 38) terms 'could-have-done-otherwise agency' but it is difficult to know what they may or may not have done were the individuals not to have been constrained by the operation of the third dimension. As Clegg (1989: 3) points out, this exposes a potential methodological weakness because 'when people say what their consciousness of something is, these accounts cannot be taken at face value nor can they function as explanations'. Any reference that is made to the actor's account will be necessarily flawed in his view. Finally, according to Haugaard (2002: 39), there is a troubling propensity in Lukes's work to distinguish knowledge as free from power in contrast to Foucauldian perspectives. In its treatment of 'false consciousness', his work assumes that power distorts knowledge without acknowledging the complex nature of the power–knowledge relationship (see below).

For Haugaard (2002: 38), one of the most notable features of Lukes's thesis is that his critique of Bachrach and Baratz precipitated a view that 'biases are not necessarily reducible to individuals' actions or deliberate non-actions but are inherited from the past in the form of structured and culturally patterned behaviour of groups'. In this respect, there are resonances between Lukes's work and the views of power contained in emergent theories of structuration (Giddens 1976). The process of structuration involves the reproduction of continually changing structures through the time–space-specific acts of agents. This was presented by Giddens (1984) as a response to the apparent determinism of structuralism and the limited account of the power of agency in Foucault's work (see below). Agents can utilise causal powers which, according to Giddens (1984), can both challenge as well as maintain structures. Thus, power is emancipatory and not simply constraining. There are two types of resources to structures of domination, namely: 'allocative resources' that include produced goods; and 'authoritative resources' that include self-expression and the body (Giddens 1984). For Haugaard (2002),

Giddens's approach represents an early attempt to synthesise the consensual and conflictual views of power but one that overemphasises the consensual aspects of power in contemporary society and underplays the contestation that is linked to structures. As an alternative approach, Haugaard (1997) seeks to build on the insights on power as constitutive provided by Foucault and the conflictual approach of Bourdieu; he argues that an understanding of power can be progressed by analysing the interactions between the goals of agents, structures, conflict and consensus.

The impact on tourism studies of theorisations rooted in Weberian, elitist, pluralist or Lukesian thinking is relatively limited. All too frequently, writings on tourism policy and development implicitly adopt some of these perspectives. For example, Reed (1997) critiques pluralist assumptions in discussion of partnership and collaboration in tourism planning (cf. Jamal and Getz 1995). Only a few writers explicitly acknowledge the influence of these power theories on their work (e.g. Hall 1994). Drawing on both pluralist and radical readings of power in her discussion of community involvement in the tourism development process in Squamish (British Columbia), Reed (1997) has identified several instances where the content of the agenda driving development forward was manipulated by the inclusion or exclusion of topics depending on the interests of those in power. Not only are power relations to be widely observed, but they also permeate through all strata of the tourism system, sometimes in manners that are either entirely unpredictable or obscure, unobserved to all but the skilled observer. A similar conclusion was drawn in Strange's (1999) case studies of tourism policy in historic towns in the United Kingdom. Despite the supposedly high profile of sustainability and conservation on the policy agenda, business and economic development goals were often far more influential in shaping tourism policy decisions. As Doorne (1998: 133) has succinctly observed in the contested transformation of the Wellington (New Zealand) waterfront, 'the latencies of power cannot be ignored'. In the chronology of this project, there are several instances where particular issues are overtly omitted from the agenda; particular stakeholder groups are excluded from decision-making or the opportunity to make decisions; and what he terms 'localised latencies' reveal themselves whereby 'the differences between the visions of oppositional groups are less distinct' (Doorne 1998: 151). This latter point is significant because it indicates the exercise of power in the (apparent) absence of conflict.

Notions of power linked to Weberian and Lukesian theories have been utilised in studies of tourism policy that have drawn on conceptual developments in the study of urban politics. The urban growth machine (Molotch 1976) and the urban regime (Stone 1989) were initially developed to examine how private business groupings shaped urban planning policies and how coalitions between politicians, bureaucrats and business representatives could coalesce into a consensual regime (see Gill, Chapter 6 in this volume). An examination of Christchurch (New Zealand) by Schöllman *et al.* (2001) claims that local place promotion strategies exhibit the characteristics of the urban growth machine model although the major developer interests central to the urban growth machine (Logan and Molotch 1987) play less of a role in the development of place marketing. Long (2000) argues that regime theory helps explain the long-standing features of a public–private partnership in the London borough of Islington. Elements of regime theory are also utilised by Thomas and Thomas (2005) to examine the role of small tourism businesses in the development of tourism policy. There are some superficial similarities with elitism and pluralism, and Thomas and Thomas (2005: 132) find that while in some locations stable coalitions may develop into regimes in others 'a quasi-pluralist landscape of shifting coalitions relating to specific issues may take hold'. They also draw on regime theory as part of an argument for a more systematic approach to understanding power relations in local tourism policy.

Tourism and Foucault: the critique, orderings and beyond

By contrast, far more explicit use has been made in tourism studies of the theorisations of power developed by post-structural writers, in particular the extensive writings of Michel Foucault. Foucault's consideration of power marked a significant departure from previous thinking that has been highly contested but has already had a significant influence on tourism studies. His numerous thoughts on power were in part aimed at challenging existing radical and liberal conceptions of power. Marxist theories that presented knowledge and ideology as repressive were critiqued for viewing knowledges, human subjects and practices as subject to the priorities of a capitalist system and the relations of production (Foucault 2002). Writings, often influenced by Weber, that emphasised the importance of authority, the state and politico-judicial systems were contested for their focus on how power was held by a group or institutions without an appreciation of the degree to which power penetrated deep into human existence, and because 'power relations are rooted in the whole network of the social' (Foucault 1982: 345).

At the heart of Foucault's early writing was a desire to understand the inseparable connections between power, knowledge and truth, and how the latter was not something simply sought but which played a central role in regulating human existence often through discourse (Foucault 1978). Consequently knowledge is at the core of what Foucault termed disciplinary power. Miller (2003) argues that this led to valuable insights into 'the various knowledges and practices which seek to transform human beings into subjects, and to generate true knowledges of them'. For Foucault, power is both productive, contributing to the collective dimensions of society, and also constitutive of subjectivity as power plays a role in developing individual identities and practices (Gordon 2002). Hirst (2005: 157) argues there is an institutional dimension to these insights on knowledge in that

> we see Foucault move from author-as-subject towards a view of the subject as agent/effect of a discursive formation. Enunciative modalities mean that only certain subjects are qualified to speak in particular ways: that certain statements cannot be made by everybody and anybody . . . So knowledges and the subjects who produce them are connected with particular institutional conditions and forms of power.

Foucault's later writing on governmentality also provided new ways of thinking about the state, political power and the dangers of power. The state was presented as having 'both an individualising and totalising form of power' but the study of power could not simply involve the study of institutions (sovereign power). Rather, it must also incorporate a study of all 'micro' governmental practices to reveal the connections between the 'political' and all the other types of power relation, practice and technologies (Foucault 1982: 332). Foucault (1982: 338) did acknowledge, however, the existence of 'power blocks' in which power relations, objective capacities and communication relations formed 'regulated and concerted systems'. Miller (2003: 205) ambitiously attempts to summarise the conceptual legacy of Foucault's changing writings on power as:

> power should not be understood according to the model of a generalised domination exerted by one group over another. Power must be understood as a multiplicity of force relations which are immanent to the domain in which they operate and are constitutive of their own organisation. Power does not derive from a single point of origin but is to be found where it operates, at the mobile and unstable interrelation of force relations at local levels. Power is neither an institution nor a structure; it is not a force that can be located. It is 'everywhere'.

Clearly, the location of power is a significant concern. However, power is not an entity solely of itself, but it is imbued in all forms of human endeavour. It is,

> the name that one attributes to a complex strategical situation in a particular society. Power relations are not external to and causally related to other types of relations such as economic processes or knowledges, but are immanent to them. Power comes from below, from the multiple force relations operating in the apparatuses of production, families and institutions which cut across the social body. Power relations are intentional, yet they are non subjective, that is to say they are marked by a calculation of aims and objectives, but do not result from the choice or decision of an individual subject. And finally, where there is power there is resistance. Power is a relational phenomenon which exists through a multiplicity of points of resistance which are present throughout the networks of power.
>
> (Miller 2003: 205)

Indeed for Miller (2003: 17), the core of Foucault's work is to understand 'the regulatory practices of the self', and he suggests that power is an inadequate term to encapsulate such a task.

Given the mobilities and multiplicities associated with tourism along with the sometimes elusive role of state tourism policy and planning (Hall and Jenkins 1995; Church 2004), it is perhaps not surprising that Foucault's considerations of power should have been influential in tourism. Urry's (2002) 'tourist gaze' was directly influenced by Foucault's examination of the regulating gaze in medicine and penal systems. The tourist gaze establishes regulating norms for doing tourism while marking the object of the gaze with meanings linked to a range of power relations and sources of authority. Urry (2002) recognised that his earlier discussions of the gaze had not been sufficiently 'embodied' to consider the way bodily enactments contribute to power relations. Undoubtedly significant are other critiques of Urry (2002) concerning the prioritisation of the visual, the static nature of the gaze and an overemphasis on the sites and governance of the gaze rather than the broader cultural processes with which tourism was intertwined (Franklin 2004). The tourist gaze, however, was key to moving forward the exploration of power in tourism so as to emphasise the centrality of the tourist rather than the state, other institutions or systemic power relations.

Beyond the gaze, Foucault's (1978) concern with the power of knowledges and discourses has also influenced studies of the 'micro' dimensions of power relations in tourism. Morgan and Pritchard (1998) directly acknowledge the influence of Foucault in their study of the semiotics and discourses (involving tone and status) that they argue are central to explaining the geographical differences in the development trajectories of resorts in south-west England. A range of other studies has indicated how the symbols, signs and sites of tourism contribute to influential discourses that can have contrasting effects on power relations by both empowering tourists while underpinning material and symbolic otherings of certain social groups, especially in post-colonial settings (see Aitchison 2001; Craik 1997; Winter, Chapter 5 in this volume). Indeed, a concern with the semiotics of tourism has provided some initial insights of the penetration of power relations deep into mundane materialities of tourism (Dann 1996; Echtner 1999; Selby 2004). However, as Franklin argues (2004), this was often by prioritising the objects and sites of tourism rather than the broader social processes involved.

If research into power relations in tourism is to be further informed by Foucauldian thinking, it will have to acknowledge, and be shaped by, the growing and forceful critique of Foucault's views on power. This has been developed from a number of theoretical, philosophical and psychological perspectives (Newman 2004). Lukes (2005: 98) argues that Foucault's later writings on governmentality and the many empirical studies by others that have followed expose tensions within his earlier work. In particular, they reveal that the notion of power as being productive in constituting governable subjects is an oversimplified 'ideal–typical' depiction which does not reveal how modern forms of power both succeed and fail. Other critiques of Foucault have focused on moral and ethical issues whereby the all-embracing, immanent conception of power leads to a view of the subject as dominated and leaves no room for human autonomy or responsibility (Hartstock 1990; Stewart 2001). Schapp (2000: 134) claims that it is the 'sense of human beings as creative agents both of the self and of the world they share that is missing in Foucault'. Not surprisingly, this has been strongly refuted by advocates of Foucault. Gordon (2002: xvii) suggests that, despite the evidence of Foucault's writings,

> one section of academia is content to this day to assert that Foucault considered truth to be no more than an effect of power, that his thought is a wholesale

> and nihilistic rejection of the values of the Enlightenment, that he and his work are incapable of contributing to any form of rational and morally responsible action.

Critiques and reactions to Foucault's writings of power of this nature have almost inevitably encouraged the development of more nuanced explorations of the creative and embodied 'power' of the tourist. Crouch (Chapter 2 in this volume) has drawn on theorists such as de Certeau (1984) to examine the enactments and negotiations in the fleeting practices, embodiments and spaces that enable tourism to be empowering. Similarly, Selby (2004) argues that the phenomenological writings of Schutz can be used to understand the relations in (co-)constructing a creative tourist experience. In other tourism writings the influence of post-Foucauldian arguments and critiques are more explicit. For example, Hartsock's (1990) feminist critique of Foucault, along with the conclusions of a range of post-structural writings, lead Aitchison (2003: 33) to argue that, for understanding gender and power relations in the context of leisure, capitalism and patriarchy must remain the key focus of examination. She maintains that it is necessary 'to provide a broad analysis of the cultural "fragments and differences" in the inter-relationships between gender and leisure while simultaneously attending to the broader structural relations of power'. As an example of this approach, Johnston's (2005) examination of gay and lesbian pride festivals illustrates the role played by bodies and performance in both challenging and reaffirming the heteronormative power relations that structure tourism and leisure spaces. Such studies are also trying to avoid the pitfalls of so-called 'resistance studies'. Influenced by Foucault's argument that resistance to power relations was endemic, these have been critiqued from a performative perspective for encouraging a domination–resistance binary, and for not appreciating 'how power is continually and creatively constructed in fleeting contextual encounters' (Rose 2002: 395).

Actor-network theory (ANT) represents a further significant post-Foucauldian perspective by which to understand power and its connectivities with tourism. While acknowledging an affinity to Foucault, proponents of ANT claim to adopt a distinctive and more empirically oriented approach to understanding agency, organisation and power (Law 1994, 2003). The emphasis in ANT is on relations and process, and how time- and space-specific translations of heterogeneous networks, agents and non-human devices produce orderings that act in and shape the social world. Law (2003: 5) argues that the perspective on power in ANT must not be associated with pluralism. Rather,

> the effects of power are generated in a relational and distributed manner and nothing is ever sown up . . . ordering (and its effects including power) is contestable and often contested . . . human beings and machines have their own preferences.

There have been strong critiques of ANT by theorists (Latour 1999). However, its emphasis on relational materialism and the role of non-human agents and devices in shaping power relations has appeal to tourist researchers seeking to understand the significance of objects involved in tourism. For example, Hetherington (2000) examines the interaction of the agency of humans and museum artefacts in constructing the tourism experience. Taking a broader perspective, Franklin (2004) argues that sociological notions of ordering influenced by ANT, Deleuzian writings on rhizonomic networks, Foucauldian thought and post-humanism have the potential for creating nothing less than a new ontology of tourism. By viewing tourism as an ordering of central importance to the current social world, Franklin (2004) outlines the possibilities to understand tourism as not other to the everyday world, as Urry (2003) has suggested. Instead, it is central to the everyday and is a key connectivity network mutually contributing to the globalising tendencies of contemporary capitalism and the associated processes such as consumerism and cosmopolitanism. The adherents of ordering theories suggest it has considerable potential for revealing the detail of how power 'works' in tourism through agents, objects, bodies and performances. Thus, it provides insights beyond more structurally oriented accounts that seek to reveal the connections between tourism and wider power relations. Franklin (2004: 297) draws on the empirical example of the growth of Thomas Cook's tourism business which he claims played a role in nationalism and nation-building. He contends that, by viewing tourism as an ordering involving work, projects, devices and governance, it is possible to show that tourism was 'something that was made to happen' rather than just a structurally linked outcome of changing technologies and demand.

Urry (2003) seeks to develop an ordering perspective of mobility in general which, compared to Franklin (2004), is more explicit in its consideration of power and draws on network metaphors and complexity theory. For Urry (2003: 112), a complexity approach involves seeing power not,

> as a thing or a possession. It is something that follows or runs. . . . It is non-contiguous. . . . Travelling light is the new asset of power. Power is all about speed, lightness, distance, the weightless, the global, and this is true both of elites and of those resisting elites.

The key transformational elements of power are mediated and informational power and their increasing 'structural power' loosens other structural elements of society and means that political and personal attempts at ordering are always challenged and disrupted (Urry 2003: 139). Such claims are not readily reconciled with increasingly draconian attempts by governments to control human movement and would no doubt be theoretically critiqued by those mentioned earlier who argue for the need to recognise autonomy and identify responsibility in the analysis of power (Schnapp 2000; Lukes 2005). Importantly, however, Urry (2003) is seeking to highlight how mobilities are central to power relations in contemporary society and the influence of mobilities extends well beyond the governance and control linked to the gaze.

Collectively, these writings would suggest the engagement of tourist studies with theories of power is more advanced than might otherwise be thought. Claims that tourist researchers are not engaged with social science theory (Selby 2004) seem less valid and the examination of power relations in tourism is certainly playing a role in correcting some of the alleged theoretical limitations of past studies of tourism (Franklin and Crang 2001). Indeed, examinations of power and tourism influenced by Foucault and other post-structural theories seem to be more developed than those that draw on radical or liberal conceptions of power. Furthermore, Foucault has had a significant methodological legacy in the social sciences by advancing the case for ethno-methodologies and it is no surprise that research influenced by post-structural theory has been central to developing qualitative methods which until recently had been marginalised in tourism studies (Jamal and Hollinshead 2001; Phillimore and Goodson 2004). However, notions of space and spatiality, despite their apparently obvious centrality to studies of tourism and power in this respect, have developed an elusive status to which we now turn our attention.

Spatialities of power and tourism

Foucault (1982: 361) argued that 'space is fundamental in any exercise of power' and unsurprisingly a number of the studies of tourism from a post-structural perspective have made space their focus of study. Edensor (2000) draws on Foucault's ideas about surveillance to explore what are permissible and prohibited practices in enclavic tourist spaces, as well as the manner in which they are practically regulated. Similarly, in addition to finding power relations at the level of the individual and the institution that constrain and manage tourist behaviour, Cheong and Miller (2000: 372) discover power relations in locations which at first inspection might apparently be unpromisingly non-political to untrained eyes: interpersonal transactions between tourists and guides; ethical codes; and the design, content and publication of guidebooks. With an almost evangelical zeal, they (2000: 371) concluded from their Foucauldian reading that power is 'omnipresent in a tripartite system of tourists, locals and [information] brokers'.

To some degree, however, the notion of the tourism gaze reified the study of the spaces of tourism and this has been critiqued by Franklin (2004). Urry (2003: 113), draws on concepts of complexity to argue for a fluid and dispersed view of the relations between power and space. He claims that, from a complexity perspective, power 'may be increasingly detached from a specific territory or space'. Despite this detachment he recognises that time–space configurations are entwined in the moorings and mobilities based around globally integrated networks and attractors which play a key role in transforming the social world.

Foucault (1978: 94) also recognised the mobile nature of power, noting that 'power is exercised from innumerable points, in the interplay of non-egalitarian and mobile relations'. For Westwood (2002: 2), such a view signifies that power should not be treated as a capacity outside or beyond social relations; rather, power is constitutive of social relations. In a classical Foucauldian reinterpretation, she notes that power is to be found everywhere and always present. Westwood sets out to document the terrains in which power is constitutive in the social. She argues that not only are there identifiable modalities of power but also distinctive sites of power. The former refer to the different forms of power and the manner in which power is enacted, while the latter are social spaces where power is exercised. Among the sites of power are: racialised power; class and power; engendered power; sexualised power; spatial power; and visual power. Modalities of power are: repression/coercion; power as constraint; hegemony and counter-hegemony; manipulation and strategy; power/knowledge; discipline and governance; and seduction and resistance (Westwood 2002: 3). For her, such a framework is useful because it moves analysis of power away from conceptualisations centred on capacity towards the specificity of strategies and tactics used in distinctive social settings (Westwood 2002: 135).

Allen (2003: 2) in his study of place, space and power adopts a similar starting point of power as a relational effect of social interaction. For him, 'people are placed by power, but they experience it at first hand through the rhythms and relationships of particular places, not as some pre-packaged force from afar and not as a ubiquitous presence'. Power is therefore not an arbitrary construct; instead, it is always of a particular kind. He notes that it is vital to differentiate because acts such as 'domination, authority, seduction, manipulation, coercion and the like possess their own relational peculiarities' (Allen 2003: 2). This is a reasonably orthodox position to assume. Where his thesis differs is in his argument that if power has a presence at all it is through the interplay of forces in place (Allen 2003: 11). This position is presented as an antidote to the increasingly accepted view that power is all around us. Among some of the more abstract theorisations of power, there is a simple propensity to assume that power is uniformly and evenly distributed across space. Indeed, Allen (2003: 3, emphasis original) argues that 'we have lost the sense in which power is *inherently* spatial, and conversely spatiality is *imbued* with power'. Power may require proximity or function through reach. For instance, how are domination, authority and seduction exercised? Are there differences in their exercise when the parties are close by or far apart? And from a perspective of tourism, perhaps a more prescient question is one of how they come into contact with one another.

Again, these might seem like obvious questions, but as Allen (2003) argues there has been practically no attention among theorists to how geography impacts on the operation of power. As an initial effort to address this shortcoming, he identifies three genres of conceptualisation in which the spatiality of power is 'considered'. First, the writings of Weber – and others for whom power is an entity – are criticised because they take for granted how power is dispersed across space. Power is produced and reproduced across space largely unproblematically from place to place, and geography is viewed by and large as only a minor disruption to the distribution of power. The second genre is associated with Michael Mann and Manuel Castells and, once more, Allen decries this group for an equally unproblematic treatment of power over space. On this occasion, such accounts portray power as navigated through complex and multiple networks in society. In terms of their pointing to the intricate organisation and orchestration of power among actors, such views are highly relevant. They are, however, problematic in so far as they adopt a metaphor of power as a flow of electrical energy through a circuit without recourse to practicalities and limitations of flow throughout the system. Rather than a uniform or continuous transmission across tracts of space and time, as the second genre would imply, Allen (2003: 8) views power as always constituted in time–space. In this respect, he identifies the work of Foucault and Deleuze as emblematic of a third genre. Power is practised, not possessed and that practice is immanent. Simply put, power is present, not a backcloth; moreover, it is not imposed from above or externally but is seen as coextensive with its field of operation (Allen 2003: 9).

Hirst (2005) is more dismissive of the potential of Deleuze for understanding space and power arguing that the notion of de-territorialisation is misplaced. For Hirst (2005) the state, the city and the building remain three key spatial scales at which power coalesces. Flyvbjerg (1998) also stresses spatialised power at the city scale. In his study of Aalborg, Denmark he draws on Bourdieu and Foucault to illustrate how the Chamber of Commerce, the conservative local media and the local police determined the urban planning agenda. Allen (2003) adopts a rather different theoretical and empirical approach in his attempt to place power as he draws on a combination of the theoretical writings of Lefebvre (1991) and a series of case studies. Rather than trying to pursue Lefebvre's (1991) elusive representations of space, representational spaces and spatial practices, Allen (2003: 162) adopts a 'practiced view of power' in keeping with Lefebvre to examine 'how space is claimed exclusively and to focus on *what* exactly is exercised in the name of power'. One of the case studies is the tourism and retail spaces of the Potsdamer Platz in Berlin, including the Sony Plaza. Allen (2003: 182) claims here that space is dominated by a 'seductive presence' so that what seem like low key, intimate consumer invitations involve effective power from a distance for the Sony Corporation. The role of distance and presence is one of the key spatial paradoxes for understanding power in relation to tourism and mobilities. As Allen (2003: 183) notes,

> it is true that power has to have a presence to be effective, the nature of that presence and its effect will vary from mode to mode. Just as there is no everywhere to power, so there is no such thing as a universal blanket presence.

Clearly space matters for tourists and all those connected to tourism. A temporary co-presence will be central to the exercise of power in the spaces of tourism, but these spaces also involve complex networks of distanciation, especially as producers need to send out their seductive invitations not just into tourism spaces but into the home and human imaginations. Distance, 'absence' and co-presence are important issues in the consideration of tourism and power to which we return in the conclusion. By studying power in the context of tourism it is possible to both 'place' and 'mobilise' power, and hence to consider both the spaces where power coalesces and also the range of implications of increasing mobilities for the contrasting theorisations of power.

Structure of the book

In the preceding sections we have explored the various ways in which power has been theorised and investigated empirically, and we have introduced the ways in which power and tourism have been connected. Power is an essentially contested concept. No single or accepted hegemonic position exists among power theorists on what power is or what power ought to be. None of the theoretical or conceptual constructs or positions introduced thus far is unproblematic. There are more popular, even more fashionable, views of how to conceptualise power but this does not necessarily mean they are any the more effective than the more unfashionable in their ability to interpret the social or political dimensions of tourism. In this collection there has been no intellectual policing; no single, major 'paradigmatic' strand of thinking prevails here. Almost inevitably, not every theoretical perspective can be covered but a diversity of approaches to power is present as a means of commending the versatility and capacity of power theory for tourism analysis. Some of the authors champion or challenge particular theoretical positions of power with, for example, Foucauldian (Winter) and Lukesian (Coles and Scherle, Hall) readings, whereas others set out to establish how the potentialities of power discourse would add to our understanding of tourism and vice versa (Shaw, Timothy, Lew).

The collection comprises three main parts, with a synthesis of the main perspectives and prospects for the future development of power discourse in tourism in the concluding chapter. Part I (chapters 2–5) considers the relationships between power, performance and practice. It explores notions of the negotiation and experience of power, with a particular emphasis on the embodiment and performance of power in tourism. The individual tourist is the predominant unit of analysis. David Crouch (Chapter 2) surveys recent arguments concerning the 'practical ontologies' of tourism and emphasises how lay geographies are used by tourists to negotiate and make sense of tourism thereby producing complex and non-linear entanglements. Crouch uses these notions to explore the significance of the Mediterranean and America in relation to British tourism and culture. These ideas are also progressed in an empirical sense in Carl I. Cater's (Chapter 3) case study of embodiment in adventurous pursuits in Queenstown, the self-proclaimed 'adventure capital of the world'. There are overlaps between Cater's work and Gareth Shaw's (Chapter 4) investigation of holiday-making among people with a disability. While Shaw does not make such an extensive use of power theory, his contribution warns of the need to reflect carefully on the nature of practical ontology in his study of holiday-making among tourists who have a disability. Rooted in recent literatures on the dominant 'social model of disability' (Gabel and Peters 2004), Shaw argues that tourism for people with a disability is empowering not just in the sense that holidays represent the opportunity to overcome immobility. Rather, through tourism there is a release from the shackles of everyday routine for people with a disability, their family and their carers. The final contribution in this part is Caroline Winter's (Chapter 5) Foucauldian reading of the Ghan Train as tourist attraction and emblem of cultural politics of Australian nation-building. Winter seeks to reveal the connections between wider social processes and the signs and symbols of the Ghan Train in Australia by working with Foucault's notions of technologies. She argues that through the train, visual representations and the landscape, the contribution of Indigenous and Afghan people towards the Australian nation has been subjugated, and creates an outcome that favours the stories and efforts of white Australians. Collectively, these chapters emphasise the role individual tourists can play in shaping the power relations of tourism.

In Part II (chapters 6–8), the connections between power, property and resources are surveyed. These essays focus on political arrangements between groups in society and focus more squarely on the technologies – or as Morriss (2002) might put it, the vehicles of power – and the role of state institutions in mediating power relations. Alison M. Gill (Chapter 6) explores the politics of property development in the year-round resort of Whistler in Canada. Whistler has been a long-term research interest for Gill and her chapter presents a retrospective of how power relations have been mediated using urban regime theory (Stoker 1996). Although the nature of the regimes and the power relations among and between stakeholders have evolved over the years, one constant has been the discourse surrounding 'bed units'. Although originally nothing more than a planning tool to regulate the load of development in a sensitive environment, 'bed units' have assumed an almost mythical status in Whistler. They have become a currency in their own right, fought over and regulated in the contested process of development.

The idea of changing regimes is picked up in Alan A. Lew's contribution (Chapter 7). New urban regeneration initiatives embody the cultural settings and mediations of power in post-socialist settings. Morriss (2002) has pointed out that the majority of debate on power seems to be transacted in the English language. Whether or not this is a misleading assertion is an altogether different issue, but his observation points to the predominance of anglophone perspectives on power and hence raises questions of their ability to interpret power relations in other cultures (see also Coles and Scherle, Chapter 10 in this volume). Lew's chapter reminds us of the distinctive networks of *guanxi* operating in and among Chinese communities around the world (Lew and Wong 2004), how power is articulated through these networks of affiliation, and how the 'informal' (as Western commentators might consider) collides with the formal. His chapter considers notions of power in a command economy. Authority and domination are identified in power discourses which mainly pertain to Western liberal democracies. China is becoming one of the main source countries for international tourism while there is considerable and rapid development of infrastructure for overseas visitors and domestic tourists, not least related to the Beijing Olympics in 2008 (Zhang *et al.* 2005). Although the Chinese economy is becoming more liberal, urban regeneration is perceived as essentially a state-centred project (Wu 2003). The formal structures described by Lew are still quite rigid but his contribution highlights the emergence of new urban forms as a manifestation of the agencies of individuals and groups of entrepreneurs in an unfolding form of structuration (Giddens 1984); informal social and cultural systems of power resonate in and interact with formal political and economic structures.

Chapter 8 by Church and Ravenscroft continues the consideration of neo-liberalism and state institutions but focuses on property rights and the legal system. An examination of the conflicts in England over access to inland water for leisure between landowners, anglers and canoeists reveals how the socio-spatial process of resource mobilisation linked to legal rights interact with state institutions and the structural principles of a neo-liberal society to produce leisure and tourism outcomes that favour those with property rights. The chapter demonstrates, however, the dependent nature of power relations so that those with authority rely on the actions and discourses of those they seek to exclude from particular spaces to justify the maintenance of their property rights.

In the final part of the book (chapters 9–11), issues of power and governance are explored. Once again, the role of state institutions features prominently but here special reference is afforded to the idea of empowerment in two respects: in a more general, relative sense associated with restructuring of power relations over time (Timothy, Coles and Scherle); and second, it focuses on empowerment as a specific ideological aspiration associated with, and articulated through, tourism and in the best interests of the communities it is intended to benefit (Coles and Scherle, Hall). Not surprisingly, Lukesian thinking has great potential in tourism studies in this respect. Dallen J. Timothy (Chapter 9) surveys recent progress by the tourism academy and in wider interdisciplinary studies of tourism. His chapter demonstrates that empowerment has become a key conceptual and practical concern for social scientists from a large variety of disciplinary backgrounds, including members of the tourism academy. Multiple, contested perspectives abound in a highly fragmented and uncoordinated corpus of studies on tourism of relevance to tourism scholars. The great challenge is to draw together common strands from this plurality of perspectives. Timothy's review stresses that the time is right to introduce theoretically informed readings into studies of tourism and empowerment; to date, our understanding benefits from a large empirical base but lacks conceptual refinement. Where theoretical frameworks have been applied they have almost exclusively viewed power through a Weberian lens (Sofield 2003), whether explicitly or implied.

Tim Coles and Nicolai Scherle adopt a different tack to issues surrounding empowerment (Chapter 10). Their chapter looks at the struggles of Moroccan tour operators to achieve more equitable commercial outcomes in their dealings with their German counterparts. Tourism is viewed as a 'passport to development' and a means by which Morocco will achieve greater economic independence over time. Coles and Scherle use perspectives from organisational studies on the power tactics within organisations (based on the work of Lukes) combined with perspectives on intercultural communications to consider how power relations are constructed and played out in extended spatial commodity chains between Germany and Morocco. These theoretical frameworks and conceptual toolboxes reveal certain contradictions and tensions in the unfolding commercial relationships between German businesses and their Moroccan counterparts, especially where conflict is involved. Power is understood and used in quite instrumental means by both the perceived powerful and powerless to secure preferred outcomes. Crude calculations of the exercise of power may not be consistent with philosophical pronouncements (Morriss 2002) but importantly reveal that the powerless can act in subtle ways to enhance their power and that both parties can feel empowerment in commercial relationships. In Chapter 11, C. Michael Hall adopts an explicitly Lukesian approach to the location of power in tourism. As a final contribution here, Hall considers the functioning of power in and through a discussion of heritage tourism, Olympic bids and multiscaled governance. This highlights the way in which key organisations, such as the World Tourism Organisation (WTO) and the World Travel and Tourism Council (WTTC), use knowledge and influence to define the 'rules of the game' concerning tourism globally, while operating through Lukes's second and third dimensions of power to ensure certain issues receive little attention in the political discourses of tourism. Both dimensions serve to defend and further substantiate the positions of such supra-national organisations as 'global' leaders.

References

09/11 Commission (2004) *The 09/11 Commission Report. Final Report of the National Commission on Terrorist Attacks Upon the United States*. Authorized edition. New York: W.W. Norton.

Aitchison, C. (1999) 'Heritage and nationalism: gender and the performance of power', in D. Crouch (ed.) *Leisure/Tourism Geographies: Practices and Geographical Knowledge*. London: Routledge.

—— (2000) 'Young disabled people, leisure and everyday life: reviewing conventional definitions of leisure studies', *Annals of Leisure Research*, 3(1): 1–20.

—— (2001) 'Theorising Other discourses of tourism, gender and culture: can the subaltern speak (in tourism)?', *Tourist Studies*, 1(2): 133–147.

—— (2003) *Gender and Leisure. Social and Cultural Perspectives*. London: Routledge.

Alexander, M.J. (1998) 'Imperial desire/sexual utopias: white gay capital and transformational tourism', in E. Shohat (ed.) *Talking Visions*. Cambridge, MA: MIT Press.

Allen, J. (2003) *The Lost Geographies of Power*. Oxford: Blackwell.

Ap, J. (1992) 'Residents' perceptions on tourism impacts', *Annals of Tourism Research*, 19: 665–690.

Arendt, H. (1970) *On Violence*. London: Penguin.

Arino, O.B. (2002) 'Sustainable tourism and taxes: an insight into the Balearic Eco-tax Law', *European Environmental Law Review*, June: 169–174.

Ashley, C. (1998) 'Tourism, communities and national policy: Namibia's experience', *Development Policy Review*, 16: 323–352.

Bacharach, S.B. and Lawler, E.J. (1980) *Power and Politics in Organizations. The Social Psychology of Conflict, Coalitions and Bargaining*. San Francisco, CA: Jossey-Bass.

Bachrach, P. and Baratz, M.S. (1962) 'Two faces of Power', *American Political Science Review*, 56: 941–952.

Bachrach, P. and Baratz, M.S. (1963) 'Decisions and nondecisions: an analytical framework', *American Political Science Review*, 57: 641–651.

Bachrach, P. and Baratz, M.S. (1970) *Power and Poverty: Theory and Practice*. New York: Oxford University Press.

Baranowski, S. (2000) 'Strength through Joy: tourism and national integration in the Third Reich', in S. Baranowski and E. Furlough (eds) *Being Elsewhere. Tourism, Consumer Culture, and Identity in Modern Europe and North America*. Ann Arbor, MI: University of Michigan Press.

Baranowski, S. and Furlough, E. (2000) 'Introduction', in S. Baranowski and E. Furlough (eds) *Being Elsewhere. Tourism, Consumer Culture, and Identity in Modern Europe and North America*. Ann Arbor, MI: University of Michigan Press.

Basch, H. (2004) 'Who's in control? Insights from the front line', in School of Tourism and Leisure Management, University of Queensland (eds) Proceedings of the 14th International Research Conference of the Council for Australian University Tourism and Hospitality Education (CAUTHE) 'Creating Tourism Knowledge'. Brisbane: University of Queensland.

Bastakis, C., Buhalis, D. and Butler, R. (2003) 'The perception of small- and medium-sized tourism accommodation providers on the impacts of the tour operators' power in Eastern Mediterranean', *Tourism Management*, 25: 151–170.

Begg, A. (2000) *Empowering the Earth. Strategies for Social Change*. Dartington: Green Books.

Bell, S. (2002) 'Only an all-out war on turrism can bring to an end this war on turrism!' *The Guardian*, 29 November 2002 (#1853).

Benton, T. (1981) '"Objective" interests in the sociology of power', *Sociology*, 15: 161–184.

Bianchi, R. (2002) 'Towards a new political economy of global tourism', in R. Sharpley and D. Telfer (eds) *Tourism and Development. Concepts and Issues*. Clevedon: Channel View Books.

Binnie, J. (2004) *The Globalisation of Sexuality*. London: Sage.

Bjorkland, E.M. and Philbrick, A.K. (1975) 'Spatial configurations of mental processes', in M. Belanger and D.G. Janelle (eds) *Building Regions for the Future: Notes of Documents du Recherche No. 6*. Quebec: Department of Geography, University of Laval.

Bramwell, B. and Lane, B. (2000) 'Collaboration and partnerships in tourism planning', in B. Bramwell and B. Lane (eds) *Tourism Collaboration and Partnerships. Politics, Practice and Sustainability*. Clevedon: Channel View Books.

Bramwell, B. and Sharman, A. (1999) 'Collaboration in local tourism policymaking', *Annals of Tourism Research*, 26(2): 392–415.

British Broadcasting Corporation (BBC) (2002) Kenya terror strikes target Israelis. Online document. Available at: http://news.bbc.co.uk/2/hi/africa/2522207.stm [last accessed 29 July 2005].

Britton, S. (1991) 'Tourism, capital, and place: towards a critical geography of tourism', *Environment and Planning D: Society and Space*, 9: 451–478.

Brown, M. (1999) 'Travelling through the closet', in J. Duncan and D. Gregory (eds) *Writes of Passage: Reading Travel Writing*. London: Routledge.

Browne, K. (2007 forthcoming) 'A party with politics?: (Re)making LGBTQ Pride spaces in Dublin and Brighton', *Social and Cultural Geographies*.

Buchanan, D. and Badham, R. (1999) *Power, Politics and Organizational Change. Winning the Turf Game*. London: Sage.

Burns, P. (1999) 'Paradoxes in planning. Tourism elitism or brutalism?' *Annals of Tourism Research*, 26(2): 329–348.

—— (2004) 'Tourism planning. A Third Way?' *Annals of Tourism Research*, 31(1): 24–43.

Butler, R.W. (1980) 'The concept of a tourist area cycle of evolution: implications for the management of resources', *Canadian Geographer*, 24(1): 5–12.

—— (2006) *The Tourism Area Life Cycle. Volume 1: Applications and Modifications*. Clevedon: Channel View Books.

Cable News Network (CNN) (2002) Al Qaeda claims responsibility for Kenya attacks. Online document. Available at: http://archives.cnn.com/2002/WORLD/africa/12/02/kenya.probe/ [last accessed 29 July 2005].

Cantu, L. (2002) 'De Ambiente: queer tourism and the shifting boundaries of Mexican male sexualities', *GLQ: A Journal of Lesbian and Gay Studies*, 8: 139–166.

Cheong, S-M. and Miller, M.L. (2000) 'Power and tourism. A Foucauldian observation', *Annals of Tourism Research*, 27(2): 371–390.

Church, A. (2004) 'Local and regional tourism policy and power', in A.A. Lew, C.M. Hall and A.M. Williams (eds) *A Companion to Tourism*. Oxford: Blackwell.

Clancy, M. (1998) 'Commodity chains, services and development: theory and preliminary evidence from the tourism industry', *Review of International Political Economy*, 5(1): 122–148.

Clegg, S. (1989) *Frameworks of Power*. London: Sage.

Cochrane, E. (2005) The United States Conference of Mayors. International travel challenges/economic impact (Washington, DC, 22 April). Online document. Available at: http://www.usmayors.org/uscm/us_mayor_newspaper/documents/04_25_05/cochran.asp [last accessed 29 July 2005].

Cocks, C. (2000) 'The chamber of commerce's carnival: city festivals and urban tourism in the United States, 1890–1915', in S. Baranowski and E. Furlough (eds) *Being Elsewhere. Tourism, Consumer Culture, and Identity in Modern Europe and North America*. Ann Arbor, MI: University of Michigan Press.

Coleman, S. and Crang, M. (eds) (2002) *Tourism: Between Place and Performance*. Oxford: Berghahn.

Craik, J. (1997) 'The culture of tourism', in C. Rojek and J. Urry (eds) *Touring Cultures: Transformations of Travel Theory*. London: Routledge.

Crase, L. and Jackson, J. (2000) 'Assessing the effects of information asymmetry in tourism destinations', *Tourism Economics*, 6(4): 321–334.

Crouch, D. (1999) 'The intimacy and expansion of space', in D. Crouch (ed.) *Leisure/Tourism Geographies*. London: Routledge.

—— (2000) 'Places around us: embodied lay geographies in leisure and tourism', *Leisure Studies*, 19(1): 63–76.

—— (2004) 'Tourist practices and performances', in A.A. Lew, C.M. Hall and A.M. Williams (eds) *A Companion to Tourism*, Malden, MA: Blackwell.

Dahl, R.A. (1958) 'A critique of the Ruling Elite Model', *American Political Science Review*, 52: 463–69.

—— (1961a) 'Equality and power in American scoiety', in P.F. Drucker, D.C. Miller and R.A. Dahl (eds) *Power and Democracy in America*. Westport, CT: Greenwood Press.

—— (1961b) *Who Governs? Democracy and Power in an American City*. New Haven, CT: Yale University Press.

—— (1968) 'Power', in D.L. Sills (ed.) *International Encyclopedia of the Social Sciences*. New York: Free Press and Macmillan.

Dann, G.M.S. (1996) 'Tourist images of a destination: an alternative analysis', *Journal of Travel and Tourism Marketing*, 5(1/2): 41–55.

de Araujo, L. and Bramwell, B. (1999) 'Stakeholder assessment and collaborative tourism planning: the case of Brazil's Costa Dourada Project', *Journal of Sustainable Tourism*, 7(3/4): 356–378.

de Certeau, M. (1984) *The Practice of Everyday Life*. Berkeley, CA: University of California Press.

de Kadt, E. (1979) *Tourism: Passport to Development*. Oxford: Oxford University Press.

Dessewffy, T. (2002) 'Speculators and travellers: the political construction of the tourist in the Kadar regime', *Cultural Studies*, 16(1): 44–63.

Doorne, S. (1998) 'Power, participation and perception: an insider's perspective on the politics of the Wellington Waterfront redevelopment', *Current Issues in Tourism*, 1(2): 129–166.

Dowding, K. (1996) *Power*. Buckingham: Open University Press.

Doxey, G.V. (1976) 'When enough's enough: the natives are restless in Old Niagara', *Heritage Canada*, 2: 26–27.

Echtner, C.M (1999) 'The semiotic paradigm: implications of tourism research', *Tourism Management*, 20(1): 47–57.

Edensor, T. (2000) 'Staging tourism: tourists as performers', *Annals of Tourism Research*, 27(2): 322–344.

—— (2001) 'Performing tourism, staging tourism: (re)producing tourist space and practice', *Tourist Studies*, 1: 59–82.

Elliot, J. (1997) *Tourism. Politics and Public Sector Management*. London: Routledge.

Fallon, F. (2001) 'Conflict, power and tourism on Lombok', *Current Issues in Tourism*, 4(6): 481–502.

Flyvbjerg, B. (1998) *Rationality and Power: Democracy in Practice*. Chicago, IL: Chicago University Press.

Forstner, K. (2002) 'Community ventures and access to markets: the role of intermediaries in marketing rural tourism products', *Development Policy Review*, 22(5): 497–514.

Foucault, M. (1978) *The History of Sexuality. Volume 1. An Introduction* (trans. R. Hurley). New York: Pantheon.

—— (1982) 'The subject and power', in L.D. Hubert and P. Rabinow (eds) *Michel Foucault: Beyond Structuralism and Hermeneutics*. London: Harvester Wheatsheaf.

—— (2002) 'Truth and Judicial Forms. Lectures delivered at the Pontifical University of Rio de Janeiro May 1973', in J.D. Faubion (ed.) *Michel Foucault. Power. Essential Works of Foucault 1954–1984*. London: Penguin Books.

Franklin, A. (2003) *Tourism: An Introduction*. London: Sage.

—— (2004) 'Tourism as an ordering. Towards a new ontology of tourism', *Tourist Studies*, 4(3): 277–301.

Franklin, A. and Crang, M. (2001) 'The trouble with tourism and travel theory?' *Tourist Studies*, 1(1): 5–22.

Fyall, A. and Garrod, B. (2005) *Tourism Marketing. A Collaborative Approach*. Clevedon: Channel View Books.

Gabel, S. and Peters, S. (2004) 'Presage of a paradigm shift? Beyond the social model of disability toward resistance theories of disability', *Disability and Society*, 19(6): 585–611.

Galbraith, J.K. (1983) *The Anatomy of Power*. London: Corgi Books.

Giddens, A. (1968) '"Power" in the recent writings of Talcott Parsons', *Sociology*, 2: 257–272.

—— (1976) *New Rules of Sociological Method*. London: Hutchinson.

—— (1984) *The Constitution of Society*. Cambridge: Polity.

Gordon, C. (2002) 'Introduction', in J.D. Faubion (ed.) *Michel Foucault. Power. Essential Works of Foucault 1954–1984*. London: Penguin Books.

Hall, C.M. (1994) *Tourism and Politics. Policy, Power and Place*. Chichester: J. Wiley.

—— (2000) *Tourism Planning. Policies, Processes and Relationships*. Harlow: Prentice Hall.

—— (2003) 'Politics and place: an analysis of power in tourism communities', in S. Singh, D. Timothy and R. Dowling (eds) *Tourism in Destination Communities*. Wallingford: CAB International.

Hall, C.M. and Jenkins, J. (1995) *Tourism and Public Policy*. London: Routledge.

Hall, C.M., Timothy, D.J. and Duval, D.T. (2004) 'Security and tourism: towards a new understanding?', in C.M. Hall, D.J. Timothy and D.T. Duval (eds) *Safety and Security in Tourism. Relationships, Management and Marketing*. Binghamton, NY: Haworth Press.

Hannam, K. (2002) 'Tourism and development I: globalization and power', *Progress in Development Studies*, 2(3): 227–234.

Hartstock, N. (1990) 'Foucault on power: a theory for women?', in L. Nicholson (ed.) *Feminism/Postmodernism*. London: Routledge.

Haugaard, M. (1997) *The Constitution of Power*. Manchester: Manchester University Press.

—— (2002) *Power: A Reader*. Manchester: Manchester University Press.

—— (2003) 'Reflections on seven ways of creating power', *European Journal of Social Theory*, 6(1): 87–113.

Henderson, J. (2003) 'Ethnic heritage as a tourist attraction: the Peranakans of Singapore', *International Journal of Heritage Studies*, 9(1): 27–44.

Hetherington, K. (2000) 'Museums and the visually impaired: the spatial politics of access', *Sociological Review*, 48: 444–463.

Hindess, B. (1996) *Discourses of Power. From Hobbes to Foucault*. Oxford: Blackwell.

Hirst, P. (2005) *Space and Power: Politics, War and Architecture*. Cambridge: Polity Press.

Hollinshead, K. (1994) 'The unconscious realm of tourism', *Annals of Tourism*, 21: 387–391.

—— (1999) 'Surveillance of the worlds of tourism: Foucault and the eye-of-power', *Tourism Management*, 20: 7–23.

Hubbard, P. (2002) 'Sexing the Self: geographies of engagement and encounter', *Social and Cultural Geography*, 3: 365–382.

Hunter, F. (1953) *Community Power Structure: A Study of Decision Makers*. Chapel Hill, NC: University of North Carolina Press.

Hunter, W.C. (2001) 'Trust between culture: the tourist', *Current Issues in Tourism*, 4(1): 42–67.

Ioannides, D. (1998) 'Tour operators: the gatekeepers of tourism', in D. Ioannides and K. Debbage (eds) *The Economic Geography of the Tourist Industry: A Supply-side Analysis*. London: Routledge.

Jamal, T. and Getz, D. (1995) 'Collaboration theory and community tourism planning', *Annals of Tourism Research*, 22(1): 186–204.

Jamal, T. and Hollinshead, K. (2001) 'Tourism and the forbidden zone: the underserved power of qualitative injury', *Tourism Management*, 22(1): 63–82.

Johnston, L. (2005) *Queering Tourism: Paradoxical Performances at Gay Pride Parades*. Abingdon: Routledge.

Judd, D.R. and Simpson, D. (2003) 'Reconstructing the local state. The role of external constituencies in building urban tourism', *American Behavioral Scientist*, 46(8): 1056–1069.

Kayat, K. (2002) 'Power, social exchanges and tourism in Langkawi: rethinking resident perceptions', *International Journal of Tourism Research*, 4: 171–191.

Keitz, C. (1991) 'Reisen zwischen Kultur und Gegenkultur – "Baedeker" und die ersten Arbeitertouristen in der Weimrer Republik', in H. Spode (Hg.) *Zur Sonne, zur Freiheit! Beiträge zur Tourismusgeschichte. Berichte und Materialien Nummer 11*. Berlin: Verlag für universitäre Kommunikation.

Klemm, M.S. and Martín-Quirós, M.A. (1999) 'Changing the balance of power. Tour operators and tourism suppliers in the Spanish tourism industry', in L.C. Harrison and W. Husbands (eds) *Practising Responsible Tourism. International Case Studies in Tourism Planning, Policy and Development*. Chichester: J. Wiley.

Koshar, R. (2002) *German Travel Cultures*. Oxford: Berg.

Laslo, D. (2003) 'Policy communities and infrastructure of urban tourism', *American Behavioral Scientist*, 46(8): 1070–1083.

Latour, B. (1986) 'The powers of association', in J. Law (ed.) *Power, Action and Belief: A New Sociology of Knowledge*. London: Routledge.

—— (1999) 'On recalling ANT', in J. Law and J. Hassard (eds) *Actor Network Theory and After*. Oxford: Blackwell.

Law, J. (1994) *Organising Modernity*. London: Sage.

—— (2003) 'Notes on the theory of the actor network: ordering, strategy and heterogeneity', Centre for Science Studies, Lancaster University, Lancaster LA1 4YN. Available at: www.comp.lancs.ac.uk/sociology/soc054jl.html [last accessed November 2005].

Lefebvre, H. (1991) *The Production of Space*. Oxford: Blackwell.

Levine, D.N. (2000) 'On the critique of "Utilitarian" theories of action. Newly identified convergences among Simmel, Weber and Parsons', *Theory, Culture and Society*, 17(1): 63–78.

Lew, A.A. and Wong, A. 'Sojourners, guanxi and clan associations: social capital and overseas Chinese tourism to China', in T.E. Coles and D.J. Timothy (eds) *Tourism, Diaspora and Space*. London: Routledge.

Logan, J. and Molotch, H. (1987) *Urban Fortunes. The Political Economy of Space*. Berkeley, CA: University of California Press.

Long, P. (2000) 'Tourism development regimes in the inner city fringe: the case of Discover Islington, London', in B. Bramwell and B. Lane (eds) *Tourism Collaboration and Partnerships: Politics, Practice and Sustainability*. Clevedon: Channel View Books.

Lukes, S. (1974) *Power: A Radical View*. London: Macmillan.

—— (1977) *Essays in Social Theory*. London: Macmillan.

—— (2005) *Power: A Radical View*. Basingstoke: Palgrave Macmillan (second edition, original text with two major new chapters).

Madrigal, R. 'A tale of tourism in two cities', *Annals of Tourism Research*, 20: 336–353.

Mann, M. (1986) *The Sources of Social Power. Volume 1. A History of Power from the Beginning to AD1760*. New York: Cambridge University Press.

—— (1993) *The Sources of Social Power. Volume 2. The Rise of Classes and Nation-states, 1760–1914*. New York: Cambridge University Press.

Markwell, K. (1996) 'Towards a gay and lesbian leisure research agenda', *Australian Leisure*, 7: 42–44.

Miller, P. (2003) *Domination and Power*. London: Routledge & Kegan Paul.

Mills, C.W. (1956) *The Power Elite*. New York: Oxford University Press (second edition 1959).

Molotch, H. (1976) 'The city as growth machine', *American Journal of Sociology*, 82: 483–499.

Morgan, N. and Pritchard, A. (1998) *Tourism Promotion and Power. Creating Images, Creating Identities*. Chichester: J. Wiley.

—— (1999) *Power and Politics at the Seaside*. Exeter: Exeter University Press.

Morriss, P. (2002) *Power. A Philosophical Analysis*. Manchester: Manchester University Press.

Mosedale, J.T. (2005) 'Capital mobility in the tourism sector: an analysis of integrated corporations', unpublished paper presented at 'The End of Tourism? Mobility and Local–Global Connections', 23–24 June, Centre for Tourism Policy Studies, University of Brighton, Eastbourne, UK.

Moubayed, S. (2001) Amin Gemayel says his family's history 'runs parallel to Lebanon's', Washington Report on Middle East Affairs. October. Online document. Available at: http://www.wrmea.com/archives/october01/0110029.html [last accessed 14 August 2005].

Mowforth, M. and Munt, I. (1998) *Tourism and Sustainability. New Tourism in the Third World*. London: Routledge.

Murphy, P.E. and Murphy, A. (2004) *Strategic Management for Tourism Communities. Bridging the Gaps*. Clevedon: Channel View Books.

Newman, S. (2004) 'The place of power in political discourse', *International Political Science Review*, 25(2): 139–157.

Parsons, T. (1963) 'On the concept of political power', *Proceedings of the American Philosophical Society*, 107(3): 232–262.

—— (1967) *Sociological Theory and Modern Society*. New York: Free Press.

Phillimore, J. and Goodson, L. (2004) *Qualitative Research in Tourism. Ontologies, Epistemologies and Methodologies*. London: Routledge.

Poggi, G. (2001) *Forms of Power*. Oxford: Polity.

Pritchard, A. (2004) 'Gender and sexuality in tourism research', in A. Lew, C.M. Hall and A. Williams (eds) *A Companion to Tourism*. Oxford: Blackwell.

Pritchard, A., Morgan, N.J. and Sedgley, D. (2002) 'In search of lesbian space: the experience of Manchester's gay village', *Leisure Studies*, 21(2): 105–123.

Pritchard, A., Morgan, N.J., Sedgely, D., Khan, E. and Jenkins, A. (1998) 'Reaching out to the gay tourist: opportunities and threats in an emerging market segment', *Tourism Management*, 19: 273–282.

—— (2000) 'Sexuality and holiday choices: conversations with gay and lesbian tourists', *Leisure Studies*, 19: 267–283.

Prus, R. (1999) *Beyond the Power Mystique. Power as Intersubjective Accomplishment*. Albany, NY: State University of New York Press.

Puar, J.K. (2002a) 'A transnational feminist critique of queer tourism', *Antipode*, 34: 935–946.

—— (2002b) 'Circuits of queer mobility: tourism, travel and globalization', *GLQ: A Journal of Lesbian and Gay Studies*, 8: 101–137.

Reed, M.G. (1997) 'Power relations and community-based tourism planning', *Annals of Tourism Research*, 24(3): 566–591.

Richter, L. (1983) 'Tourism, politics and political science: a case of not so benign neglect', *Annals of Tourism Research*, 10(3): 313–335.

Rose, M. (2002) 'The seductions of resistance: power, politics and a performative style of systems', *Environment and Planning D: Society and Space*, 20: 383–400.

Ryan, C. (2002) 'Equity, management, power sharing and sustainability – issues of the "new tourism"', *Tourism Management*, 23(1): 17–26.

Schapp, A. (2000) 'Power and responsibility: should we spare the king's head?', *Politics*, 20(3): 129–135.

Scheyvens, R. (1999) 'Ecotourism and the empowerment of local communities', *Tourism Management*, 20: 245–249.

Schöllmann, A., Perkins, H.C. and Moore, K. (2001) 'Rhetoric, claims making and conflict in touristic place promotion: the case of central Christchurch, New Zealand', *Tourism Geographies*, 3: 300–325.

Scott, J. (1991) *Power*. Oxford: Polity.

Selby, M. (2004) *Understanding Urban Tourism. Image, Culture and Experience*. London: I.B. Tauris.

Selin, S. (1999) 'Developing a typology of sustainable tourism partnerships', *Journal of Sustainable Tourism*, 7(3/4): 260–273.

Selin, S. and Chavez, D. (1995) 'Developing an evolutionary tourism partnership model', *Annals of Tourism Research*, 22(4): 844–856.

Semmens, K. (2005) *Seeing Hitler's Germany. Tourism in the Third Reich*. Basingstoke: Palgrave Macmillan.

Shaw, G. and Williams, A.M. (2004) *Tourism and Tourism Spaces*. London: Sage.

Simmons, D.G. (1994) 'Community participation in tourism planning', *Tourism Management*, 15(2): 98–108.

Simpson, K. (2001) 'Strategic planning and community involvement as contributors to sustainable tourism development', *Current Issues in Tourism*, 4(1): 3–41.

Sofield, T.H.B. (2003) *Empowerment for Sustainable Tourism Development*. Oxford: Pergamon Press.

Stewart, A. (2001) *Theories of Power and Domination*. London: Sage.

Stoker, G. (1996) 'Regime theory and urban politics', in R.T. LeGales and F. Stout (eds) *The City Reader*. New York: Routledge.

Stone, C. (1989) *Regime Politics: Governing Atlanta 1946–1988*. Lawrence, KS: University Press of Kansas.

Strange, I. (1999) 'Urban sustainability, globalisation and the pursuit of the heritage aesthetic', *Planning Practice and Research*, 14(3): 301–311.

Swain, M.B. (1995) 'Gender in tourism', *Annals of Tourism Research*, 22(2): 274–266.

Teo, P. (2002) 'Striking a balance for sustainable tourism: implications of the discourse on globalisation', *Journal of Sustainable Tourism*, 10(6): 459–474.

Thomas, R. and Thomas, H. (2005) 'Understanding tourism policy-making in urban areas, with particular reference to small firms', *Tourism Geographies*, 7(2): 121–137.

Thurlow, C. and Jaworski, A. (2003) 'Communicating a global reach: inflight magazines as a globalizing genre in tourism', *Journal of Sociolinguistics*, 7(4): 579–606.

Timothy, D.J. (1997) *Tourism and Political Boundaries*. London: Routledge.

—— (1998) 'Co-operative tourism planning in a developing destination', *Journal of Sustainable Tourism*, 6(1): 52–68.

Trist, C. (1999) 'Recreating ocean space: recreational consumption and representation of the Caribbean marine environment', *Professional Geographer*, 51(3): 376–388.

Urry, J. (2002) *The Tourist Gaze*. London: Sage.

—— (2003) *Global Complexity*. Cambridge: Polity.

Veijola, S. and Jokinen, E. (1994) 'The body in tourism', *Theory, Culture and Society*, 11: 125–151.

Vogler, C. (1998) 'Money in the household: some underlying issues of power', *Sociological Review*, 46(4): 687–713.

Wearing, B. (1995) 'Leisure and resistance in an aging society', *Leisure Studies*, 14: 263–279.

Weber, M. (1978) *Economy and Society: An Outline of Interpretive Sociology*. Berkeley, CA: University of California Press.

Westwood, S. (2002) *Power and the Social*. London: Routledge.

Williams, D.R. (2002) 'Leisure identities, globalization, and the politics of place', *Journal of Leisure Research*, 34(4): 351–368.

Wu, F. (2003) 'The (post-) socialist entrepreneurial city as a state project: Shanghai's reglobalisation in question', *Urban Studies*, 40(9): 1673–1698.

Zhang, H.Q., Pine, R. and Lam, T. (2005) *Tourism and Hotel Development in China*. Binghamton, NY: Haworth Press and International Business Press.

Part I

Power, performance and practice

2 The power of the tourist encounter

David Crouch

Introduction: power and the non-representional

Recent contributions to the debates on tourism and leisure geographies have started to articulate discourses of so-called 'non-representational' ways of encountering space (Coleman and Crang 2002; Cant 2004; Kayser-Nielsen 2003; Larsen 2004; Pons 2003; Urry and Sheller 2004; Crouch 2001; Crouch and Desforges 2003). This chapter builds on these insights in order to try to progress the 'structure of power' debate considered through this volume. This particular power debate tends to have elucidated successfully the institutional frameworks through which tourism and leisure operate, and are operated (Oakes 1997, 2004; Meethan 2001, 2003; Britton 1991). Yet the debate has tended to open up the complexity of institutional participation with much room left to explore another layer of complexity interwoven across it. Rather than argue against a version of power situated in institutions (and for example a 'mediated culture'), this chapter calls for an engagement of a more institutional process of power and the potentialities for the participation in power processes by individuals. Here the possibilities of participation by individuals are considered from debates concerning the encounter. In the concluding commentary directions and concerns that such an engagement might involve are sketched.

'Non-representational' geographies have sought to reposition the individual, the human, the subject into the making of space and hence into the making of tourism. The individual emerges as 'doing' tourism, even as 'making' tourism (Coleman and Crang 2002; Crouch 2001). Crucially, this approach engages the individual in multiple relations with culture (Nash 2000) through tourism. Key components of this approach that are explored here include the reflexive and embodied character of the tourist encounter; that is, practice, or performance of tourism, is essential and it engages emerging debates of 'practice as practical ontology' whereby the individual may engage in potential interventions and interruptions in prevailing (and multiple) discourses of tourism and leisure in contemporary cultural life. In this respect, space is considered as a complex conduit through which the individual may participate in these processes of producing the world; that is,of making sense of it, of making tourism. The active participation of the individual in processes of tourism positions the individual as a key player in the exercise of power in tourism.

In one particular dimension, this is because the tourist makes sense of the world through doing tourism; the tourist does not use (nor need) it pre-packaged by tourism institutions (although these packagings may offer useful reference points along the way) in order to engage selectively in the sense of making encounters. In particular, this requires direct engagement with a conceptualisation of power in terms of the construction, production and circulation of meaning. Thus, making sense of the world and circulating its meanings is a significant intervention in the processes of power; it influences the ways in which individuals may choose to act and what they do.

The capacity of individuals to act themselves has been a feature of diverse literatures: from Bakhtin to Foucault, Gramsci to de Certeau, and in recent cultural geographies of resistance (cf. Crang and Thrift 2000; de Certeau 1984; Pile and Thrift 1995). However, as Thrift (1997) has argued, resistance is not the only way through which the individual may be considered to encounter the world and make sense of it. Individuals make their own contributions to what the world means to them. Taking Shotter's (1993) notion of everyday life encounters with the world, through such actions the individual constructs his or her world; the world is given meaning. Thus, doing tourism may reasonably be considered as one clear example of such action (Crouch 1999).

Individuals as tourists may, then, have power over what the world means, how actions are given meaning, what is done and so on. The individual emerges as less conditioned and determined by surrounding contexts, but engaged in taking those contexts and acting, thinking and feeling in relation to them. Thus, the world may be less prefigured for the tourist than is often presumed (Crouch 2004a). Power over the giving of meaning – or Shotter's 'ontological knowledge' – may not be trivial. Instead, these dimensions of being a tourist contest familiar presumptions of power in tourism; that is, they complicate – rather than argue against – the power of institutions. Thus, this component of power is not one of individuals working in response to cultural influences and significations produced beyond their own lives, but one of production and circulation among wider contexts and their own life experiences.

The next section outlines key aspects of recent discussions concerning space in contemporary cultural frameworks and change. The dominance of theories privileging the power of the context outwith individuals' lives is contested through a consideration of arguments concerning everyday practices. It considers what Ingold (2000) has termed 'dwelling' and it takes an important cue from Miller (1997) on consumption. Following this, the relevance of debates on the body as involved in life encounters is highlighted. The world emerges as accessed and understood in more complex ways. Such a reconceptualisation of the encounter and space stimulates a profound rethinking of not only what the tourist does, but of how the tourist may be engaged in processes of power in what tourism and its spaces may signify. To reiterate, in no way does this rethinking negate the power of institutions and more traditional readings of tourism power, for instance, on marketing and commodifications, or of the erosive power of global investment (Meethan 2001). Rather, such rethinking enables us to re-engage the often disenfranchised individual (tourist) in processes of making tourism; that is, of power over what tourism means, and how it may refigure the world, as well as contemporary lives.

Ideas of the power of culture and the making of space

This chapter seeks to offer a discussion of the individual in tourism and leisure as involved in a negotiative process. The individual's life is in tension between holding on to identity, ontological security, and a grasp of 'position in the world'; it is simultaneously in a process of 'going further' whereby new, revised and exploratory identities may be forged, even if temporarily. The cultural (and societal) significance of this negotiation, and its consequences in terms of processes of power, are considered in relation to prevailing contexts in which tourism and leisure happen. As developed here, the individual mediates his or her life in complexities of action, practice and performance. The individual negotiates the diverse components of his or her life. This may include negotiation between individuals, and between the self and broader, often mediated, cultural contexts and influences. Through numerous and unevenly related practices and performances, the individual mediates what his or her life means and how it makes sense. What happens in terms of, for instance, going further and holding on is not a single polarity, but a case of multiple, asymmetrically interrelated components. Tourism and leisure offer examples of this complexity of holding on and of going further: they exemplify the distinctive ways in which life is made sense, meaning is produced and circulated, and the individual is engaged in processes of power. The ways in which the individual engages in producing meaning, and value, through what is done in temporary moments of being a tourist, work not only in actions and their significance during those moments, but in relation to other experiences of being a tourist, and other components, practices and performances surrounding what is done in what might be called 'home life' leisure. Tourism and leisure deliver an array of components through which life is made sense, negotiated, progressed (or not). Again, the making and circulation of meaning is an active process in which individuals are involved, through which they can participate in an empowering process in relation to other contexts of contemporary power.

Tourism and leisure happen in space, or rather in relation to space. Space is not to be fetishised but is an inescapable means through which encounters in the world occur. Space is a key component of how meaning and value are produced. Space can be important in power processes in ways that acknowledge both contexts of investment, institutions, the media and the like, as well as in the actions and making sense that individuals do themselves. Of course, these components of process are not isolated. Nor do they work in linear relation. Tourism and leisure are no exceptions. Space is metaphorical and material; metaphor is constructed and constituted by individuals and other components of cultural life. Thus through metaphor, space is made powerful as metaphor itself, and much geographical, cultural and media as well as lay discourse is maintained (Lash and Urry 1994).

Several threads in recent debate that have significant influence on our understanding of space have argued for the increasing detachment of meaning and value production from the individual or that the individual's sense of the world is increasingly abstracted in wider flows and affects beyond their control, over which they have no power. For example, notions of time–space compression have suggested that increasing technologised speed and virtual access alongside processes of globalised complexity accelerate the circulation and distribution of meaning in contemporary culture, and 'real distance' diminishes the significance of particular spaces (Harvey 1990). Cultural de-differentiation is presented as providing a weakening of the identification that individuals hold in relation to spaces and their lived experience, rendering spaces alienating from human life and emptied of meanings previously held. Human action becomes marginalised in the production of meaning. New spaces are produced that are thus effectively 'non-spaces' as they have not been constituted through human activity and its lent cultural meaning and value (Auge 1995).

The significance of this space-creation process is, arguably, experienced more widely in the outsider-ness of human life in its playful, post-identity activities of gazing rather than engagement (Urry 2002). It is in mobilities signified in the fleeting and implicitly unstable that thereby determine the further superficiality of contemporary life, requiring a ready-made reference point from elsewhere outside the individual's life practices (Urry 2002). Thus the world is engaged and consumed as meaningful through processes of detached consumption where the producers of objects of consumption and their advertising increasingly produce meaning, reference, context and thereby operate as the key players in dynamic cultural economies. Similar complexity and time–space compression as well as their effects on investment and circulation affect the production sphere (Lash and Urry 1994).

Baudrillard (1981: 85) argued for the importance of 'strategies of desire' through which consumers' needs are mobilised, and their nascent interests are captured in a process of consumption before consumption. These strategies, he argued, consist of the signs on which the value of products are conveyed in the process of seduction. Baudrillard's strategies *for* consumption are crucial points in his version of cultural economies. The effective and affective power of signs are displayed and systematised through their display and communication. The power shifts from the objects themselves – and the subjects of their consumption – to their circulation in representations, their fuller consumption dominated by their sign-value, their value invested in anticipation. Cultural industries, the media and so on reframe life's meanings and produce new cultural economies which become constructed and constituted whereby individuals merely 'sign up' (*sic*) to abstracted bundles of meaning.

A position emerges of a world constructed and constituted through representations, produced in the contexts of products and media, and combined with contemporary social and cultural detachment. Individual human experience is at the margins of the contemporary world and its processes. Through various strands of these debates in uneven combination, there emerges a prevailing discourse on space, place and geography. A particular English version of landscape has conceptualised landscape as text and, like the notion of tourism *destination* held in representations produced elsewhere from the lives of individuals who may use or visit, it is characterised by perspective and surveillance by 'others' yet where the individual may also be spectator and interpreter appropriated in the gaze (Ringer 1999; Macnaghten and Urry 1998). Space is arguably produced and thereby its meaning constructed through media and product design, identification and promotion (Crouch and Lubbren 2003). More recent discussion on mobilities, including specifically tourism, have compounded these various processes (time–space compression, complexity, globalisation, the gaze of/on space, and space's increasing power as represented through mediated culture) into an idea of multiple and complex long-time/distance mobilities that increasingly make spaces significant, and implicitly (threaten to) become overwhelming – over*power*ing – of other grounds and practices through which individuals may have some power of making meaning in the spaces of their lives (Urry 2003). These processes, mobility and so on, hold the key power in the way life is given meaning and spaces are made sense.

Space that bears significance is presented as largely outside individual influence, power and processes. The potential for virtual 'contact' releases the need to position individuals in actual material spaces. In general terms, space becomes something that may offer objects of play as detachment and without responsibility. Identities are 'bought into'; they are constituted neither in continuities of social distinction nor through individuals' lives and life practices. Similarly, the body in space emerges as an objectified component of this play, object of interest, curiosity, desire (Featherstone and Turner 1995). The emerging emphasis on contemporary cultural economies tends to work with a prioritised and privileged understanding of vision as detachment, the power of representations, and the alienated power and the marginal role of individuals exist practically as support cast.

However, there are other complexities emerging from different discourses. An emphasis on the institutional in tourism and leisure has a long history: from a commercially driven presumed framework of how tourism works to a worthy effort to handle such discourses critically (Meethan 2001; Britton 1991). In each of these there is a focus on the individual, also as members of social categories, acting in relation to structure. That structure is the framework through which this interaction occurs; the interaction is implicitly presented as linear, consumers working in relation to frameworks. While this discourse contains the potential for resistance, it does not tend to acknowledge a great complexity. For this we may turn to Miller's (1998) creative work on material culture.

In the question 'why some things matter?' Miller identifies a significantly emerging trend from the ways in which individuals work meaning into material things, products, artefacts, in ways that function neither to prescribe meaning or obviously resistant ones. Elsewhere Miller has responded to overemphasis on linear power from production to consumption in his work both on shopping in the UK (1998) and in buying Coco-Cola in Trinidad (1997). The overarching power of global companies, and of globalisation in the enforcement of meaning, is unsettled, if not dispelled. In tourism it is familiar to understand a world in which globalisation and the economic exercise of financial power and branding change lives and values. Our individual tourist is not heroic in subverting acknowledged power thus exercised, nor necessarily intentionally resistant (Thrift 1997). Yet it may be that in working inter-subjectively, individuals negotiate their lives and intervening cultural power in more complex and uncertain ways; that is, less in ways that intentionally resist than otherwise also work in other cultural processes.

In a reference to elements recognisably touristic, of adverts and films, spectacles and events, Taussig (1992: 141) cautions us over the dominant modes of constructing and constituting that are familiar in tourism discourse:

> The reading of ideology into events and artifacts, cockfights and carnivals, advertisements and films, private and public space, in which the surface phenomenon . . . stands as a cipher for uncovering horizon after horizon of otherwise obscure systems of meanings . . . [S]uch a mode of analysis is simple minded in its search for 'codes' and manipulative because it superimposes meaning on the 'natives'.

It is often in these modes of surfing tourism evidence in brochures, in what agencies do and how individuals 'fall in line' in their acts of purchase, that tourism has been analysed and understood (Williams 2004). In the forms of cultural praxis briefly outlined, new meanings emerge. The ways in which this works are identified through an acknowledgement of practical ontology and our working out what matters; of identifying value through work done through, in part, an encounter with the spaces in which action happens. Action is placid and active; it is acutely transacted and observed in relaxation. In identifying the latter, we seek to move well beyond a Foucauldian notion of the gaze now archetypal in much tourism thinking (Urry 2002). While shifting tourism thinking towards the individual, the gaze implies too much detachment, dis-involvement. This is not withstanding it makes powerful insight into the active pace of the 'consumer', tourist or individual, as reflexive, multiple and diverse, perhaps less as intentional thinker than as 'dwelling' (Lash and Urry 1994; Edensor 1998).

Ingold (2000) progresses an interpretation of the individual in the world as engaged by working things from experience as well as circumstances, and in relation to contexts. As individuals work their lives through, they may work out some components for themselves. Shotter (1993) called this a practical ontology of active and physical engagement in the world. The individual works not merely in an abstract mental process but also in physical action. Physical action informs feeling and thinking; physical action expresses attitude and feeling.

In the next sections the fragments of practical ontology are sketched through processes of making sense of the world, figuring and refiguring what is there, what is imagined, what is prefigured and proscribed in material objects and available cultural metaphors through doing, through encounters with the world, in spaces, and as a tourist. Burkitt (1999) discussed the taking up of these components of practical, embodied action as informing not only the meaning of things, but the meaning of our place in the world, in negotiating our own identity. A way of thinking through this process of negotiation is explored in terms of the individual in action, as a 'tourist', working out how they find themselves in the world and in relation to the world, as identities, in particular through the ideas, developed from discourses on performance, of 'holding on' and 'going further' (Crouch 2003). Working our lives out as a tourist – and from having been a tourist – we can make new knowledge, refigure what is there, and the emerging knowledge can contribute to popular discourses of everyday life, through which in turn ideas, feelings and events in the world may be renegotiated.

The body debate

Merleau-Ponty's (1962; Radley 1995; Crossley 1995, 1996) interpretations of the relation between space and the individual offers a way of rethinking how individuals engage and encounter their surroundings in terms of the embodied character of what they do. Crucially these recent developments of Merleau-Ponty's work, aligned critically with current work on performance, have developed explanations of the individual engaged in the world iteratively, with the world inscribed through cultural mediation (Crang and Thrift 2000; Nash 2000). The notion of embodied practice as expressive provides a useful orientation for thinking through relations between touch, gesture, haptic vision and other sensualities and their mobilisation in *feelings of doing* (Harre 1993). The word 'doing' distinguishes what people may do (without particular practical outcome) from that oriented around tasking. By expressively encountering, the individual engages in body-practice as an everyday activity of living (Radley 1995). Burkitt (1999) has drawn together ideas of body-practice with social constructionist notions of identity, and in particular Shotter's notion of 'ontological knowledge', to argue that embodied practice may be important in working identity. Similarly, Ingold (1995) identifies a process of *dwelling* whereby encounters with objects, individuals, space and the self direct the progression of life, including the way it gains meaning (cf. Harrison 2000).

These ideas can be mobilised towards thinking of space in terms of negotiating life in the act of spacing. In recent debates on performance and its performativities, the possible constituents of spacing have offered a further step. In particular, Deleuze and Guattari's (1988; Grosz 1999) attention to the uncertainties of flows and momentary character of performativity elaborate the uncertainties and complexities of spacing (and its potential as part of performativity) to reconstitute life. Take, for instance, cultural mediation which provides frameworks through which the individual in relation to the world is inscribed (Taussig 1992): in the contemporary period this occurs through official and institutional protocols. Performativity is profoundly bodily (Dewesbury 2000) and the body is expressive (Radley 1995, Tulloch 2000). Performance is being theorised in ways that provide ground for the active working of the individual in her/his life. The individual's performance is proscribed by protocol, in ritualised practice, working to pre-given codes, habitually repeated, and conservative (Butler 1997; Carlsen 1996). Yet inherent in performance is the possibility that relations with contexts may be reconfigured, broken, adjusted or negotiated (Lloyd 1996; Thrift and Dewesbury 2000).

While performance can emphasise the framework of everyday protocols, the performative errs to the potential of openness. The reconstitutive potential of performance is increasingly modulating life and discovering the new, the unexpected, in ways that may reconfigure the self in a process of 'what life (duration, memory, consciousness) brings to the world: the new, the movement of actualisation of the virtual, expansiveness, opening up' (Grosz 1999: 25), enabling the unexpected. Like practice, performance is a 'cultural act, critical perspective, a political intervention' (Roach 1995: 46), and so may contain the transformative. Radley (1995) argues that individuals create a potential space in which they can evolve imaginary powers of feeling. Performativity articulates practices (Schieffelin 1998: 195, quoted in Tulloch 2000: 4) and it enables closer interrogation of practice. Performativity opens up the individual as engaged in a process of 'becoming' in ways that are only incompletely proscribed. These ideas share character with Ingold's (2000) notion of doing in terms of 'dwelling'. He distinguishes between ideas for things, space and so on, as prefigured and determinate, while the motor of 'dwelling' that sustains the present and future from which contemplation and new possibilities of reconfiguring the world in flows can occur.

Both performative and embodied practice are characterised in doing. Each is articulated for the individual in (multiple) terms of doing as constituting, their own significations, as material or embodied semiotics, and each may respond to other representations of the world (Game 1991; Crouch 2001). Performance as performativity is taken further as ongoing and multiple interrelations of things, space and time in a process of becoming, in engaging the new that may be, like Radley's consideration of embodied practice, unexpected and unconsidered, not only prefigured, but also suggestive of a similar performative shift beyond mundane, routine habituality. 'Going further' may emerge from exactly those apparently momentary things (Dewesbury 2000). Moreover, the borders between 'being' as a state reached and becoming are indistinct and constantly in flow (Grosz 1999), although they may be focused in *the event* (Dewesbury 2000: 487–489). In the present discussion, becoming is distinguished from being in the sense of Grosz's *becoming* as 'unexpected' where performativities may open up new, reconstitutive possibilities. It is in the notion of multiple routes of 'becoming' that the discourse on performativity is particularly powerful.

To pursue a performance approach to tourism and to tourists in terms of the tourist experience presents considerable potential for our making sense of what happens in tourism, including the significance of tourism in contemporary culture and how places and experience are communicated between and amongst people. There are further possible insights in terms of the role of the tourism industry, the knowledge and value of touristic environments and cultural heritage, and tourism's role in the making and negotiation of contemporary identities.

Tourists' lay geographies and tourism's practical ontologies

Recent contributions have acknowledged insights from philosophical discussions concerning the body (Crang 1999; Coleman and Crang 2002; Crouch and Desforges 2003; Crouch *et al.* 2001). In this section the particular focus is on the gathering acknowledgement of the importance of the individual encountering the world bodily, through which distinctive, complex and fluid practical ontologies may negotiate both individual life and relations of power in the ways in which the world, tourism destinations and experiences are produced in meaning and circulated.

The self and the world are engaged bodily through doing; that is through touch, body-tactics and other sensuous and inter-subjective encounters. Space is constructed and constituted bodily through doing. Performance enacts a multiple lay geography through which the world, the self and others are figured in the feeling of doing (Crouch 2001). Prefigured meanings and significations are reflectively worked into this constitution. Emerging practical geography is a patina of performance. Destinations are given meaning through performance as well as prior significations. In her autobiographical accounts of performing Bondi beach and the British Pennines, Game (1991) discusses her bodily encounters as reaching things that cannot be seen and making material semiotics 'in the doing'. Rather than its ostensible features pre-constructing the contours of experience, body-practice constructs and constitutes space. The individual tourist enacts space-ing, in an ongoing and fluid time–space encounter. Shotter's (1993) discussion of practical ontology focuses upon the ways in which people construct and constitute the world through doing. For him, this is essentially (inter-subjectively) through practical work. Burkitt (1999) has developed this in terms of bodies of thought, with individuals acting bodily. In a comparative way Ingold (1995: 76) distinguishes two perspectives, of 'building' and of 'dwelling':

> People do not simply import their ideas, plans or mutual representations of the world, since their very world, to borrow a phrase from Merleau-Ponty, is the homeland of their thoughts. Only because they dwell therein can they think the thoughts they do.

That process of 'making sense' is arguably more fluid, vague, unordered and non-linear than familiar notions of ontology. It may also be more complex in terms of time and the ways in which influences are encountered, engaged and negotiated (Crang 2001). This complexity through doing may be extended to a consideration of identities and identity-formation. Burkitt (1999) argues for the significance of embodied practice, in terms of the ways in which people make sense of their identities, of who they are and how they relate to others. To adopt such a view utilises fluid notions of identities constantly negotiated in relation to other contexts (Oakes 1997), constantly fluid in and through individuals' practices, perpetually and reflexively refigured, multiple and complex, rather than fixed. For Burkitt (1999), identities are contingently and fluidly constituted through and by embodied engagement in the world. Performance in particular activities and moments may be used to modulate or intensify this instability of identities, subject to the fleeting and irrational character of performance (Dewesbury 2000).

Alongside and amongst numerous other life-practices tourism may be used to escape, to confront the unexpected, to change and to disrupt. It may be used to hold on, and to steady life and its identities (see Crang 1997 for the example of how tourists use photographs). Moments of being a tourist are not contained, but rather they flow over, amongst, in and out of other parts of individuals' lives. In a similar vein, the increasing complexities of local performance and performative flows around the world complicate the polarisation of hosts and guests as they do of being a tourist and not-tourist (Clifford 1997; Crouch *et al.* 2001; Larsen 2004). The phenomenon of being a tourist is therefore relational. However, it serves to conceptualise the ongoing flow of events and significations with particular moments, nodes, knots of collision along the flows where the particular character of performance becomes felt significant (Harrison 2000). The next section briefly explores these possible knots through a reflection on space.

Identity, and the power of being a tourist

'Performance as habitual' and 'performance as becoming' are powerful informative ideas for considering tourism, what the tourist does, and what the tourist makes sense of in that doing. The polarisation of tourism from everyday life implies that everyday life has no room for the unexpected, and that being a tourist cannot be mundane. Tourism is familiarly analysed as a distinct life-zone where the power of the tourism industry's global reach is predominant. In this, perhaps theory-making is too easily informed by the literature of the tourist industry. Increasingly, a conceptualising of life in separate compartments in and through which the world is encountered, valued and understood may be critiqued by an analysis that envisages multiple and circuitous flows through which the individual encounters the world. As performative encounters are significant in constituting the world, the self and each other, the familiar compartmentalisation of tourism misses its complexity. For example, Crang (1997) identifies the importance of holiday snaps in developing an ongoing and non-linear flow of personal and shared heritage into which moments of encounter are injected.

Tourism is often imagined in the production of images and brands as an appeal to a tabula rasa, the notional innocent individual, as 'fresh territory', awaiting signs to seduce (Crouch 2004a, 2004b). Yet by far the majority of individuals going to do tourism have been before; their experience as a tourist is engaged in their life course. In a more subtle process, the particular instance or action of being there and doing is in itself new, or not fully framed for them. As things are done, other 'events' are remembered and re-placed into the present. Memory is also temporalised and can reinvigorate what one is doing now; however, it is also reinvigorated and may be re-routed in the 'now' (Crang 2001). Spacing in memory is engaged afresh. Memories colour performativity in the now, but neither in a rerun of the past nor as a linear consequence of something prepared earlier by someone else.

Performing time/spacing appears to be more than a linear 'moving on' from ideas, and memory is operated as an active character of performativity. As Bachelard (1994: 57) argues, we have 'only retained the memory of events that have created us at the decisive instants of our pasts'. These are drawn into a focus through the character of performativity in nodes and knots of what they notice. When individuals speak of what and how they 'do', they compile events reduced to an instant where things matter. Although we 'may retain no trace of the temporal dynamic of the flow of time' (Bachelard 1994: 57), moments of performance (can be revealed) when, and through which, things are remembered as significant. In the doing, moments of memory are recalled, reactivated in what is done, and thus, drawn upon to signify. It is less that memory is performed than it is 'in performance'. Thus, memory and the immediate are performed as complex variants of time. Informants do not simply remember by picking the memory up momentarily, rather they return to it through performance repeatedly and re-form it. Time, too, is performed repeatedly, differently and embodied there.

The several strands of happening are negotiated in the performance 'made sense' in complex time as body-thinking beings (Burkitt 1999). This process of negotiating leaves room for the individual to 'take it' where s/he likes. Harvey (1996) typologised human action in an effort to identify where power lies. He distinguished between 'routine' and transformative action. I want to suggest that in the apparently routine there can be the transformative. Domosh (1998) observes that routine practices can generate their own micro-politics and Thrift (2003) has more recently identified that politics is not only 'Olympian', or even too evidently negotiative with structure. Tourist performances can be significantly in excess of prefigured meanings, brand significance and mediated cultural frameworks. In what they do, tourists make a sketch (Radley 1995). It is this capacity of performativity to be productive that characterises becoming. It produces feeling. It changes the intensity and pitch of how things are signified and what things mean. Yet that becoming is not isolated from other parts, or contexts, of life. It is negotiated. As Crossley (1995, 1996) argued with reference to 'culture' and embodied practice, people do not act independently of culture, rather culture is felt bodily (Csordas 1990).

The complex flows of workings of frameworks and action may be briefly considered in terms of the available cultural contexts of the Mediterranean and of America in England, in an exploration of the discourse argued (Crouch and Selwyn 2003; Campbell 2004). It is familiar to assert the power of distinctive commercialised identities in the mediation of America through holidays and the designs of destinations, and the signification of their globally available products. Similarly, the Mediterranean is a bundle of significances in contemporary Britain. These significances too have different temporalities. Crucially, each of these two examples may be regarded as powerful components of contemporary global economies, more 'obviously' mediated cultural economies and the spaces that may convey them. The following explores reflections on the complexities of place-construction and representation, and the value and meaning given to practised, performed spaces. Two kinds of geography are considered, briefly, here. One is the cultural economy through which the Mediterranean is constructed and constituted, and thus prospectively consumed. The other is the cultural economy through which 'American landscapes' are produced for consumption in Britain (Selwyn 1996; Campbell 2004).

The Mediterranean has been significant in diverse ways in British culture for over a century (Inglis 2000), not least in terms of its artistic and commodified representations. The Mediterranean may be consumed through the prefigured circulation of iconographies of culture and of religion, each of which has deep, long-term histories. More recently, especially in the artists' depictions of the early decades of the twentieth century, these often contested histories have been combined with new layers of the exotic, sexual and sensual temptation, and of risk borne on depictions of harems and their bodily display (Lubbren 2003). Yet, since the mid-twentieth century newer versions of heat, risk, the body and the exotic have been inserted in more popular discourses and appeals to diverse kinds of tourists. The purpose has been to suggest they engage the recent delights of their bodies on beaches and elsewhere, very recently emerging in versions of clubbing life.

However, destinations of tourists such as the Mediterranean may be given meaning through performance as well as through their prior significations. For example, as Game (1991) shows, the projected imagery does not hold intact, but the individual makes the trip, encounters and makes sense in a new compilation. Through reflection on the imagery that promotes Bondi beach, she identifies in her own experience something very different. For her it is the intimate engagement with the features of the beach, the elements, a book to read, her own memories of doing similar things on this beach years ago, that constitute her encounter. Rather than its ostensible features pre-constructing the contours of experience, body-practice can be significant in the way space is constructed and constituted through bodily, imaginative encounters that may also make playful use of memory and time (Crang 2001).

In a different way but still suggestive of this interplay, Andrews's (2004) investigation of charter tourists in Magaluf and Portonova (Majorca) unpicks individual and group efforts to reconstruct an idea of Britishness – racist, sexist, militarily patriotic, alienating. However, while the destination she examines is known for components of Britishness, English breakfasts and enclaves of British, especially English, ex-patriots and the like, it was significantly in the doing through embodied encounters in clubs, on the beach, in bars and the like that the idea of Britishness is mobilised. Mobilisation happens in body-actions, both as display, but also in the physicality of often very expressive, heavily inter-subjective and 'social' encounters. The individual tourist is reinterpreted as performing, and performance produces an embodied semiotics.This would suggest that practical ontology can refigure prefigured cultural economies. Moreover, the values and meanings through which relations are constituted through visits such as these may be at variance with what are often considered to be prevailing meanings and their significations.

In a similar way to the commercialised cultural projections of the Mediterranean, this kind of explanation works with respect to the mediation of America. It, too, functions in a linear fashion to produce an experience of consumption, experience and place that is projected through global commodification and politically constructed meanings and values. The American Dream, western ideals, freedom and superiority, global commercial and lifestyle power are arguably signified in landscapes that are intended to contain these representations in a simulacrum of power, product and positioning (Campbell 2004). Yet how do these supposed projections 'work' in the experience, in the embodied encounter that individuals may make in the spaces that are enveloped (in design terms) with American stylisation? Campbell speculates on the limits of how we understand the cultural circulations of America thinking through the encounter the individual makes. Dewesbury (2003) interrogates a diversity of meanings and identities engaged in a visit to London Bridge, Arizona. It is difficult to be secure in an argument that presumes linear follow-through of the construction of such a cultural economy as 'America'.

Projected identities of the Mediterranean and of America only go so far in making offerings of cultural identity and identification; the proscribed and prefigured may only be incompletely grasped through a discourse that regards the consumer, the tourist, the individual as 'in receipt of', rather than as part of, power in the processes of contemporary cultural economy components. Power operates in terms of the influences involved in identity-processes, and embedded in these processes are the important components of making knowledge as discourse and the circulation of that knowledge. Embedded in their processes are the important components of how knowledge is made as discourse and its circulation in and through what individuals do, feel and think. A process of lay, or everyday, geographical knowledge happens that may be able to work its own signification, to work in ways different from a projected cultural economy 'from without' and, as Crang (1997) argues, which resembles the making of lay iconography (Crouch and Parker 2003). In varied ways individuals can contest and negotiate the tensions of the unexpected and the habitual (Grosz 1999; Carlsen 1996). There is a pervasive tension – and not a singular polarity – in the narratives between 'holding on' and 'going further' that many respondents appear to negotiate, and these can be played out in their encounters with space. Their apparent 'holding on' suggests a safeguarding of parts of their lives which they feel provide identity, control, and being able to engage repetitively on the ground they know, while perhaps at the same time, perhaps for others, this performance emerges as something new.

Through its ability to figure and refigure, to make distinctive sense of what tourism and its spaces signify, 'doing tourism' becomes affective in both processes of identity and in the exercise of power over what tourism signifies, its values and meanings in contemporary life. Burkitt (1999) argues its importance in the refiguring of negotiations of the complexities and fluidities of contemporary identities and values, and argues that 'the bodily' is significant in the ways that identities are constituted. The reconstitutive suggests potential refiguring of identities. Identities may be characterised in practice and performativity negotiatively with contexts. Through our bodies we expressively perform who we are. The fluidity and openness of performativity may be used to refigure identities, working alongside (other) contexts in the way that Barnett (1999) has argued towards the 'flattening out' of contexts and practices. The body in its performance signifies, as well as is signified, by our identities and, as Butler (1997) argued, affects and is an affect of our identities. Identity can be significant in the inter-subjective character of performance, in its embodied semiotics and in being and becoming 'in relation to others', in structuring feeling through doing (Harrison 2000). The identities of individuals are not constituted as tourists. Although they can define themselves in relation to signs of the industry they self-define as people through their idiosyncratic performativities. They thus negotiate, unevenly, routes and roots (Bell 1996: 9). Performativities can distinctively colour geographical knowledge.

> [T]he ability to reflect consciously on thought and sensation, which are initially spatially located, comes through this symbolic dimension. This dimension is blended with space and time, for symbols are used for a means of communicating with others . . . a means of communicating with ourselves.
>
> (Burkitt 1999: 77)

These events are engaged in multiple flows, across events and amongst contexts in plural ways, as practical or performative ontology (Shotter 1993; Dewesbury 2000: 477). There are multiple tensions and negotiations, therefore, at play. Contexts can serve as part components of the world in which the tourist engages in physical, mental and contemplative action. The lay knowledge that evolves through the encounter the individual makes is available as a resource through which the individual can negotiate the multiple contexts not only of tourist life but everyday life. Encountering the world through the temporary, informed and informing moments of being a tourist can shape the ways in which the individual understands the world, and his/her place in it. This is activated through a process of mediation, of holding on and going further. The individual engages in a practice of identification and of power, shaping, reshaping and making sense. Meanings, values, relations and identities are open to negotiation, to being sustained, and to being reworked.

Concluding observations

The positioning of embodiment, performance and ontological, practical knowledge as important in processes of power contributes greater complexity to the ways in which power in tourism may be apprehended and understood. It does not argue against the importance of investigating, interpreting and engaging the work, nor the power, of institutions, cultural industries, cultural mediations and so on. Rather, it points towards a mutual and multiple engagement of these diverse components, and also to the necessity not to work, or to presume, linearity from contexts outwith the individual to the ways in which the individual may be presumed to act in, or to make sense of, the world.

Some prevailing discourses in tourism may assert their own power over ideas, feelings, attitudes and identification. Another set of, or other, discourses circulate through lay geographies. In no sense does this identification of the importance of lay geographies intend to ignore the pernicious consequences of tourism's economic power (Meethan 2001). However, the various consequences of such a straitjacketed view of power may not be so directly transferred to its cultural effect on how the world is understood and lives are identified (Oakes 2004). Protocols of how to think, and value, which are inscribed through the tourism 'industry' onto consumers, may only go so far. The industry already acknowledges the complexity, diversity, nuanced character of 'the consumer' (*sic*). Their protocols may not work. Together, and in flows, these plural discourses interact, collide and are negotiated messily. The consequent presumptions on the globalisation of 'culture', with the embedding of projected touristic ideals and identities into everyday lives, are therefore open to contestation, as Miller (1997) exposed in relation to the cultural globalisation of soft drinks.

Mobilising the potential insight from the narrative developed in this chapter offers a number of research concerns and, implicitly, methodologies (Crouch 2004b). To understand these lay geographies offers an opportunity to make sense of the power of tourism as investment, effort and discourse in contemporary culture. It is timely to develop investigative platforms that can unpack the complexities argued here. There have been several contributions already (Crang 1997; Crouch 2003). Clifford's (1997) arguments about routes and roots also points to entanglements rather than linearities and prior prescriptions in tourism. A much greater investigation of the qualitative character of tourism is required. This should carefully push beyond the limits of prior presumptions and reconsider presumptions of what tourism is in order to enable a deeper reflexivity on the subjectivities of being a tourist; that is, of the tourist more as individual who happens to be engaged in tourism temporarily.

Indeed, in recent discussions on mobilities, it is interesting that two key components of mobilities remain as yet unsaid (Sheller and Urry 2004). The meaning of spaces that tourists identify and articulate need to be tracked back through diverse and multiple influences, including brochures but also life relationships, other space-knowledges. They should not be read off like litmus tests of what may seem to be obvious influencers which may be merely the easiest to find. There is no evidence that justifies the power of the gaze (i.e. as sign reading) in relation to other life-practice components; there is little evidence, for that matter, of the power of any tourist brochure over decisions, meanings, feelings, of any tourist. There is so much yet to unpack here but in another way, the individual 'move/s' temporally between being a tourist and not being so. These are not polarities but fluidities, nodes or points in a life, highly influenced by temporalities of life itself. Further, the spaces of life overlap; relationships, for example, are not forgotten – far less escaped – 'on holiday' but taken with us. Being a tourist is not so different as it is often, even conceptually, claimed. Being a tourist, doing tourism, and doing leisure merge; they are not separate bubbles as the advertisements may luxuriously claim. These are fantasies that to date tourism studies has found seductive. The power of institutions over what people do, or the power of advertising over decisions made, remains unchecked or at best hardly contested. Knowledges, the decisions and desires overlap. Spaces known in one part of the patina of encounter merge and (dis)colour the other. Finally, mobilities need to be considered in the wide spectrum in which they occur, in order to make sense of the way in which different influences may (not) work in tourism and for the tourist. It is important to make sense of the non-linear, multiple entanglements, to make better sense of institutional and culturally mediated meanings and their workings of power, and inside all of this, the tourist. With attention to these threads of possible investigation, it may be that we can make far better sense of the complexity of power that makes tourism work through the production, enaction and circulation of meanings and values associated with the performance of tourism.

References

Andrews, H. (2004) 'Escape to Britain: the case of charter tourists in Mallorca', unpublished PhD thesis, London Metropolitan University.

Auge, M. (1995) *Non-Places: An Introduction to a Theory of Super-modernity*. London: Verso.

Bachelard, G. (1994) *The Poetics of Space*. Boston, MA: Beacon Press.

Barnett, C. (1999) 'Deconstructing context: exposing Derrida', *Transactions of the Institute of British Geographers New Series*, 24: 277–294.

Baudrillard, J. (1981) *For a Critique of the Economy of the Sign*. St Louis, MO: Telo.

Bell, V. (1996) 'Performativity and belonging: an introduction', in V. Bell (ed.) *Performativity and Belonging*. London: Sage.

Britton, S. (1991) 'Tourism, capital, and place: towards a critical geography of tourism', *Environment and Planning D: Society and Space*, 9: 451–478.

Burkitt, I. (1999) *Bodies of Thought*. London: Sage.

Butler, J. (1997) *Excitable Speech: A Politics of Performance*. London: Routledge.

Campbell, N. (2004) 'Producing America: re-defining global tourism in a post-media age', in D. Crouch, R. Jackson and F. Thompson (eds) *Convergent Cultures: The Media and Tourist Imaginations*. London: Routledge.

Cant, S. (2004) 'The tug of danger with the magnetism of mystery: descents into the comprehensive, poetic-sensuous appeal of caves', *Tourist Studies*, 3(1): 67–82.

Carlsen, M. (1996) *Performance. A Critical Introduction*. London: Routledge.

Clifford, J. (1997) *Routes: Travel and Translation in the Late Twentieth Century*. Cambridge, MA: Harvard University Press.

Coleman, S. and Crang, M. (eds) (2002) *Tourism: Between Place and Performance*. Oxford: Berghahn.

Crang, M. (1997) 'Picturing practices: research through the tourist gaze', *Progress in Human Geography*, 21(3): 359–373.

—— (1999) 'Knowing, tourism and practices of vision', in D. Crouch (ed.) *Leisure/Tourism Geographies*. London: Routledge.

—— (2001) 'Rhythms of the city: temporalised space and motion', in J. May and N. Thrift (eds) *Time/Space: Geographies of Temporality*. London: Routledge.

Crang, M. and Thrift, N. (eds) (2000) *Thinking Space*. London: Routledge.

Crossley, N. (1995) 'Merleau-Ponty, the elusory body and carnal sociology', *Body and Society*, 1: 43–61.

—— (1996) *Inter-subjectivity: The Fabric of Social Becoming*. London: Sage.

Crouch, D. (ed.) (1999) *Leisure/Tourism Geographies*. London: Routledge.

—— (2001) 'Spatialities and the feeling of doing', *Social and Cultural Geography*, 2(1): 61–75.

—— (2003) 'Spacing, performance and becoming: the tangles of the mundane', *Environment and Planning A*, 35: 1945–1960.

—— (2004a) 'Flirting with space', in C. Cartier and A. Lew (eds) *Seductions of Tourism*. London: Routledge.

—— (2004b) 'Tourism, research practice and tourist geographies', in B. Ritchie, P. Burns and C. Palmer (eds) *Tourism Research Methods: Integrating Theory with Practice*. Wallingford: CAB International.

Crouch, D. and Desforges, L. (2003) 'The sensuous in the tourist encounter: introduction to the power of the body in tourist studies', *Tourist Studies*, 3(1): 5–22.

Crouch, D. and Lubbren, N. (eds) (2003) *Visual Culture and Tourism*. Oxford: Berg.

Crouch, D. and Parker G. (2003) 'Digging-up Utopia? Space, place and land use heritage', *Geoforum*, 34(3): 395–408.

Crouch, D. and Selwyn, T. (2003) 'The Med in mind', unpublished paper available from the author, University of Derby.

Crouch, D., Aronsson, L. and Wahlstroem, L. (2001) 'The tourist encounter', *Tourist Studies*, 1(3): 253–271.

Csordas, T.J. (1990) 'Embodiment as a paradigm for anthropology', *Ethos*, 18: 5–47.

de Certeau, M. (1984) *The Practice of Everyday Life*. Berkeley, CA: University of California Press.

Deleuze, G. (1993) *The Fold: Liebniz and the Baroque*. London: Athlone Press.

—— (1994) *Difference and Repetition*. London: Athlone Press.

Deleuze, G. and Guattari, F. (1988) *A Thousand Plateaus: Capitalism and Schizophrenia* (trans. R. Hurley). London: Athlone Press.

Dewesbury, J.D. (2000) 'Performativity and the event', *Environment and Planning D: Society and Space*, 18: 473–496.

—— (2003) 'Tourist: pioneer: hybrid: London Bridge, the mirage in the Arizona Desert', in D. Crouch and N. Lubbren (eds) *Visual Culture and Tourism*. Oxford: Berg.

Domosh, M. (1998) '"Those gorgeous incongruities": polite politics and public space on the streets of nineteenth-century New York', *Annals of the Association of American Geographers*, 88: 209–226.

Edensor, T. (1998) *Tourists at the Taj*. London: Routledge.

Featherstone, M. and Turner, B.S. (1995) 'Body and society: an introduction', *Body and Society*, 1(1): 1–12.

Game, A. (1991) *Undoing Sociology*. Buckingham: Open University Press.

Grosz, E. (1999) 'Thinking the new: of futures yet unthought', in E. Grosz (ed.) *Becomings: Explorations in Time, Memory and Futures*. Ithaca, NY: Cornell University Press.

Harre, R. (1993) *The Discursive Mind*. Cambridge: Polity Press.

Harrison, P. (2000) 'Making sense: embodiment and the sensibilities of the everyday', *Environment and Planning D: Society and Space*, 18: 497–517.

Harvey, D. (1990) 'Between space and time: reflection on the geographic information', *Annals of the Association of American Geographers*, 80(3): 418–434.

—— (1996) *Justice, Nature and the Geography of Difference*. Oxford: Blackwell.

Inglis, F. (2000) *The Delicious History of the Holiday*. London: Routledge.

Ingold, T. (1995) 'Building, dwelling living: how animals and people make themselves at home in the world', in M. Strathearn (ed.) *Shifting Contexts: Transformations in Anthropological Knowledge*. London: Routledge.

—— (2000) *The Perception of the Environment: Essays in Livelihood, Dwelling and Skill*. London: Routledge.

Kayser-Nielsen, N. (2003) 'New Year in Nampnas: on nationalism and sensuous holidays in Finland', *Tourist Studies*, 3(1): 83–98.

Kristeva, J. (1996) *The Portable Kristeva*. New York: Columbia University Press.

Larsen, J. (2004) 'Families seen sightseeing: performativity of tourist photography', *Space and Culture*, 8(4): 416–434.

Lash, S. and Urry, J. (1994) *Economies of Signs and Space*. London: Sage.

Lloyd, M. (1996) 'Performativity, parody and politics', in V. Bell (ed.) *Performativity and Belonging*. London: Sage.

Lubbren, N. (2003) 'North to South: paradigm shifts in European art and tourism, 1880–1920', in D. Crouch and N. Lubbren (eds) *Visual Culture and Tourism*. Oxford: Berg.

Macnaghten, P. and Urry, J. (1998) *Contested Natures*. London: Routledge.

Meethan, K. (2001) *Tourism in Global Society*. Basingstoke: Palgrave.

—— (2003) 'Mobile cultures? Hybridity, tourism and cultural change', *Tourism and Cultural Change*, 1(1): 11–28.

Merleau-Ponty, M. (1962) *The Phenomenology of Perception*. London: Routledge.

Miller, D. (1997) 'Coca-Cola: a black sweet drink from Trinidad', in G. McKay (ed.) *Consumption and Everyday Life*. London: Sage.

—— (1998) *Material Culture: Why Some Things Matter*. London: Routledge.

Nash, C. (2000) 'Performativity in practice: some recent work in cultural geography', *Progress in Human Geography*, 24(4): 653–664.

Oakes, T. (1997) 'Place and the paradox of modernity', *Annals of the Association of American Geographers*, 87(3): 509–531.

—— (2004) 'Modernity and tourism', in C. Cartier and A. Lew (eds) *Seductions of Tourism*. London: Routledge.

Pile, S. and Thrift, N. (eds) (1995) *Mapping the Subject*. London: Routledge.

Pons, P-O. (2003) 'Being-on-holiday: tourist dwelling, bodies and place', *Tourist Studies*, 3(1): 47–66.

Radley, A. (1995) 'The elusory body and social constructionist theory', *Body and Society*, 1(2): 3–24.

Roach, J. (1995) 'Culture and performance in the cirum-Atlantic world', in A. Parker and E. Sidgewick (eds) *Performativity and Performance*. London: Routledge.

Ringer, G. (1999) *Destinations: Cultural Landscapes of Tourism*. London: Routledge.

Schieffelin, E.L. (1998) 'Problematising performance', in F. Hughes-Freeland (ed.) *Ritual, Performance, Media*. London: Routledge.

Selwyn, T. (1996) *The Tourist Image*. Chichester: John Wiley.

Sheller, M. (ed.) (2004) *Tourism Mobilities: Places to Stay, Places in Play*. London: Routledge.

Sheller, M. and Urry, J. (eds) (2004) *Tourism Mobilities: Places to Play, Places in Play*. London: Routledge.

Shotter, J. (1993) *The Cultural Politics of Everyday-life*. Cambridge: Polity Press.

Taussig, M. (1992) *The Nervous System*. London: Routledge.

Thrift, N. (1997) 'The still point: resistance, expressive embodiment and dance', in S. Pile and M. Keith (eds) *Geographies of Resistance*. London: Routledge.

—— (2003) 'Performance and . . .', *Environment and Planning A*, 35: 2019–2024.

Thrift, N. and Dewesbury, J-D. (2000) 'Dead geographies – and how to make them live', *Environment and Planning D: Society and Space*, 18: 411–432.

Tulloch, J. (2000) *Performing Culture*. London: Sage.

Urry, J. (1990) *The Tourist Gaze*. London: Sage.

—— (2002) *Sociology beyond Societies*. London: Sage.

—— (2003) *Global Complexity*. Cambridge: Polity.

Williams, S. (2004) *Tourism and the Social Sciences*. London: Routledge.

3 Adventure tourism

Will to power?

Carl I. Cater

The popularity of adventurous activity as a leisure and tourism pastime continues to grow among both established Western markets and emergent global ones. The Adventure Travel Society (1999), a US industry organisation, reports that people throughout the world will continue to leave the beaten path in record numbers; even in a soft economy, active outdoor recreation will increase in the years to come. Adventure travel and its related expenditure contributes $220 billion annually to the US economy alone (ATS 1999). Although this figure results from a broad and perhaps less than rigorous calculation, it is an indication of how significant these practices have become over the last decade or so. Other commentators suggest that the adventure tourism industry grew at an estimated 10 to 15 per cent every year during the 1990s (Hawkins 1994). An explosion in adventure travel shows, magazines and television programmes is also indicative of this trend. Adventure tourism, along with ecotourism and cultural tourism, has become part of a spectrum of 'new' tourist practices with clearly different ethics to traditional 'mass' tourism. New destinations, environments and activities have been developed to fuel this demand, with previously inconceivable locations, such as the polar ice caps, the depths of the oceans and even outer space, becoming readily accessible to those with the will and the money. This is not only significant from the importance of the size of the industry, but also in its broader societal impact, particularly as many of these adventurous activities carry an entire lifestyle with them.

It has been long suggested that one of the primary motivations for such human endeavour is self-actualisation (Maslow 1987). Indeed the power over self is certainly a persistent theme in all leisure activity. The caveat to this, however, is an increased understanding that the body itself has power such that even at the level of the individual, contested practices may take place. These dynamic negotiations of embodied action are a cornerstone of the adventurous practices this chapter seeks to uncover. A corresponding trend – to meet the demand for these desires – has been the increasing commercialisation of adventurous activity. In rural locations in particular, adventure tourism has reinvigorated flagging peripheral economies (Cloke and Perkins 1998a). This has a twofold implication in understandings of power. First, increasing commercialisation of adventure greatly complicates the simpler relationships that may occur between individual and activity during independent adventure participation. Adventure providers add levels of power relationships that contextualise the experience and greatly increase the performative aspects to that experience. Second, the development of adventure pursuits may have significant potential to alter place identity. However, these new meanings for rural tourism locations may not go uncontested.

This chapter seeks to examine the way in which the growth of adventure tourism demonstrates these complex articulations of power, both at the level of the individual and tourist place. Adventurous spaces are often inherently unliveable, and this is undeniably part of the attraction of momentarily dwelling within them, reinforcing their touristic character (Rose 1999: 248). We cannot exist within the raging whitewater rapids or in the space between bridge and canyon. The journeys into this unfamiliar terrain are exploratory and touristic. However, their extreme nature places them on the margins of tourist and body-space. These assertions make it apparent that experiences are simultaneously negotiated at several scales. Recent examinations on the scalar nature of power, by for example Herod and Wright (2002) and Allen (2003), stress that there is no certainty to power or the scales at which it manifests itself. According to Allen (2003: 190), 'we need to be a little more curious about powers' spatial constitution in a landscape that does not assume fixed distances, well defined proximities and effortless reach'.

Indeed, the drive to adventure is fundamentally about the disruption of such neat preconceptions to which Allen refers. In the adventure experience, the self attempts to show power over the body, and over the natural spaces in which activity takes place. However, nature within and without is not always as pliable as desired, inscribing its own authority on proceedings. The situation is further compounded by the commercial nature of the adventure product, whereby these barriers are in part negotiated by external providers, in particular through their embodiment, the guides. These individuals etch out their own power relations through the mediation of experience. Moreover, these performances of power take place within a variable catalogue of prior experiences, circulated images and tourism marketing that compounds the situation further. Consequently, it is useful to adopt a topographical perspective. The discussion will highlight these interactions at levels of participant, provider and place. Correspondingly, it also calls for a more nuanced appreciation of experience that has relevance beyond touristic and geographical inquiry.

'The adventure capital of the world'

A destination that has been enormously successful in tapping into the increased interest in adventure tourism is that of Queenstown in New Zealand, which has marketed itself as 'the adventure capital of the world' (Figure 3.1). Through this branding, the destination has positioned itself among a band of global locations that package a range of adventure activities in spectacular natural settings. Among these are Victoria Falls (Zimbabwe), Bend (Oregon, USA), Voss (Norway) and Interlaken (Switzerland). It is important to recognise that Queenstown has always been a tourist destination, primarily as a result of its scenic splendour, but its character has been changed and supplemented by the growth of adventure tourism which is represented by a plethora of activities ranging from bungy jumping to

Figure 3.1 Queenstown – 'the adventure capital of the world'

Source: reprinted courtesy of Destination Queenstown.

canyoning. The town and its environs function as a huge adventure playground, with paragliders and aerobatic aircraft whirling above, mountain bikes speeding past and jetboats speeding across the lake. It is for this reason that the town itself is dominated by booking agents, shops, bars and restaurants that extol this adventurous ethic. The latter are particularly important for that 'après adventure' celebration, and they have names like 'the edge' or 'surreal', reinforcing the percieved association with the sublime.

The destination is not large by international standards, with a resident population of approximately 16,000 in the Queenstown regional tourism organisation (RTO) region (TRCNZ 2004). Visitors are split fairly evenly between the international and domestic markets, with a total of 433,000 day and overnight visits to the region from the former in 2003 and 594,000 from the latter (TRCNZ 2004). Tourism dominates employment, with 46 per cent of the workforce directly employed in the tourism sector (TRCNZ 2004). The majority of material used in this chapter comes from research conducted in the resort over six months in 1998. Semi-structured in-depth interviews were conducted with approximately 100 participants in a range of adventure activities from November 1998 to April 1999.

Performing power

If we are to explain how adventurous activity prioritises particular notions of space, first we need to consider how adventure manifests itself at the level of the individual. To partake in adventure is basically to act performatively. Performing various roles in public space is one of the more significant ways that individuals make sense of their worlds and especially of their own bodies. It is apparent that 'bodies are used to act out roles in various settings, which confirm and resist (at different times and places and sometimes simultaneously) wider sets of expectancies' (Cresswell 1999: 176). Conspicuously then, bodies have the ability to perform (or not) in any given situation. As Denzin (1997: 97) suggests, 'performance is interpretation . . . a performance is a public act, a way of knowing, and a form of embodied interpretation'. Furthermore, these embodied performances shape, and are shaped by, the spaces that they inhabit. As Rose (1999: 258) puts it,

> The body is entangled with fantasy and discourse; fantasy mobilizes bodies and is expressed through discourse; and discourse, well, discourse is disrupted by fantasy and interrupted by the body. And all of these relations are articulated spatially; their performance articulates space.

It appears that we are increasingly moving towards what Turner (1996: 1) has called a 'somatic society', in which 'the body is now part of a self-project within which individuals express their own personal emotional needs through constructing their own bodies'. One only needs to take a glance at the Sunday newspaper supplements to see how prevalent identities of body are at a societal level. As a result,

> the individual body is connected into larger networks of meaning at a variety of scales. [Embodiment] refers to the production of social and cultural relations through and by the body at the same time as the body is being 'made up' by external forces.
>
> (Cresswell 1999: 176)

A significant reason for this development has been the change in the place of the body in society more generally. While in early capitalism there was a close connection between the body and work, this has been eroded especially in recent times with the reduction of bodily work leading to 'an entirely different and corrosive emphasis on hedonism, desire and enjoyment' (Turner 1996: 4) as the focus for embodied concern. Drawing on earlier work by MacCannell (1976, 1992) and others, the body is an authentic site of consumption in a way that solely visual experience is not. As Urry (1999: 41) suggests, the other senses may offer us a 'more direct and less premeditated encounter with the environment' that contrasts with 'the abstractive and compositional characteristics of sight'. The construction of the recreational body is thus important in understanding participation in adventure activities.

While grounded in a long history of adventurous tourist practice, adventure tourism is clearly about an active participation with the body. The inability of tourist metaphors such as Urry's (1990) 'gaze' to capture this enacting body has been highlighted in previous work (cf. Veijola and Jokinen 1994). Game (1991: 167) suggests 'that there is a way of "being in" that refuses an objectifying gaze and inscribes a different desire. This desire is not primarily spectacular, but relates to touch'. These haptic experiences are undoubtedly what participating in adventure tourism is all about (Lewis 2000). Playing and performing with the body clearly applies in the arena of adventure sports, and confirms Poon's (1993: 90) assertion that touristic practice is progressively more about *being* than *having*. Despite, or perhaps as a result of, the fact that we live in a world of an 'increasingly pronounced visual character . . . made with the visible in mind' (Lefebvre 1991, quoted in Urry 1999: 35), there is a *move* to experience with the lived body as a whole. The Cartesian idea of a disembodied visual gaze sits ill at ease with the subject's belonging to an embodied, material world.

Power over the body

This acknowledgement does not diminish the desire to demonstrate power and control over the body. A frequent comment from participants in adventure tourism was the desire to achieve the perfect form, particularly where the activity is captured on film or camera. In bungy jumping this may be a perfect dive (Figure 3.2); in other activities, the look of confidence and domination in the frame. The look is one of harmonic conquest of the environment, embodying the adventurous ethic itself. As Schiebe (1986: 142) notes, 'the *right stuff* as a concept is bound up with vertigo and transcendence in the escape from earth's bondage. If you have the right stuff you are in harmony with the sublime.' Respondents are undeniably aware of the desire to bring this expression of having the 'right stuff' into their performances:

> Absolutely totally focused to get it right, I wanna go and see it, look at my style. I wanted that good spread thing going, nice little y-shaped descent and that.
>
> (Gareth, Pipeline Bungy)

> I was trying to act quite cool about it. I must have done, because the bloke said to me 'you're not scared are you?' and I lied and said 'nah', and he said 'say goodbye to your mates' and I said 'see you later lads' on the video, and then when I jump off the bridge I wink at the camera, sound stuff. To be honest, if other people are scared it makes me feel a bit better, cause they are so . . . like her and her mate were really, really scared, so that made me feel a bit more in control. If I was the only one who was scared I would have felt pretty fucking panicky. Just before I jumped though, there was a point where I thought this is a little bit stupid, paying to jump off a bridge. When you get up there you have a sudden realisation that you have paid to jump off a bridge and everyone is watching you do it, so you are resigned to the fact, but you do think it is pretty silly!
>
> (Anthony, Pipeline Bungy)

This discussion fits neatly within the motivational ideas of self-actualisation, as suggested by Maslow (1987). Participants temporarily give up their comfort zone

Figure 3.2 The body in action

Source: Carl I. Cater.

in order to achieve a higher level of needs satisfaction. At these higher levels sublime experiences are representative, as 'the values that self-actualisers appreciate include truth, creativity, beauty, goodness, wholeness, aliveness, uniqueness, justice, simplicity and self-sufficiency' (Maslow 1987: 147). Participants undoubtedly demonstrate a desire to enforce their will on their bodies to achieve successful completion of the experience. Furthermore, through this very individualised will to power tourist narratives are often constructed. This is clearly demonstrated in the stories that adventure tourists use to regale others and construct their own identity on returning home (Desforges 1998).

The powerful body

However, in this recognition of performing bodies it is important not to construct them as somehow passive. Indeed, bodies have an undeniable material existence and as Grosz (1990: 72) has argued 'what is mapped onto the body is not unaffected by the body onto which it is projected'. Moreover for Shields (1991: 269), *performing* subjects are

> not just a body being acted upon, but a fractured/multiple agent which is possibly contradictory but always actualising and deforming structural codes; hence elaborating a performance supported by social rituals and exchanges which confirm different personae.

As Cloke and Perkins (1998b: 189) point out, 'adventure tourism is fundamentally about active recreational participation, and it demands new metaphors based more on "being, doing, touching *and* seeing" rather than just "seeing"'.

Clearly, the body has significant power in and of itself to shape the experience. What it is important to recognise is that the body is *the* location for the negotiation of the experience (Figure 3.2). Once again this challenges a supposition that consciousness is the locus for intentionality (Merleau-Ponty 1962). One of the most interesting observations came from one of the bungy jump guides for AJ Hackett, who highlighted how he sees bodies moving in every participant, narrating the example of a person who was about to jump:

> You can see with a lot of people, yeah, you see she has done a lot of it because she has let go and she is standing upright on the edge you know, so she has done that step and all she has to do is go forward. And you see her body is slightly leaning back, well that is negative, and what you can actually do is feel the body as you talk to them actually start to move forward, or you are talking to them and their body is leaning back but their feet are moving forward, and you can see that – this person is going, they are going.
>
> (Stu, jumpmaster, AJ Hackett)

> He just said 'lean forward'. I had hold of his waist so it felt okay. He said 'lean forward' and I said 'I can't. I physically can't throw myself off', but then as he counted down I just went 'oh well', leant forward a little bit, and then before I knew it I was off. I just didn't think that I would be able to go, but you just go forward that much and that's it then.
>
> (Becky, UK, 22, Kawarau Bungy)

Moving forward from these observations, it becomes clear that the body is possessed of the ability to act in a knowing manner. Drawing on Merleau-Ponty's observations on the body subject, there is a challenge here to the phenomenological preoccupation with consciousness as the locus of intentionality, for example as proposed by Husserl: consider, 'the plunge into action is, from the subjects' point of view, an original way of relating himself to the object, and is on the same footing as perception' (Merleau-Ponty 1962: 110). Mobility, therefore, is the most basic form of intentionality, and this is crucial in any understanding of the way in which new forms of embodied action develop. As Cresswell (1999: 176) contends,

> Just as abstract space can be transformed into social space (or place) by taking power seriously, so it is possible to think of human movement as a social phenomenon, as a human geographical activity imbued with meaning and power.

The intentional body is vital to understanding how we relate to, and construct, geographical space, especially in tourism studies given the focus on inherently mobile corpora. However, we should not see the struggle for embodied power as necessarily being one of foregone dominance. For Nietzche (1968: 636), all bodies are continually asserting their power;

> every specific body strives to become master over all space, and to extend its own force (its will to power) and to thrust back all that resists its extension. But it continually encounters similar efforts on the part of other bodies and ends by coming to an arrangement 'union' with those that are sufficiently related to it: thus they conspire together for power.

Clearly there is a method whereby wilful forms are regulated by processes both internal and external. As Harrison (2000: 510) notes, 'as practices become habitual and strata are laid down . . . a relatively standard distribution of organs occurs, one that is socially acceptable'. Recognition of the ambition of the body is paralleled by a concession that it must also be a site of resistance. As Grosz (1990: 64) suggests,

> as well as being the site of knowledge power, the body is thus also a site of resistance, for it exerts a recalcitrance, and always entails the possibility of a counter-strategic reinscription, for it is always capable of being self-marked, self-represented in alternative ways.

By definition, adventure tourism offers a direct challenge to acting as the tourist body 'should', even though, as will become apparent, the situation is considerably more complicated than this when we examine the 'places of adventure'. As Johnston (1989) points out, the social organisation of adventure, mediated by various subcultures, affects the actions and views of participants, and their seeking and acceptance of risk. However, this is not to view the adventure tourist as some porous canvas; 'individuals are active, as well as reactive, able to interpret and alter their circumstances, subcultures and institutions' (Johnston 1989: 43). Furthermore, the body may occasionally refuse to perform as is expected or required of it, even at the individual level. Feelings may be different, utterances may be surprising, and behaviours may be unpredictable. Sometimes the body may fail altogether in achieving what was set out, demonstrating a clear power of its own. Fear, which is constituted in this setting as a distinctively embodied emotion, may have more power than rational decisions based on cognition or memory.

Excluded bodies

In any consideration of power, one also needs to identify processes of marginalisation and exclusion. The practice of adventure tourism certainly creates such geographies. At sites of adventurous activity it is usual to observe a placard that sets out an embodied ideal for the participants (Figure 3.3). Nevertheless, participation in adventure tourism is not restricted to the young, white males epitomised by 'Pepsi-Max' advertisements (cf. Cater 2000). In fact participation is much broader than popular opinion seems to believe, with all ages, abilities and classes

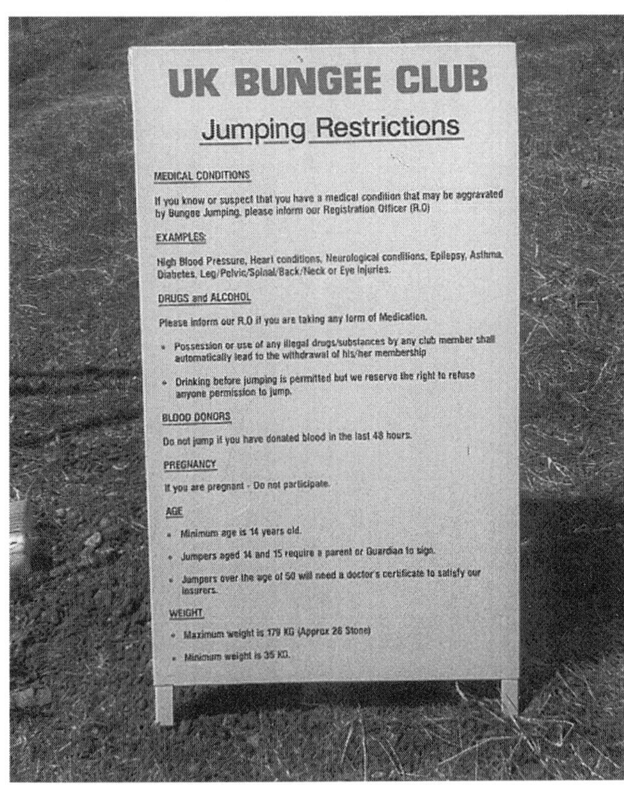

Figure 3.3 An ideal of embodied perfection for participation in adventure

Source: Carl I. Cater.

participating. Although there are clear masculinities to the promotion of adventure, the performance is just as popular with both sexes, and the average adrenaline seeker is just as likely to be Korean as Californian. However, the growth of adventure activities in less developed locations has the potential to expose significant disparities in the manner in which tourists and locals experience those environments (Mowforth and Munt 2003).

Without a doubt, there are also significant embedded gender roles in adventure that have potential to influence power relationships. These may be roles that are merely played, as shown by Hirschmann (1984), but actual bodily gender is also important, particularly in the marketing of the industry. Swain (1995: 249) points to the 'predominant tourism brochure representations of men associated with action, power and ownership, while women are associated with passivity, availability and being owned'. There is a machismo element to the practice of adventure tourism, which Turner (1996: 3) explains as arising out of the reorganisation of labour and thus leisure and consumption;

> Young, working-class men have become a surplus population, whose machismo image of toughness no longer has a direct functional relevance to their enforced leisure. The labouring body has become a desiring body.

One should be wary of attributing the growth in adventure purely to such a phenomenon, but there is some explanation here, as illustrated by Bordieu (1993: 354), who shows how the popularisation of certain sports may be parallel to an increase in the willingness to 'gamble with the body itself', replacing the earlier elite ethic of 'the body as an end in itself'. The hedonistic body is a clear figure within the promotion of adventure.

Provider power

Hence our scalar-based appreciation of power must recognise that there are external factors that the adventurous corpora must negotiate. In commercial adventure tourism, these performances of embodied power are heavily mediated through the provider. In particular, this results from the handing over of a significant part of the responsibility to the adventure company in question. One commentary on the 1999 Swiss canyoning disaster, in which twenty people died, observed;

> Most people don't have the time or want to reach a level of independent competence in many adventure activities, so in effect they go shopping for the expertise, buying in experience, handing the duty of care to someone who does have the right certificate or training.
>
> (*The Guardian*, 29 July 1999, G2: 3)

Consequently, the performance and conduct of guides are integral parts of the experience, and thus have important ramifications for any consideration of power. This power takes two forms, the most important of which is that of physically ensuring that the activity is completed successfully and without any harm coming to the client. Often more visible, however, are the guides' roles in creating what the adventure will be like as a tourist product. As such, adventure tourism guides are primary gatekeepers in the mediation of experience. The participants have an expected idea of the person who will lead them through the activity, typically active, outdoorsy, knowledgeable and larger than life, acting as the 'exemplars of the bodily habitus expected of and desired by tourists' (Crang 1997: 151). This is obviously enhanced by the company advertising, as shown for a jetboat ride (Figure 3.4) which positions their drivers as being 'personality plus', a clear reference to their distinctive personalities. The guides themselves are perhaps even more aware of the roles that they are expected to play than the participants:

> How important do you think your role is in providing a positive experience? It's very important, it's of the upmost importance, you've gotta be leaping around and make sure they are happy.
>
> (Mark, driver/guide, Kawarau Bungy)

This is not to say that all of the guides are from one mould, as the presence of a variety of personalities is important to the effective operation of the activity. One of the guides for AJ Hackett, the principal bungy jumping company, detailed how,

Figure 3.4 Dart River Safari drivers and guides are sold as being 'personality plus'
Source: reprinted courtesy of Dart River Safaris.

while his colleagues shared attributes such as being 'outdoorsy types', they also had to have varied personalities in order to keep people jumping. To illustrate this, he offered the example of how some participants responded well to being treated more abruptly, practically being told to jump, while others required more gentle persuasion. A range of characters who could fulfil these roles was necessary. Thus, communication skills are highly prized by the adventure operators:

> If they have experience on ropes and karabiners and that, we would like that, but we choose a person on how they communicate at the end of the day, and if they are on to it. But communicating with people if they are liked, just the look, it plays a huge factor in it, because if you are tying a person up then you have got to look confident. They can't be way over the top though, so we have to be selective on appearance and how they communicate with people.
>
> (Matt, marketing manager, Pipeline Bungy)

What is interesting, however, is the manner in which all this 'fun' may be at odds with any form of reality. Many of the commentaries on whitewater rafting or jetboats were an interesting blend of truth, myth and jokes, with no acknowledgement of where the boundaries lie. As well as demonstrating the position of power held by the guides, this confirms an assumption that the tourists do not, in fact, care how authentic the commentary is, just as long as the performance is authentic only in its enjoyment (Ritzer and Liska 1997: 102). As Cloke and Perkins (2002: 12) have noted, 'tourists are often told stories of place which blur the boundaries between history and fiction'. An example of this process of mystification can be seen in the trips with Queenstown Rafting:

> Boarding the bus the guide started a rapport by identifying the various nationalities on the bus and making fun of various national traits. He started by telling the Japanese clients that he didn't like eating in Japanese restaurants because the fish was never cooked properly, and pointing out to the Australians on board that the safety leaflet had pictures especially for them. The ride up Skippers Canyon continued on a humorous theme, making jokes about the drive and the rafting, and often making the experience sound more dangerous than it was, but in a comedic fashion. Thus we were told that the electric pylons hanging precariously on the side of the canyon were in fact a net to catch the bus if our driver, who had only driven this far three times, made a mistake. Historical sites were pointed out, but again humour involved so that the boundaries became blurred. A tale emerged about the Skippers pub ruin, where barmaids were employed for such a short period of time that the landlord hired the ugliest barmaid he could find, naturally from Australia, who was the great grandmother of our driver, and couldn't we see the family resemblance? This theme continued on the trip itself, with gold sluicing equipment becoming a Kiwi version of a bread making device, and a tale about the river guide's grandfather mining for 'leaveitrite' (leave it right there). When people fell in, the others were told that they are merely sacrifices to the water gods.
>
> (Personal account, Queenstown Rafting)

Often this picture is enhanced by the very organisation of the adventure trips, with military-style organisation often being the norm. Participants are usually kitted up at an operations headquarters, and then transported via minibuses, masquerading as troop carriers, into the arena of adventure whilst being briefed by the guides. Clearly this enhances the feeling of entering *into* an adventure, and suggestively draws interesting parallels with an earlier age of wars as occupying the niche that adventure now serves.

The guides undoubtedly enhance the desire to 'look good' during the practice of the activity. For example, on a river surfing trip, 'participants were told that the most important thing was to *try to look cool* whilst walking down to the river suited up in a wetsuit' (personal account, Serious Fun river surfing). The last thing that most participants hear before a bungy is the jumpmaster telling them to look out and give a good swan dive (Figure 3.2). Of course, it is worth entertaining that these notions are not solely performative, since concentrating on one's performance may also have the result of increasing the concentration on the task at hand which increases safety. A good arcing dive in a bungy also happens to be the safest manner of jumping. In addition, the adventure companies must perform to specific criteria to support their adventurous credentials, although this may be somewhat at odds with commercial practice. The founders of Pipeline Bungy left AJ Hackett precisely because they felt that the company had moved too far away from its adventurous, 'fun' and personal roots to follow an overly corporate ethos:

> We wanted to get back to the fun on the bridge you know, none of this big corporate international scene with management structures and that. Just a good team around us, who we knew were the right kind of people, and focused on keeping it simple and giving the customers a good time and developing more of a relationship with them basically.
>
> (Andrew Brindsley, founder, Pipeline Bungy)

It is interesting to consider the ramifications of such an acknowledgement for the expansion of adventure tourism operations. Is it possible, at least outwardly, for adventure tourism operators to exist as part of a large tourism operation? From a principled viewpoint, commercial interest and adventure may be opposed, and careful masking of such relationships takes place. This leads to the situation where 'operators of alternative adventurous sites will *both* wish to distance themselves from the hegemonic imagery and style of Disney *and* adopt tried and tested techniques of branding, marketing, pricing, ancillary product sales, safety, staff performance and the like' as described by Cloke and Perkins (2002: 6). Thus, in visiting the operations of any of the adventure companies, it is notable that one is not made aware of any of their ownership details or 'sold' other trips that may be offered by these bigger organisations. Certainly large tourism operations have made forays into the adventure tourism market, for example First Choice's £9 million acquisition of Exodus in 2002. However, the importance of the construction of adventure as a personalised, embodied experience leads them to often mask their presence where necessary.

Power, place and discourse

From a spatial perspective, individual negotiations of power manifest themselves in a distinct manner. Quite apart from the provision of spectator platforms and viewing areas that create an audience for adventurous experiences, it is clear that the sites of activities are carefully constructed in order to maximise the performance. This display of adventure is an integral part of the experience, and it is perhaps this that is most influential in the new rural character of Queenstown. Although probably less than a third of the visitors to Queenstown will actively participate in an adventurous activity (Berno *et al.* 1996), almost all will spectate, and thus form part of the experience. For example, most of the coach tours make a point of stopping at the Kawarau bungy jump on their route into the region, and large purpose-built spectator platforms provide for this. The carefully staged nature of adventurous spaces is apparent in this description of Goldfields Jet:

> On the trip itself we are going to enhance the relics that we have got on the side, like the Swiss guns and all the things that the guys have to stop and do a talk for, to look at. There are a few of these props now, but we are going to drop in a lot more props. Currently they are spread around the hills as ruins, so we put them down by the river so we will really enhance the trip.
>
> (Nick Flight, marketing manager, Shotover Jet)

Theming the various activities is recognised as a highly important process, particularly in an environment that has become as competitive as Queenstown. A clear theme helps to identify an activity from its competitors, and this can be seen in the success of 'Catching the Canyons' with Shotover Jet, or the clear branding of the Pipeline. Where the latter is concerned, the rebuilding of a fake pipeline across the gorge (to hark back to a sluicing pipe that was in the area a century before) defines the frontier gold *rush* element to the Pipeline experience. So successful has this piece of adventure architecture become that few participants realise the actual inauthenticity of this icon. Such a process may be presented as a grassroots example of the invocation of what Hollinshead (1998) terms a 'distory' in tourist practice, the selective presentation of history in a commodified and manipulated fashion, characterised by the processes of historical appropriation by the Disney corporation. Such practices enable,

> a postmodern seizure of useable storylines from the past which are decontextualised and romanticised and thereby turned into nostalgia which can decorate, and be purchased, but are frequently much diminished in connected meaning.
>
> (Hollinshead 1998: 80)

Whilst it is important to recognise that this process is unlikely to have the rigidly authoritarian delineation seen in McDisneyization (Ritzer and Liska 1997), the close personal networks detailed previously and below, combined with a hegemonic touristic focus in Queenstown, may propagate a similar phenomenon, albeit of a less observable nature. Moreover, Weber (2001: 374) suggests that careful diversification of the product may allow access to previously untapped markets without alienating operators' existing customer base, and this strategy can be clearly already observed on the ground. For example, AJ Hackett's instigation of historical tours of the Kawarau bridge and provision of related information clearly captures a wider market, as well as extending the attraction for the existing customers. Likewise, Shotover Jet's 'history-added' Goldfield Jet and 'nature-added' Dart River Jet may be observed as part of the same process. Such dominant meanings are broadly circulated, as the performance is not solely confined to the geographical boundaries of Queenstown. Operators ensure that the experience is available in a number of 'take away' media formats through which the consumption of performance can be continued. An account of an individual's first bungy jump reveals the importance of the video as a souvenir:

> The video has captured a moment of pure open-mouthed terror. But from the perspective of my safe little flat in London, it is the screams of pure admiration from the onlookers that now gives me the biggest thrill.
>
> (*Weekend Telegraph*, 12 January 1998: 24)

However, we should not accept these spatial delineations as being the only meaning that Queenstown carries. Nevertheless, the power of this identity means that there are distinct exclusions as regards the types of country visions that are entertained in the resort. The spaces of adventure in which the activities are actually performed are remarkably confined. Locations are carefully sited (and sighted) in order to maximise the potential for scenic backdrops, while maintaining just the right level of access (seemingly adventurous, but not inconvenient). Less spectacular or inaccessible landscapes may be completely excluded because they do not conform to the ideals required by the adventure tourism industry. Undoubtedly, representations of the landscape associated with adventure tourism are dominated by those produced by the Pakeha (white) imaginary, whereby the New Zealand landscape was seen as virgin territory. This land was one to be conquered and tamed, and brought within the rationalising gaze of the new settlers. Such a rationale clearly reinforces representations of adventure in Queenstown today. As discussed above, one is encouraged to jump off a bungy constructed along the route of a restored gold mining sluice pipeline; the gold *rush* might be physically different today, but it still contains such motifs as exploiting the rural environment for the fulfilment of personal gain. Aboriginal ways of natural interaction, with a more harmonious view of the environment, are conspicuously absent. A reluctance to engage with alternative representations in tourism marketing may be identified as an issue at the national level as well as at the destination scale (Aitken and Hall 2000).

The power of nature

Despite this acknowledgement, nature does have an undeniably powerful role to play in the context and meaning of adventure tourism. While it is recognised that contemporary theory would deconstruct the notion of a single 'Nature', and certainly the study in question would agree with the concept of multiple 'natures' (Franklin 2002), there is clearly an identified external environment in which adventure activity takes place. A natural playground like that of Queenstown is vital to the whole adventure experience. Not only do the surroundings provide a ready-made environment for the practice of adventure (for example whitewater rivers, deep canyons, mountains, etc.), it has been hypothesised that a ruggedly beautiful setting is crucial to the attractions of the adventure itself. In interviews, participants certainly responded positively when asked about the environment in which the activity is performed, with phrases such as 'it's beautiful' or 'it's gorgeous', and they were undeniably quick to point out the attractions of undertaking those pursuits in a rural setting:

> Nobody would do this up a river with condos down the side of it, you have got to do it in a natural setting.
>
> (Charles, Canada, over 55, Shotover Jet)

> Beautiful man, I wouldn't do it if there wasn't any water, I wouldn't do it. Its beautiful, it helps you do the thing. If it was in a car park I would be like no way.
>
> (Anthony, UK, 15–25, Kawarau Bungy)

However, respondents seemed less sure about the actual impact landscape has on the activity itself:

> Its nice, nice scenery, but half the time you don't notice do you?
>
> (Scott, UK, 22, River Surfing)

> Ah yeah, things like that, beauty, all that, it depends on who you get on the day. I think that's a really small group of people who are interested in that, a lot of people who come, you could put this place in a dump, they wouldn't care you know. It's a big bridge and they are going to jump off it, you know?
>
> (Stu, jumpmaster, AJ Hackett)

It would appear, then, that landscape aesthetics are important to the practice of adventure, but their relation is peculiar. During the practice of the adventure activity itself, concerns with aesthetics are removed almost entirely. As such, it is almost like these attributes are a prerequisite for adventure, but they go no further than this; the conditions of a beautiful setting are fulfilled, 'the box is checked', and the activity can proceed. There is a seeming contradiction in the heightened moments of adventure sports whereby one nature is powerfully negotiated through the body, while at the same time there is a process of distanciation from the external environment. These views would concur with the claims made by Hall (1992: 144) that, in the practice of adventure tourism, the environmental setting takes on a subordinate role. While this setting does form an important backdrop for the practice of the activity, it is the performance of the activity itself that remains the primary attraction.

Perhaps such an acknowledgement is not so surprising if we consider the place of the natural setting within the performance of adventure. As highlighted previously, a significant part of the adventure ethic, at least in historical terms, was the domination of landscape. Paradoxically then, to overstate the importance of nature to the experience would actually negate the passive role that this conquest demands. Such an observation is confirmed by examination of the cognitive maps completed by some of the participants. When respondents were asked to draw a map of Queenstown, scenic backdrops were commonly represented, illustrating the importance of these to the adventure tourist experience. However, several of them made a clear effort to place the self within the landscape. This clear identification of the individual performing actively within the landscape refutes a notion of a romantic gaze.

However, the founder of one of the major bungy companies suggests that, while the scenic location of adventure activities may have minor importance in the actual practice, it does have a vital role in the memory of that practice. In a way, the landscape is something that can be taken away in terms of the whole experience. In effect, the landscape delivers a value-added component;

> I think that a lot of the people who come here to bungy and it's their first jump. They come up to Skippers or Pipeline, they are so focused on that, that

> they don't actually think about where they are going or what they are going to experience along the way. But when they think back, they have the whole scenery thing, its all value-added stuff.
>
> (Andrew Brindsley, founder, Pipeline Bungy)

This sentiment is clearly echoed by the participants, and this is, of course, far more acceptable to adventurers, because nature in memory or on a videotape is bounded and commodified:

> I mean they have got bungy jumps back in England, but they are off cranes and that, it isn't the same thing is it? You want videos with like the scenery like this.
>
> (Shaun, UK, 25–35, Kawarau Bungy)

Here adventure tourism operators have been very successful at commodifying the adventurous landscape in terms of a product that can be sold and circulated, thus reinforcing the adventurous myths and ideologies that I hinted at earlier. To date, this commodification of nature through such media has received only minor attention (Franklin 2003; Thrift 2001; Cloke and Perkins 2002).

Of course, one also needs to consider the fact that on rare occasions nature does demonstrate its true power in the case of serious or fatal accidents. Most accidents in adventure tourism are attributable to a severe underestimation of natural conditions and inadequate risk assessments. Cultural approaches would identify through this the agency of nature as an actor in the adventure experience. For all the providers' efforts, the outdoor environment is not a theme park, and it cannot be predicted with such accuracy. On a far more mundane level, natural conditions can still determine the enjoyment of the experience. Although weather conditions in particular may dominate such experiences, there are numerous other ways in which nature can influence outcomes. Interactions with recalcitrant wildlife, which are often sold as part of the adventure product, are frequently not up to the expectations of participants. These scenarios demonstrate that nature has an agency manifest in the ability to 'push back' (Thrift 2001; Franklin 2003). Consequently, it becomes apparent that while the landscape is an important part of the adventure experience, its relationship is not as simple as it might at first appear. Mirroring the activities themselves perhaps, the place of a natural setting is highly dynamic in the negotiation of the experience, and this needs to be acknowledged in any consideration of power.

Concluding remarks

Contemporary adventure tourism is emblematic of more symbiotic association between individual and environment than one of domination by either party. Consistent with Nietzsche's view above, to a large degree, individuals conspire with nature, both within and without, for the experience that results. Nietzsche's *Übermensch* is also a character of some relevance here, as the individual who seeks to uncover place in the surrounding environment through expressions of power.

> The *Übermensch* is the man who overcame himself and grows each day. He is the one who reveals his will to power on the environment that surrounds him. He lives in the moment of each and every second. He is the man who has organised his passions of chaos and is disciplined.
>
> (Nietzsche 2002: 259)

What this does not go on to say is that the *Übermensch* may discover other bodies, personal, provider- or place-related which are also seeking to exert their own power relations. This highlights the dynamic nature of adventurous interactions which are in practice more likely to take on a conciliatory tone. These complex articulations of power have been demonstrated and this account has been assisted by recognising scalar tendencies in power at the level of participant, provider and place.

The negotiation of equilibrium takes place as much through the body as it does via any notion of a gaze. In exploring the negotiation of power in adventurous tourism places, this chapter also moves towards addressing the lacuna identified by Veijola and Jokinen (1994: 149) who bemoan '*the absence of the body* from the corpus of the sociological studies on tourism'. As Macnaghten and Urry (1993: 268) suggest, discourses of rural tourism frequently contain 'a number of problematic assumptions based on conceptualising individuals partaking in countryside leisure which produces passive and docile bodies'. On the contrary, the commentary has illustrated the undeniable power of the tourist body, and advocates a more active, embodied and sensuous understanding of the consumption of the countryside. We need therefore to uncover further the textures of this embodied interaction with the landscape if we are to seek to understand more fully the geographical processes behind adventure tourism.

At the same time, Nietzsche suggests that the individual needs to avoid complexity in order to maintain power. The struggle for adventure, and indeed for tourism, is increasingly led by the desire to escape the complexities of the modern world. In an unpredictable and partially commodified natural environment, however, the discussion has shown that these reified dreams are unlikely to be fulfilled, especially where power relations are fluid in the manner highlighted by Allen (2003) and discussed above. Of course, this does not prevent contemporary adventurers from repeatedly returning to these spaces in order to renegotiate these expressions of power. Indeed, the growth of adventure tourism perhaps demonstrates that many amoung us are modern-day Zarathustras seeking to gain power in the mountains.

References

Adventure Travel Society (ATS) (1999) The Importance of Adventure Travel and Ecotourism. Online document. Available from: http://www.adventuretravel.com/seminar_home.htm (last accessed 8 August 1999).

Aitken, C. and Hall, C.M. (2000) 'Migrant and foreign skills and their relevance to the tourism industry', *Tourism Geographies*, 2 (1): 66–86.

Allen, J. (2003) *Lost Geographies of Power*. Oxford: Blackwell.

Berno, T., Moore, K., Simmons, D. and Hart, V. (1996) 'The nature of the adventure experience in Queenstown', *Australian Leisure*, 81: 21–25.

Bordieu, P. (1993) 'How can one be a sports fan?', in S. During (ed.) *The Cultural Studies Reader*. London: Routledge.

Cater, C. (2000) 'Can I play too? Exclusion and inclusion in adventure tourism', *North West Geographer*, 3(2): 50–60.

Cloke, P. and Perkins, H.C. (1998a) '"Pushing the limits": place promotion and adventure tourism in the South Island of New Zealand', in H.C. Perkins and G. Cushman (eds) *Time Out: Leisure Recreation and Tourism in New Zealand and Australia*. Auckland: Longman.

Cloke, P. and Perkins, H.C. (1998b) 'Cracking the canyon with the awesome foursome: representations of adventure tourism in New Zealand', *Environment and Planning D: Society and Space*, 16: 185–218.

Cloke, P. and Perkins, H.C. (2002) 'Commodification and adventure in New Zealand tourism', *Current Issues in Tourism*, 5(6): 521–549.

Crang, P. (1997) 'Performing the tourist product', in C. Rojek and J. Urry (eds) *Touring Cultures: Transformations of Travel and Theory*. London: Routledge.

Cresswell, T. (1999) 'Embodiment, power and the politics of mobility: the case of female tramps and hobos', *Transactions of the Institute of British Geographers* NS, 24: 175–192.

Denzin, N.K. (1997) *Interpretative Ethnography: Ethnographic Practices for the 21st Century*. London: Sage.

Desforges, L. (1998) 'Checking out the planet: global representations/local identities and youth travel', in G. Valentine and T. Skelton (eds) *Cool Places: Geographies of Youth Cultures*. London: Routledge.

Franklin A. (2002) *Nature and Social Theory*. London: Sage.

—— (2003) *Tourism: An Introduction*. London: Sage.

Game, A. (1991) *Undoing the Social: Towards a Deconstructive Sociology*. Buckingham: Open University Press.

Grosz, E. (1990) 'Inscriptions and body-maps: representations of the corporeal', in T. Threadgold and A. Cranny-Francis (eds) *Feminine/Masculine and Representations*. Sydney: Allen & Unwin.

The Guardian (1999) 'Criticism grows of disaster gorge trip', *The Guardian*, 29 July, G2: 3.

Hall, C.M. (1992) 'Adventure, sport and health tourism', in B. Weiler and C.M. Hall (eds) *Special Interest Tourism*. London: Belhaven.

Harrison, P. (2000) 'Making sense: embodiment and the sensibilities of the everyday', *Environment and Planning D: Society and Space*, 18(4): 497–518.

Hawkins, D.E. (1994) 'Ecotourism: opportunities for developing countries', in W.F. Theobold (ed.) *Global Tourism: The Next Decade*. Oxford: Butterworth-Heinemann.

Herod, A. and Wright, M.W. (eds) (2002) *Geographies of Power: Placing Scale*. Oxford: Blackwell.

Hirschmann, E.C. (1984) 'Leisure motives and sex roles', *Journal of Leisure Research*, 16(3): 209–223.

Hollinshead, K. (1998) 'Disney and commodity aesthetics: a critique of Fjellman's analysis of "distory" and the "historicide" of the past', *Current Issues in Tourism*, 1 (1): 58–81.

Johnston, M.E. (1989) 'Peak experiences: challenge and danger in mountain recreation in New Zealand', unpublished PhD thesis, Lincoln University, New Zealand.

Lewis, N. (2000) 'The climbing body, nature and the experience of modernity', *Body and Society*, 6(3–4): 58–80.

MacCannell, D. (1976) *The Tourist*. London: Macmillan.

—— (1992) *Empty Meeting Grounds: The Tourist Papers*. London: Routledge.

Macnaghten, P. and Urry, J. (1993) 'Constructing the countryside and the passive body', in C. Brackenridge (ed.) *Body Matters: Leisure Images and Lifestyles*. Brighton: LSA (Leisure Studies Association publication 47).

Maslow, A. (1987) *Motivation and Personality* (third edition). New York: Harper and Row.

Merleau-Ponty, M. (1962) *The Phenomenology of Perception*. London: Routledge.

Mowforth, M. and Munt, I. (2003) *Tourism and Sustainablity. New Tourism in the Third World* (second edition). London: Routledge.

Nietzsche, F. (1968) *The Will to Power*. New York: Vintage.

—— (2002) *Beyond Good and Evil: A Prelude to the Philosophy of the Future*. Cambridge: Cambridge University Press.

Office of National Statistics (2000) *Travel Trends. A Report on the International Passenger Survey 1999*. London: HMSO.

Poon, A. (1993) *Tourism, Technology and Competitive Strategies*. Wallingford: CABI.

Ritzer, G. and Liska, A. (1997) '"McDisneyization" and "Post-Tourism": Complementary perspectives on contemporary tourism', in C. Rojek and J. Urry (eds) *Touring Cultures: Transformations of Travel and Theory*. London: Routledge.

Rose, G. (1999) 'Performing space', in D. Massey, J. Allen and P. Sarre (eds) *Human Geography Today*. Cambridge: Polity.

Schiebe, K. (1986) 'Self narratives and adventure', in T. Sarbin (ed.) *Narrative Psychology: The Storied Nature of Human Conduct*. New York: Praeger.

Shields, R. (1991) *Places on the Margin. Alternative Geographies of Modernity*. London: Routledge.

Swain, M.B. (1995) 'Gender in tourism', *Annals of Tourism Research*, 22(2): 247–266.

Thrift, N. (2001) 'Still life in present time: the object of nature', in P. Macnaghten and J. Urry (eds) *Bodies of Nature*. London: Sage.

TRCNZ (2004) *New Zealand Regional Tourism Forecasts 2004–2010: Queenstown RTO*. Wellington: Tourism Research Council New Zealand.

Turner, B. (1996) *The Body and Society* (second edition). London: Sage.

Urry, J. (1990) *The Tourist Gaze*. London: Sage.

—— (1999) 'Sensing leisure spaces', in D. Crouch (ed.) *Leisure/Tourism Geographies: Practices and Geographical Knowledge*. London: Routledge.

Veijola, S. and Jokinen, E. (1994) 'The body in tourism', *Theory, Culture and Society*, 11: 125–151.

Weber, K. (2001) 'Outdoor adventure recreation. A review of research approaches', *Annals of Tourism Research*, 28(2): 360–377.

Weekend Telegraph (1998) 'Throwing caution- and myself- to the wind', *Daily Telegraph*, 12 January: 24 (Weekend Telegraph Section).

4 Disability legislation and the empowerment of tourists with disabilities in the United Kingdom

Gareth Shaw

Introduction: tourism and disability in context

In Britain, around 40 per cent of individuals do not participate in holidays during any one year (Shaw and Williams 2002). Of course, such general figures conceal a range of reasons for not taking a regular holiday, including illness, lack of money and pressure of work (English Tourist Board 1989). A small but growing literature has started to explore these people who are excluded from the holiday process. In a pioneering study, Haukeland (1990) working in Scandinavia categorised such people as either 'constrained' or 'unconstrained', with the former wanting to travel, but being unable to do so, whereas the latter chose not to go on holiday (see also Davidson 1996; Deem 1996). It is the former group that further research has focused on, with initial attention being directed to the study of socially disadvantaged families (Hughes 1991; Smith and Hughes 1999). Smith and Hughes (1999) have shown in a somewhat limited study that, for these people, the meaning of holidays remains distinctive. Such meanings encompass: 'escape from normal routines', 'the strengthening of family relationships' and 'the improvement of general well-being'. Of course, these motives can also relate to other holidaymakers, but Smith and Hughes found that amongst socially disadvantaged families, patterns of holiday consumption were also different from the 'norm', in that destination choice and frequency were both greatly limited because of economic constraints.

Such disadvantaged families represent one element within the so-called 'constrained' travellers group. There may also probably be a relatively large proportion of disabled people within this group, although comparatively little is known about their holidaymaking patterns, with most views being based on limited information, mainly derived from recent industry commissioned reports (see Shaw *et al.* 2005). What we do know is that there are an estimated 8.6 million disabled people in Britain with a variety of disabilities who are drawn from a range of family and social circumstances. These figures are based on Department for Education and Employment (DfEE) estimates and relate to adults only (aged 16 and over).

The passing of the Disability Discrimination Act (DDA) 1995 has brought increased political and economic attention to people with disabilities, especially from service providers and the tourism industry. The DDA makes it illegal for service providers (e.g. attractions and accommodation units) to discriminate against people with disabilities. The DDA itself is in three main parts, phased in since the legislation became law in 1996. In this context, part three holds the major implications for the tourism industry. This states that service providers should have taken reasonable measures to alter their premises to accommodate people with disabilities, and this became law in October 2004. The response from government has been to highlight such needs for change, some of which were in *Tomorrow's Tourism* which included notions of widening access to the elderly and disabled (Department of Culture, Media and Sport 1999).

We have, therefore, a growing interest in people with disabilities as potential tourists but, significantly, this tends to be centred around a fairly narrowly defined agenda. In turn, this is strongly associated with the requirements of the DDA and, more especially, the circumstances surrounding service providers within the tourism industry. Set aside from this there is an area of recent research focusing on tourists with disabilities, which is attempting to uncover the needs, motives and behaviour of these tourists. This chapter seeks to explore the emerging literature, focusing particularly on tourists with disabilities alongside other more general studies of the disabled. The main aim of the chapter is to uncover the relationship between disability and holidaymaking in the UK by covering three main themes: the participation of people with disabilities in holidaymaking; the meaning of holidays to such people and their families; and the nature of holiday decision-making by people with disabilities in terms of perceived and real levels of accessibility. These are set within a general framework of legislation and the nature of empowerment. The chapter begins with a broad ranging review of current research on tourists with disabilities.

Disability legislation and tourism research: extending the debate

Increasing campaigns to provide equality to people with disabilities and, more importantly, the introduction of legislation to halt discrimination against this group, has brought a range of responses. These include not only a change in the activities of service providers and the tourism industry generally, but also a growing research agenda within tourism studies. The passing of disability discrimination legislation in the UK has mirrored developments in Australia (Darcy 2002), Canada (Canadian Tourism Commission 1997) and the USA (Burnett and Baker 2001). However, as Darcy (2002: 61) points out, disability and tourism research 'have largely remained separate areas of study, with little substantial research being undertaken' on combining such issues.

Despite such criticisms, the situation is changing, especially in Australia and North America where a number of theses have initiated research on the constraints and behaviour of tourists with disabilities (see, for example, Daruwalla 1999; Foggin 2000). In the USA, for example, Ray and Ryder (2003) have explored the motivations of disabled travellers in a preliminary study that makes interesting comparisons between able-bodied tourists and those with physical impairments within the context of 'nature' tourism. This also raises the importance of marketing the visitor experience, a theme touched upon by Salt (2004) within the UK.

Similarly, within the UK a succession of recent contributions have at least drawn attention to the plight of people with disabilities wanting to engage in the holiday process (Goodall 2002; Phillips 2002; Veitch and Shaw 2004a, 2004b).

As yet, there has been an overemphasis on the legislation and problems of physical access which have served, according to Shaw and Coles (2004), to narrow down the research agenda. Within the UK this has been further compounded by the publication of a range of reports by tourism policymakers. There are gradual signs of the debate widening in the UK, in part following studies in Australia by Darcy (2002) and Daruwalla (1999). For example, recent attention has been given to understanding all the barriers conditioning tourists with disabilities, including the communication of holiday information to such people (Veitch and Shaw 2004b).

Such progress stands in marked contrast to more general research of disability or, indeed, within the geographies of disability which is attempting to engage with many of the mainstream debates in disability studies (Park *et al.* 1998). Within Britain, an early core debate has been constructed around the 'social model of disability', which emerged following the activities of the Union of the Physically Impaired Against Segregation (1976). The model, which has become an important, if recently contested, paradigm (Shakespeare and Watson 2001), argues that 'people with impairments are disabled/excluded by a society that is not organised in ways that take account of their needs' (Tregaskis 2002: 458). More especially, Barnes (1991) has highlighted some significant distinctions within the model between impairment and disability, namely, that:

- impairment is the functional limitation within the individual caused by mental or sensory impairment
- disability is the loss or limitation of opportunities to take part in the normal life of the community on an equal level with others, due to physical and social barriers.

Associated with these ideas are what Chalmers (1978) has called a series of 'protective belt' theories that expand upon the basic social model, the most influential of which is the modernist, materialist one. This views exclusion as arising under modern capitalism. More recently, others have stressed the importance of culture and the media within society in indirectly justifying and maintaining the exclusion of disabled people (Tregaskis 2002; Hughes and Paterson 1997). Within this context Barnes (1991: 5) has argued that it is necessary to view 'current social responses to impairment as the cultural product of the interaction between means of production and central societal values'. Such ideas may be particularly important within the cycles of production and consumption that occur within tourism. Of equal importance within tourism are the ideas within disability studies that call attention to the growing need to study the experiences of people with disabilities relative to abled-bodied people (Hughes 1991). Only by doing this can the experiences of people with disabilities be fully appreciated. Despite such viewpoints, Gleeson (1997: 181), amongst others, has characterised the disability debate as suffering from 'the legacy of theoretical deprivation' due to its failure to engage with major theories of society.

From the perspective of research within tourism, the debate is certainly still somewhat limited. Indeed, few, if any, of the social constructs of disability and their related ideas have permeated tourism studies. Researchers working within tourism have shown fairly limited interest in exploring the experiences of the disabled traveller although, as was pointed out, the research agenda is starting to move on. Certainly, tourism geographers can link with debates within mainstream geography, where discourses on the geographies of disabilities have been more progressive. Park *et al.* (1998) have provided an earlier review of progress, while Gleeson (1997) has sought to outline various perspectives on the geographies of disability. Within both contributions there are some strong echoes of the central debates in mainstream disability studies. Certainly, this is the case with what Park *et al.* (1998: 211) term the 'interpretative geographies of physical disability'. In this context, notions of the social construction of disability as a state of marginalisation have been especially highlighted by Imrie (1996) in terms of the 'disabling city'. Earlier work by Imrie and Wells (1993) suggested that, within Britain, the planning system was linked to the creation of a segregationist ethos which has contributed to the physical marginalisation of people with disabilities.

Clearly, spatial structures are critical, material variables in the social construction of disability, as are temporal changes brought about by the increased pace of life in much of post-modern society (Freund 2001). Of course, both spatial and temporal trends are important within tourism and provide some key linkages with circuits of production and consumption. These links can be extended, particularly in terms of what Glenn Smith (1999) and others (see, for example, Freund 2001) have called a post-modern view of space. From this perspective, disabling space can be reformulated as semiotic space as represented by 'a disabling space of values' (Glenn Smith 1999: 63).

Such a perspective represents a much deeper understanding of the relationship between people with disabilities and space. For example, although particular public and commercial spaces may be made accessible in physical terms through increased legislation, experientially such spaces may still exclude people with disabilities. Part of this is the stereotypical role that is expected by society of people with disabilities. Within this context, Butler (1999) has researched the roles youths with disabilities are expected to follow in society by exploring discourses on the body to illustrate the construction of such social images (Park *et al.* 1998). This socio-cultural and political reproduction of disabling environments, especially within the post-modern city, has also been explored by Hahn (1986) in the context of Los Angeles. While particular attention was focused on the barriers of distance for the disabled, Hahn also stressed the significance of power relationships in both constructing and maintaining certain aspects of disabling environments.

This recognition of the importance of power relationships leads to a more careful consideration of notions of power and more especially the issue of spatialised power (Massey 1999). In terms of tourists with disabilities, there are two key dimensions to this idea. One is related firmly to what Massey (1999) and Panelli (2004) discuss as the power relationships in the construction of various barriers that may serve to exclude or marginalise certain groups, for example the disabled. The second dimension relates much more to how power is obtained, brokered and used by different groups within society. Within a geographical context perspectives on power include Allen's (1997) categorisations, encompassing power as an inscribed capacity, a resource and as technology. Alongside, Sharp *et al.* (2000) view power through domination, resistance and entanglements. As Panelli (2004: 163) points out, there are parallels between such views, and certainly notions of capacity and domination help focus attention on the 'power of the institution to "control" certain spaces'. Moreover, people with disabilities have little perceived economic power to help exercise any control since they are marginalised both in the workforce and as consumers. Of course, increasingly power is mediated by the state especially through the early ideas of social services (Woods 1997) and more recently due to the state's interests in creating an (apparently) 'inclusive society'. As will be argued in this chapter, the disability discrimination legislation forms an important part of such an agenda.

The development of notions of inclusive tourism in the UK should also be seen within the context of the European Union (EU). As the European Economic and Social Committee (2003: 10) explained:

> Removing and lessening these barriers [to disabled access] is not only a must on grounds of equal rights and opportunities and non-discrimination, as championed by the EU and its Member States, but also on contributing to the growth of an economic sector.

At the European level, the state's mediation seems to be based on two very different criteria: namely, a platform of social justice and inclusion alongside strong economic motives. Such complexities fit neatly with the idea that 'power' is best conceived as a circulation or a web or net (Panelli 2004: 164). In terms of the empowerment of tourists with disabilities in the UK, the circulation of power within and between the public and private sectors of the tourism industry represents a complex web of conflicting interests.

Such discussions are particularly relevant to disability research within tourism, given the significance of perception and image. It is the links between the social organisation of space, along with its meanings, that are of critical importance to understanding the degree to which people with disabilities are excluded. Clearly, there is some considerable way to go on developing a comprehensive research agenda within tourism that will integrate with broader debates within disability studies.

At the outset this agenda needs to embrace the social model of disability and bring the debates on the social constructs of disability into tourism research. In addition, there is a need to incorporate spatial and temporal structures which are significant material factors in the social barriers of tourists with disabilities. As Wendall (1996: 38) points out, as post-modern society becomes increasingly centred on the presumptions of rapid travel, 'the more disabling are those physical additions that affect movement and travel'. Furthermore, such time–space compression is a significant feature in much of post-modern tourism and one that holds significant implications for travellers with disabilities. Similarly, the social construction of public and private spaces, especially through the processes of commodification, have important impacts on tourists which have not been investigated. Such processes have involved the increasing theming of leisure spaces specially around powerful commercial themes (Shaw and Williams 2004; see Cater, Lew, Hall – chapters 3, 7 and 11 in this volume). In turn, this has led to more exclusive spaces and the marginalisation of certain groups (Gotham 2002; see Cater, Hall – chapters 3 and 11 in this volume). As was argued earlier, tourism spaces may be more physically accessible but, because of the symbolic meanings of certain spaces, they may simultaneously be restricted to the experiences of people with disabilities. This links with Imrie's (1996) argument that such people are often forced to live, and it could be added holiday, in very different spheres of space than the able-bodied. Before turning attention to aspects of the motives, meanings and experiences of holidays by people with disabilities, we consider the DDA in terms of its impact on the tourism industry and the empowerment of the disabled.

The Disability Discrimination Act: empowering tourists with disabilities

The Disability Discrimination Act 1995 that passed into British law in 1996 is a complex piece of legislation that has been brought into force in a series of stages. The DDA makes it unlawful to treat people with disabilities less favourably than the able-bodied because of their disability. Since its second stage, introduced in October 1999, service providers have had to make reasonable adjustments for people with disabilities, such as providing extra help or making changes to the way they provide their services. The introduction of the final part of the DDA in 2004 obliges service providers to consider making 'reasonable changes to the physical features of their premises to overcome physical barriers to access' (Department for Education and Employment 1999: 2). At the time of writing, the DDA is enforced by the Disability Rights Commission. Within tourism a watching brief is provided by Holiday Care which was renamed 'Tourism for All' in 2004. This non-government pressure group however is poorly funded but nevertheless attempts to act as a focal point for information provision, as well as liaising with the EU.

Within the tourism sector, the DDA is seen as a final, legislative response to earlier calls to make tourism much more inclusive. Such ideals stem from debates within the 1970s, which became more fully articulated during the 1980s, with the publication of English Tourist Board's (1989) *Tourism for All*. This reported on a number of key constraints including the fact that many tourism operators felt they were limited in their response to the needs of tourists with disabilities, mainly due to a lack of information, and it articulated the problems encountered by the tourism industry across the UK. Equally, many tourists claimed that accommodation providers very often made unrepresentative claims about the suitability of the facilities for the disabled. In response to such findings, the report presented a range of recommendations, urging businesses to improve access and promote the development of a voluntary national standard of accessibility. The report suggested this be achieved through the Hotel and Holiday Consortium, and that the tourism industry together with the national tourist boards cooperate on definitions and standards (Veitch 2000). The National Accessible Standard was finally agreed in 1995 as a mechanism for assessing levels of accessibility mainly for wheelchair users. Unfortunately, the scheme proved limited and by 1999 only 1.2 per cent of all serviced accommodation operators in England had joined the scheme (Veitch 2000). The scheme not only suffered from a low take-up but was also too narrow in its focus with its overemphasis on wheelchair users.

To this end, such voluntary schemes, while attempting to ease the access difficulties of certain people with disabilities, in reality did little to raise awareness amongst the majority of service providers. As the chairman of the National Disability Council stated on the general response of businesses to people with disabilities, for most firms 'disability was not really on the radar screen' (NDC 1999 quoted in Veitch 2000). In this context, well meaning reports and voluntary schemes did little to empower people with disabilities when it came to holiday-making. One key question is therefore: has the introduction of the DDA changed the attitude of service providers and, in doing so, helped to improve the plight of tourists who have disabilities? To investigate this we need to consider two strands of evidence, the first of which concerns the attitudes of policymakers within tourism, especially the national tourist boards, whereas the second relates more to the direct response of service providers within the tourism industry.

In terms of the first strand of evidence, numerous reports, both published and unpublished, have been produced which, together, have sought to convey the message of making holidays more accessible to people with disabilities. The message of inclusivity was clearly raised by government in *Tomorrow's Tourism*, which highlighted widening access for the elderly and disabled (Department of Culture, Media and Sport 1999). More specifically, it is possible to recognise a series of reports produced by the English Tourism Council (ETC), often in association with the national tourist boards of Northern Ireland, Scotland and Wales, that have stressed the economic potential of the market for tourists with disabilities. These include a joint study between ETC (2000) and the Joint Disabilities Charities Research Group (1999) which revealed, from a survey of people with disabilities that 75 per cent of respondents considered holidays to be important, whilst 66 per cent of the sample had taken a holiday in the previous year. These findings were based on some 254 questionnaires completed by adults with disabilities along with 18 in-depth interviews. This was followed by an ETC (2000) report which argued that at least some 2.7 million people with disabilities in England have a propensity to take domestic holidays. Within a broader context, the report also estimated that there are around 9.4 million disabled adults in the UK and, if children are included, the figure rises to 10 million. As Shaw and Coles (2004) have argued, the primary aim of the ETC in selecting such relatively high estimates is to send a message to businesses that this market segment holds good potential. Unfortunately, such reports fail to highlight the rather fragmented nature of this market, in that many people with disabilities are unable to take holidays because of a range of barriers. Of particular note are economic constraints, either because people with disabilities are unable to work, have lower status jobs or tend to receive lower salaries on average compared to the able-bodied. Such characteristics have been identified within the UK as, for example, in the survey conducted for the Department for Education and Employment (1998). This covered some 2,000 people with disabilities of working age and found that only 46 per cent of the sample were in work and that these were most likely to have lower skilled occupations.

Evidence from the reaction of service providers to both these publicity campaigns and to more direct advice on the significance of the DDA also illustrates the scale of the problem. As Phillips (2002) suggests, there is a belief amongst tourism agencies that many parts of the industry are reluctant to change in order to be more inclusive and meet the needs of people with disabilities. Successive surveys of service providers have drawn attention to the low levels of understanding of what the 1995 Act means to the way they operate their businesses. Thus, early work from the Department for Education and Employment (1998) found significant differences between public and private sector organisations, as well as differences in size of firm, as to how the legislation was understood.

Establishments within the public sector tended to have more awareness and a better record of action than private businesses. The dominant reason for not taking action by most businesses in the private sector was that 60 per cent already believed their services were fully accessible. Similarly, 73 per cent of establishments provided no staff training on the service provision to customers with disabilities, and 70 per cent were not covered by any formal (written) policy on the provision of services to the disabled (DfEE 1998). Since this survey, tourism policymakers have been active in promoting the DDA through a range of information guides and meet-ings. ETC issued a series of business information leaflets including 'Welcoming Disabled Travellers' (ETC 1999). This gave specific examples as to how businesses could and should improve access for the disabled. Similarly, the Disability Rights Commission (2003) launched a 'New Mini Guide' to provide information on accessible tourism. In this document, the partnership between tourist boards, Holiday Care (now renamed Tourism for All) and the Disability Rights Commission was highlighted as the 'Inclusive Tourism Partnership'. In the same year, government Transport Minister John Speller drew attention to the need for the UK aviation industry to focus on the needs of travellers with disabilities (British Tourist Authority 2003).

However, some recent surveys suggest that many service providers within the tourism industry still lack an understanding, both of what the 1995 legislation means and how to become more inclusive. For example, the BMRB Social Research (2003) survey of service providers responses to the DDA based on 2,022 telephone interviews conducted in 2003 revealed that there was still the need for substantial improvements. As Table 4.1 shows, only 24 per cent of businesses had embarked on the training of staff in disability issues, whilst 3 per cent had no knowledge of the issue. As the survey reported, staff training of disability awareness is at best patchy, although some 13 per cent of organisations did claim to have some form of internal communication mechanisms for staff feedback on dealing with customers with disabilities. The same survey also asked organisations to highlight the main barriers to making adjustments for people with disabilities in order to

Table 4.1 Existing and planned adjustments for customers with disabilities (2003)

	% of businesses			
	Existing	*Planned*	*Not undertaken*	*Don't know*
Modifications to improve physical accessibility	45	9	45	1
Changes to the provision of services	32	3	61	4
Staff training in disability issues	24	10	63	3

Source: modified from BMRB Social Research (2003).

accommodate the DDA. These fell into three main types: a lack of knowledge and awareness of legal requirements, in particular among small and medium-sized enterprises; problems of adjusting the physical environment in terms of the design difficulties to be overcome; and, finally, cost factors, which were seen by most small organisations as a major constraint.

Significantly, none of the organisations within the survey perceived the lack of staff training or awareness as a major barrier to improving accessibility for people with disabilities. This is a special cause for concern given the importance of service quality within the tourism industry and, indeed, in the tourist experience. Associated with this is the mindset and attitudes of many able-bodied people towards people with disabilities. Thus, Veitch (2000) quoting work by the Leonard Cheshire Foundation, a charity representing disabled people, found strong evidence of social exclusion (Knight and Brent 1999). In this context, people with disabilities often felt isolated from normal activities in all aspects of their lives. However, one of the most significant barriers to inclusion was seen as the attitude of other people. As Veitch (2000: 17) exemplifies, people with disabilities commonly complain that the able-bodied treat them 'as children or half wits', or 'as a nuisance and an inconvenience when they go out'. Knight and Brent (1999) also found that few able-bodied people really understood the problems faced by the disabled, with almost 25 per cent of able-bodied members of the general public feeling self-conscious if they encounter people with disabilities.

Clearly, the available evidence suggests that whilst the DDA has brought disability and access into political and, to some extent, economic debates, there is still much to be done. Furthermore, within the tourism industry, in spite of efforts by policymakers, many service providers have still to grasp the ideas of inclusive access. In this context also, the empowerment of tourists with disabilities appears to have been limited and, at best, somewhat narrow. The limitations relate to the attitudes of the industry and the failure to overcome the problems of negotiating different spaces by people with disabilities on holiday. Similarly, the narrowness concerns the overemphasis on physical barriers at the expense of the total 'visitor experience' as argued by Veitch and Shaw (2004b).

'Travel is life': the meaning of holidays to people with disabilities

Krippendorf (1987) argued that for many 'travel is life'; the phrase expresses both the cultural significance of holidays but, more importantly, it highlights the need for more inclusive forms of tourism. Other authors have expressed such inclusivity in terms of the rights to rest and holidays as laid down in the United Nations 1948 Universal Declaration of Human Rights (Shaw and Williams 1994). Murphy (1985) drew early attention to the discussion of rights to holidays under the term social tourism. More recently, Darcy (2002: 1) has revisited such ideas, arguing that much of the later twentieth century was 'characterised by campaigns for equal rights and social justice'. As was pointed out earlier, disabled groups were part of this agenda for social inclusion and rights to holidays. It is therefore somewhat surprising that, until fairly recently, few studies had been directed at the meaning of holidays to the disabled. This stands in contrast to other excluded groups, such as single-parent families (Smith and Hughes 1999) and low-income households (Williams and Windebank 2002).

Despite the limited number of past studies on tourists with disabilities, recent research has started to redress this situation, especially in Australia (Darcy 2002; Darcy and Daruwalla 1999). Progress in North America is also worthy of note (see for example Canadian Tourism Commission 1997; Burnett and Baker 2001). However, some of the most comprehensive contributions have been from Darcy (2002) who has researched the tourism participation of people with impairments. Like Shaw and Coles's (2004) preliminary study of tourists with disabilities in the UK and the meaning of holidays, Darcy also utilises, in part, the social model of disability. Following the arguments of Shakespeare and Watson (2001), Darcy (2002: 15) 'calls for the reconsideration of the personal' within the study of socially constructed constraints and barriers. Such an 'embodied ontology' brings into consideration the personal and psychological characteristics of tourists with disabilities. This is a far cry from viewing all tourists with disabilities as the same, with similar problems and needs.

By understanding what Darcy (2002) terms 'embodied ontology' it becomes possible to understand the experiences and diversity of tourists with disabilities. In this context, the 'Travel is Life' project established at the University of Exeter (see Shaw and Coles 2004) aims to incorporate the social model of disability into tourism studies. This chapter now proceeds to report on two very different surveys on tourists with disabilities. The first is a survey undertaken for Visit Britain (2003) and conducted by NOP, whilst the second is a pilot study carried out by the author. Taken together, they provide an important, if still partial, insight into the world of tourists with disabilities.

The background and objectives of the NOP (Visit Britain 2003: 1) survey was to 'explore the needs, expectations and experiences of disabled people in selecting serviced accommodation for a short break holiday'. In this sense, the study is very specific and based on five mini-group discussions and 25 in-depth interviews. Respondents were selected in different categories of impairment: visually, hearing,

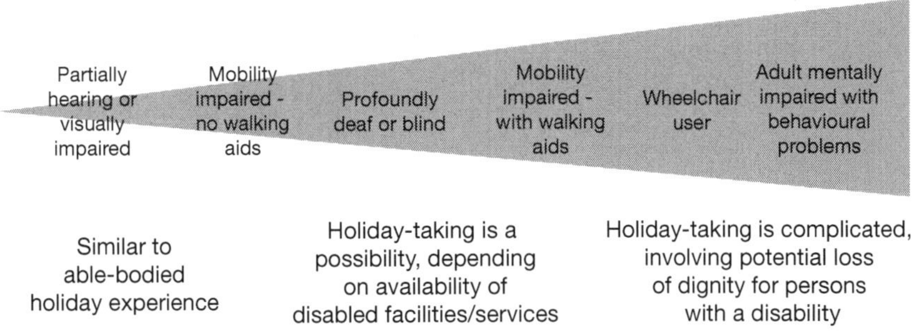

Figure 4.1 A continuum of impacts of disability on holidaymaking

Source: author, based on Visit Britain (2003).

mental, and mobility (both wheelchair and non-wheelchair users). This survey highlighted a number of key issues, including the potential relationship between type of disability and its impact on holidaymaking (Figure 4.1). The continuum of impacts suggests that certain types of disability lead to more problems of holiday-making. Of course, while this is true in relation to certain constructed barriers, it ignores the ideas of the social model of disability. In other words, it partly legitimises the socially constructed barriers to certain types of disability – especially adults with mental impairments. In terms of the latter group there has been little research on their experiences, although we do know that only 21 per cent of adults with long-term mental illness are in employment (Shaw 2004). This automatically imposes strong financial barriers on holiday-taking alongside other difficulties.

Within the context of the mentally impaired, the NOP (Visit Britain 2003) survey found a range of significant barriers to all forms of holidaymaking (Table 4.2). However, when holidays could be taken, they were regarded as significant in providing respite for other members of the household and pleasure for the disabled person.

The NOP survey also revealed a range of usually negative experiences held by tourists with disabilities, especially of holidaymaking in the UK. Given the objectives of this survey, the emphasis is very clearly on the impact of serviced accommodation on such experiences and, whilst this may be somewhat narrowly focused, as with all holidaymakers, accommodation is a critical factor (Table 4.3). Clearly, such experiences suggest that there is a considerable level of adjustment needed in the serviced accommodation sector in order to meet just the basic

Table 4.2 Barriers to holidaymaking by the mentally impaired

- Need carer assistance at all times
- Unpredictable behaviour limits scope of travel and use of accommodation
- For some (e.g. Down's syndrome) a break in routine can lead to a lack of co-operation
- Public attitudes to the mentally impaired.

Source: modified from Visit Britain (2003).

Table 4.3 Experiences of tourists with disabilities in the UK (focus on accommodation provision)

Physical barriers

- Accommodation often failed to meet needs, i.e. lack of facilities for people with disabilities, including:
  - too few disabled parking spaces
  - no policing of disabled spaces, so used by able-bodied
  - entrances too narrow
  - access only by steps
  - only a temporary wheelchair ramp available
  - lifts too small
  - insufficient room in public areas
  - bed inaccessible
  - insufficient washing facilities.

Attitudes of staff and owners

- Some degree of insensitivity
- Refuse to engage people with disabilities in discussion
- Levels of inflexibility in meeting requests of people with disabilities
- Feeling that staff are poorly trained to deal with visitors with disabilities.

Source: modified from Visit Britain (2003).

requirements of people with disabilities. The general message from this survey was that visitors with disabilities are subject to a 'lottery-style' experience when booking and using unknown serviced accommodation. As a consequence, those who do take holidays often tend to minimise the risk of having a bad experience by returning to known accommodation (Visit Britain 2003).

The second survey utilised in this chapter serves to highlight both the meaning of holidays to people with disabilities and the experiences of holidaymaking. In contrast to the NOP survey, the work undertaken by the author has more broader based objectives, although it is limited by being based on a small sample. This relates to a survey carried out in 2002 with respondents being selected from contacts made through a UK-based national disability magazine, *Disability View*. Some 30 individuals responded but, after initial contact, only 24 respondents participated in the study. Information was collected by a detailed postal questionnaire which was followed up by in-depth telephone interviews of five respondents selected randomly. The survey was aimed at uncovering the nature, meaning and experiences of holidays by people with disabilities. Unlike the NOP survey, no mentally impaired respondents were included as the emphasis was on forms of physical disability. As Table 4.4 shows, information was collected on three key themes. In addition, the final part of the questionnaire collected information on the respondents, their disability and family circumstances – including employment status. These variables are important, as they provide a series of internal barriers, which more general studies have shown link disabilities with the behaviour of the family (Altman *et al.* 1999).

Table 4.4 Major themes explored in the 'travel is life' survey

- Experience of holidays over last three years. Focus on gaining an understanding of types of holidays taken in terms of
  - travel patterns
  - organisation of travel arrangements
  - type of accommodation
  - length of stay.
- Information from non-holiday takers. This focused attention on the main reasons for not taking a holiday.
- Meaning of holidays, both to the individual as well as to other family members. Information was collected for both holiday takers and non-holiday takers within the survey.

Source: author.

Unlike previous surveys, this case study also collected information on the attitudes of those people with a disability unable to participate in holidaymaking, forming part of Haukeland's (1990) 'constrained' group of tourists. To a large extent, these people are the ones ignored by the various officially sponsored surveys produced by the UK tourism industry. Within the small case study there are five respondents (representing almost 20 per cent of the sample) who had not taken a holiday in the three years covered by the questionnaire. As Table 4.5 shows, they have a number of common characteristics. First, they are all over 56 years; second, none had worked for some time; and, finally, all have mobility difficulties. In all but two cases, financial problems were rated as 'very important' or 'important' reasons for not taking a holiday. The two respondents, 4 and 5, who were not hampered by financial constraints, gave their reasons for not holidaying as mobility difficulties. As respondent 4 explained, 'my husband has a mobility scooter and although our daughter lives in Prestatyn (Wales) we cannot get there because there are no guards vans on the trains where we could put the scooter'. In the case of respondent 5, it was the experience of attempting a holiday in an unsuitable hotel that had left a lasting, negative, impression.

For those respondents highlighting financial constrains, no amount of improvements in physical access could solve their problems. They need not only empowerment through the type of changes being promised by the DDA, but also a social tourism scheme that would help pay for holidays. In the case of all 5 respondents, the benefits of holidays were clearly recognised. Holidays meant 'a time to be completely relaxed' (respondent 1), 'an important time of rest for my wife and son' (respondent 2), and 'a time of enjoyment and change of scene' (respondent 3).

The meaning of holidays for all 24 respondents in the survey is shown in Table 4.6, which clearly identifies two main themes. The first is that holidays do have similar meanings to people with disabilities and their families as they do to the able-bodied (see Smith and Hughes 1999). For example, over 54 per cent of the sample rated holidays as a time for new experiences as 'very important', with a similar percentage rating for 'seeing new places'. These are similar to meanings rated by

Table 4.5 Characteristics of non-holidaymakers: case study respondents

Respondent	*Age*	*Gender*	*Employment*	*Disability*	*Family circumstances*
1	66–75	M	Retired from semi-skilled occupation	Stroke (had condition 15 months)	Lives with wife who acts as carer
2	56–65	M	Retired (formerly engineer)	Series of mini-strokes over 12 years	Lives with wife and son, has full-time carer
3	56–65	F	Unemployed	Unable to walk for 13 years	Lives with husband and daughter
4	76–85	M	Retired	Old age and stroke	Lives with wife, who is full-time carer
5	66–75	F	Retired	Not specified	Lives with daughter, has full-time carer

Source: author.

Table 4.6 The meaning of holidays to people with disabilities

Meanings	*Level of importance (%)*		
	Very important	*Important*	*Not important*
Meeting new people	33.3	29.2	37.5
Seeing new places	54.2	29.2	16.6
Achieving a sense of freedom	37.5	37.5	25.0
Time of rest	50.0	29.2	20.8
Get away from everyday pressures	45.0	41.6	12.5
Time for the family	45.9	37.5	16.6
Time for new experiences	54.2	33.3	12.5

Source: author.

Note: sample = 24.

other respondents without disabilities. Second, within this survey, it seems clear that the idea of holidays representing a 'time of rest' and a 'time for the family' are of increased importance for people with disabilites (Table 4.6). Indeed, in many of the detailed statements given by the respondents, such meanings were emphasised. As argued elsewhere (Shaw and Coles 2004), such findings seem to show important similarities with the findings of Smith and Hughes (1999) on disadvantaged families. Given the economic marginalisation of many people with disabilities and their families, it is hardly surprising that such parallels exist.

Conclusion

The Disability Discrimination Act 1995 has acted as a catalyst to focus attention on the needs of people with disabilities within the UK. This legislation, in all its different parts, has given rights to people with disabilities and put in place a legal requirement for service providers to cater for such consumers. Within tourism studies, it has also acted as a spur for research on the demand of tourists who have disabilities and the responses of the industry. This has led to a small but growing literature on tourists with disabilities and their families. However, as this chapter has set out to show, there are a number of significant limitations, both with the DDA itself, its impact on the industry and, finally, the way that it has conditioned perspectives on tourists with disabilities. These different issues have formed the core of key debates explored in this chapter, and each are interlinked.

In terms of the DDA, this appears to have promoted a somewhat narrow research agenda as well as a fragmented response from the tourism industry. Indeed, the main surveys point to a high degree of confusion as to the legal requirements of the 1995 legislation, especially its final part. The confusion amongst tourism and hospitality businesses is matched by an unequal response. Of equal importance is the evidence which suggests that attitudes towards consumers and tourists with disabilities remain difficult, in that such people are still viewed as problematic, and expensive to cater for. Set against these attitudes, various tourism organisations and policymakers have attempted to promote the economic benefits to be gained by businesses that focus on the market surrounding people who have disabilities. These views have stressed the untapped potential of this market sector to a rather sceptical industry. In doing so, they have neglected the social dimensions of tourism access, preferring to see only economic potentials.

This chapter has attempted to expose such weaknesses and focus on the complex set of issues which still constrain people with disabilities and their access to tourism. Of course, there is much concern about mobility problems and to this extent the DDA should see an empowerment of some tourists with disabilities. However, as the different surveys reported in the final part of this chapter show, concerns stretch beyond mobility. Of particular note are the attitudes of service workers to the disabled, alongside those of other tourists. Added to these are the financial constraints faced by many people with disabilities. This group does not represent one viable commercial segment, but rather a series of sub-groups, including those marginalised both by income and disability, and it is these people for whom the changes in disability legislation have done little to help. In this context further empowerment of people with disabilities is likely to come not from increased legislation on anti-discrimination but from a comprehensive policy of social tourism.

References

Allen, J. (1997) 'Economics of power and space', in R. Lee and J. Willis (eds) *Geographies of Economies*. London: Arnold.

Altman, B.M., Cooper, P.F. and Cunningham, P.J. (1999), 'The case of disability in the family: impact on health care utilisation and expenditures on non-disabled persons', *Millbank Quarterly*, 77(1): 39–75.

Barnes, C. (1991) 'Theories of disability and the origins of the oppression of disabled people in Western society', in L. Barton (ed.) *Disability and Society: Emerging Issues and Insights*. New York: Longman.

BMRB Social Research (2003) *Employers and Service Providers' Responses to the DDA*. London: Centre for Research in Social Policy.

British Tourist Authority (2003) 'Government news: the sky's the limit for disabled air travellers', *Westminster E-Bulletin*, 21 March, London: Government Policy Consultants.

Burnett, J.J. and Baker, H.B. (2001) 'Assessing the travel-related behaviors of the mobility-disabled consumer', *Journal of Travel Research*, 40(1): 4–11.

Butler, R.E. (1999) 'Rehabilitating the images of disabled youths', in T. Skelton and G. Valentine (eds) *Geographies of Youth Cultures*. London: Routledge.

Canadian Tourism Commission (1997) *Disabled Persons and the Tourism Industry: A Bibliography*, CTC Series No. 4. Ottawa: CTC.

Chalmers, A.F. (1978) *What is this Thing called Science? An Assessment of the Nature and Status of Science and its Methods*. Buckingham: Open University.

Darcy, S. (2000) *Anxiety to Access: The Tourism Patterns and Experiences of New South Wales People with a Physical Disability*. Sydney: Tourism New South Wales.

—— (2002) 'Marginalised participation: physical disability, high support needs and tourism', *Journal of Hospitality and Tourism Management*, 9(1): 61–72.

Darcy, S. and Daruwalla, P.S. (1999) 'The trouble with travel: people with disabilities and tourism', *Social Alternatives*, 18(1): 41–46.

Daruwalla, P.S. (1999) 'Attitudes, disability and the hospitality and tourism industry', unpublished PhD thesis, University of Newcastle, Australia.

Davidson, P. (1996) 'Holiday and work experiences of women with young children', *Leisure Studies*, 15(2): 89–103.

Deem, R. (1996) 'Women, the city and holidays', *Leisure Studies*, 15(2): 105–119.

Department of Culture, Media and Sport (1999) *Tomorrow's Tourism*. London: DCMS.

Department for Education and Employment (1998) *Employment of Disabled People: Assessing the Extent of Participation*, Research Briefs, Report No. 69. London: DfEE.

—— (1999) *Disability Discrimination Act 1995. An Introduction for Small and Medium-sized Businesses, Rights of Access to Goods, Facilities, Services and Premises*. London: DfEE.

Disability Rights Commission (2003) Attitudes and awareness survey. Available from: www.drc.org.uk (accessed January 2005).

English Tourism Council (1999) 'Welcoming Disabled Travellers', Business Information Leaflet No. 5. London: ETC.

English Tourist Board (1989) *Tourism for All*. London: ETB.

—— (2000) *People with Disabilities and Holiday Taking*. London: ETC.

European Economic and Social Committee (2003) *Tourism for Everyone*. INT/173, Brussels: EESC.

Foggin, E.S. (2000) 'The experience of leisure tourism of people with disabilities', unpublished PhD thesis, Université de Montréal.

Freund, P. (2001) 'Bodies, disability and spaces: the social model and disabling spatial organisations', *Disability and Society*, 12(2): 172–184.

Gleeson, B.J. (1997) 'Disability studies: a historical materialist view', *Disability and Society*, 12(2): 197–202.

Glenn Smith, R. (1999) 'Reflections on a journey: geographical perspectives on disability', in M. Jones and L. Marks (eds) *Disability, Divers-Ability and Legal Change*. London: Routledge.

Goodall, B. (2002) 'Disability discrimination legislation and tourism: the case of the United Kingdom', in E. Arola, J. Karkkainen and M-L. Siitari (eds) *Tourism and Wellbeing*. Conference Proceedings of the second Tourism Industry Symposium, Jyvaskyla, Finland.

Gotham, K. (2002) 'Marketing Mardi Gras: commodification, spectacle and the political economy of tourism in New Orleans', *Urban Studies*, 39(10): 1735–1756.

Hahn, H. (1986) 'Disability and the urban environment: a perspective on Los Angeles', *Environment and Planning D: Society and Space*, 4: 273–288.

Haukeland, J. (1990) 'Non-travellers: the flip side of motivation', *Annals of Tourism Research*, 17(2): 172–184.

Hughes, B. and Paterson, K. (1997), 'The social model of disability and the disappearing body: towards a sociology of impairment', *Disability and Society*, 12(3): 325–340.

Hughes, H. (1991) 'Holidays and the economically disadvantaged', *Tourism Management*, 18(1): 3–7.

Imrie, R. (1996) *Disability and the City: International Perspectives*. London: Routledge.

Imrie, R. and Wells, P.E. (1993), 'Disablism, planning and the built environment', *Environment and Planning C: Government and Policy*, 11: 213–231.

Joint Disabilities Charities Research Group (1999) *Holiday Provision for Disabled People*. London: Charities Research Group.

Knight, J. and Brent, M. (1999) *Excluding Attitudes: Disabled People's Experience of Social Exclusion*. London: Leonard Cheshire.

Krippendorf, J. (1987) *The Holiday Makers*. London: Heinemann.

Massey, D. (1999) 'Spaces of politics', in D. Massey, J. Allen and P. Sarre (eds) *Human Geography Today*. Cambridge: Polity Press.

Murphy, P.E. (1985) *Tourism: A Community Approach*. London: Routledge.

Panelli, R. (2004) *Social Geographies*. London: Sage.

Park, D.C., Radford, J.P. and Vickers, M.H. (1998) 'Disability studies in human geography', *Progress in Human Geography*, 22(2): 208–233.

Phillips, D.I. (2002) 'The Disability Discrimination Act revisited: implications of 2004 for tourism and hospitality', *Insights*: A45–A55.

Ray, N.M. and Ryder, M.E. (2003) '"Ebilities" tourism: an exploratory discussion of the travel needs and motivations of the mobility-disabled', *Tourism Management*, 24: 57–72.

Salt, C. (2004) 'Interactive barriers and visual representation: a study of tourists with disabilities', unpublished MSc dissertation, Dept. of Geography, University of Exeter.

Shakespeare, T. and Watson, N. (2001) 'The social model of disability: an outdated ideology?', in S.N. Barnartt and B. Mandell Altman (eds) *Exploring Theories and Expanding Methodologies* (Volume 2). Stamford, CT: JAI Press.

Sharp, J.P., Routledge, P., Philo, C. and Paddison, R. (2000) 'Entanglements of power: geographies of domination/resistance' in J.P. Sharp, P. Routledge, C. Philo and R. Paddison (eds) *Entanglements of Power: Geographies of Domination/Resistance*. London: Routledge.

Shaw, G. (2004) 'Space the final frontier? Disability and barriers to holidays', unpublished paper, International Geographical Union Pre-Congress Meeting on Tourism and Leisure, Loch Lomond (August).

Shaw, G. and Coles, T.E. (2004) 'Disability, holiday making and the tourism industry in the UK: a preliminary study', *Tourism Management*, 25: 397–403.

Shaw, G. and Williams, A.M. (2002) *Critical Issues in Tourism: A Geographical Perspective*. Oxford: Blackwell. (First edition published 1994.)

Shaw, G. and Williams, A.-M. (2004) *Tourism and Tourism Spaces*. London: Sage.

Shaw, G., Veitch, C. and Coles, T. (2005) 'Access, disability and tourism: changing responses in the UK', *Tourism Review International*, 8(3): 1–10.

Smith, V. and Hughes, H. (1999) 'Disadvantaged families and the meaning of holiday', *International Journal of Tourism Research*, 1: 123–133.

Tregaskis, C. (2002) 'Social model theory: the story so far', *Disability and Society*, 17(4): 457–470.

Union of the Physically Impaired Against Segregation (1976) *Fundamental Principles of Disability*. London: UPIAS.

Veitch, C. (2000) *Why Join the National Accessible Scheme*? London: English Tourism Council.

Veitch, C. and Shaw, G. (2004a) 'Access and tourism: a widening agenda', *Insights*, January: A121–A129.

Veitch, C. and Shaw, G. (2004b) 'Understanding barriers to tourism in the UK', *Insights*, May: A185–A192.

Visit Britain (2003) *Holiday-taking and Planning amongst People with a Disability*, London: NOP.

Wendall, S. (1996) *The Rejected Body*. London: Routledge.

Williams, C. and Windebank, J. (2002) 'The "excluded consumer". A neglected aspect of social exclusion?, *Policy and Politics*, 30(4): 501–513.

Woods, M.J. (1997) 'Discourses of power and rurality: local politics in Somerset in the 20th century', *Political Geography*, 16: 453–478.

5 Tourism, nation and power

A Foucauldian perspective of 'Australia's' Ghan Train

Caroline Winter

Introduction: tourism, technology and power

In western culture travel has long been associated with the collection of knowledge, the discovery of Self, the Other, faraway places and exotic environments. In the eighteenth century, knowledge created by scientific expeditions to distant lands was dispersed to the citizens of Europe by travellers who accompanied the researchers (Pratt 1992). Since then, the practice of travel has been transformed by the operation of power; this has created and differentiated a number of modes of travel. Tourism and the massive industry behind it is just one major type of travel with a different relationship to power than other forms of human movement and mobility (Coles *et al.* 2004; Sheller and Urry 2004; Hall 2004). Tourism is notable for its capacity to operate as, and be the subject of, a technology of power, and in so doing it conspires with technologies of consumption towards development of national identity. A technology of power as described by Foucault (1990, 1991) is a sophisticated arrangement of procedures which influence the behaviour of individuals towards specific ends. Contemporary examples of such a technology include marketing and advertising, and the operation of tourist business (Cheong and Miller 2000; Rooney 1997). This chapter examines how power relations articulated through tourism and technologies of power help to create, refine and communicate the identity of a nation. It discusses the use made of Indigenous Australians and Afghan cameleers to construct spaces named 'Outback', in order to promote an image of nation-state. Foucault's theorisations of power are envoked as an interpretative framework through which to address these phenomena.

Tourism is a highly productive and growing system, which incorporates multiple activities directed towards the creation of diverse social, cultural, ecological and economic outcomes. Its complexity and breadth are shown in Weaver and Opperman's (2000: 3) definition as,

> the sum of the phenomena of relationships arising from the interaction among tourists, business suppliers, host governments, host communities, origin governments, universities, community colleges and non-governmental organisations in the process of attracting, transporting, hosting and managing these tourists and other visitors.

Rather than being highly structured, tourism is a highly fragmented industry and according to McKercher (1993: 11), it is 'a multi-jurisdictional nightmare' where no one is in charge. These characteristics would suggest it is virtually impossible to identify tourism with a specific institution or control mechanism. Any analysis of tourism demands flexible thorisations such as those developed by Foucault (1990, 1991) who argues that knowledge is intimately connected with power, and it is directed through an apparatus that is dispersed throughout society. This chapter argues that tourism is both a part of, and an effect of, a system of power. One feature of tourism is its interaction with the consumerism of contemporary society, and in this respect it expands the system of power. The first part of this chapter describes Foucault's theory of power-knowledge, and how it may be used to unpack some of the effects of tourism in creating knowledge about the nation Australia. The tourism industry is argued to operate as a technology which is positively focused on creating and promoting the nation. The sign for the nation is a fusion of Indigenous Australians and their culture, the history of the Afghan cameleers, and the desert environment which together mediate the idea of the Outback, the spiritual heart of the nation. The second part presents a reading of the Ghan Train as an attraction within the tourism system which travels the Outback. The Ghan is a physical object which combines imagery from its own past and combines it with those of Indigenous Australians and Afghan cameleers to help produce an attraction, or sign for the nation, which can be physically experienced. The effect of these practices is to position Indigenous Australians and Afghan people in the past and to obscure them in the present.

Foucault and power-knowledge

Foucault (1991) argues that power and knowledge are linked in a relationship in which each is not only unable to exist without the other, but where each continuously creates the other. A discussion of knowledge implies power. According to Foucault (1991: 27),

> We should admit rather that power produces knowledge . . . that power and knowledge directly imply one another; that there is no power relation without the correlative constitution of a field of knowledge, nor any knowledge that does not presuppose and constitute at the same time power relations.

Foucault's various works describe the development of fields of knowledge including sexuality, madness, criminality and medicine. Within these fields, power is formed as a network of relations between people acting in various social roles, rather than through a 'top down' structure. In *Discipline and Punish*, Foucault (1991) demonstrates that this system is more efficient than the traditional sovereign-based system because it transfers power throughout the social body. Individuals operating within a number of 'technologies' regulate and discipline themselves in the belief that they are under the continual surveillance of a gaze which has the power to punish them.

Power classifies all ways of knowing the world, but only some forms of information are privileged and given credibility and status as 'knowledge'. The rest is discarded as 'untruth' which Foucault calls 'subjugated knowledge' (Foucault 1980: 81). In describing the development of knowledge about the natural environment, Foucault (1994: 140) notes that power selects only a few elements upon which the basis of identity and difference are to be examined. These elements act as signs to determine the relative importance of objects such that those which are unrelated to the sign are considered irrelevant and become virtually non-existent.

In many nations the notion of race has been used as an element by which to demarcate commonality/difference and establish national identity. Langton (2000: 28) argues that in the western world 'during the period 1890 to 1915 race as an organising idea claimed precedence over all previous formulations of nation and state'. The classification of people as belonging to a 'race' is based on a visual assessment of physical characteristics and it persists despite scientific genetic evidence to the contrary. Langton points out that following the Second World War, the United Nations through UNESCO sought to depose racism through a process of education, legislation and research. This programme largely failed and Australians, in particular, were unaffected by the efforts of UNESCO. Langton (2000: 30) argues that the education system's failure to teach modern scientific knowledge has allowed the myth of Indigenous Australians as 'some pre-human type' to survive. The analysis of the Ghan Train below illustrates the differential treatment of Indigenous Australians and Afghan people in the construction of tourism attractions and national identity.

Tourism can be regarded as a particular type within the broad knowledge of travel. The ways through which tourism operates as a technology of power-knowledge is discussed in the next section. A second body of knowledge is that relating to the notion of national identity which is a specific form of social identity created through discourse and disseminated through education, mass communication and other institutions (De Cillia *et al.* 1999). The way in which national identity is created and reproduced is also the result of power, and its operation can be described according to Foucault's theory. In this chapter, national identity is discussed from the perspective of its articulation with the power-knowledge network established by tourism. In other words, both tourism and national identity can each be perceived as distinct manifestations of power-knowledge which interact and link to expand and distribute their ideas more efficiently throughout society. Thus, tourism plays an important role in the creation of the body of knowledge concerning the nation and national identity, which in turn provides part of the attraction of 'Australia' for tourists.

Technologies of power

The myriad processes involved in the practice of tourism can be conceptualised as a 'technology', a term used by Foucault to encompass all of the techniques used in the practical operation of power, including the relationships between individuals, the organisation of knowledge and the 'mastery of its forces' (Foucault 1991: 27).

Foucault argued that each body of knowledge is achieved through a particular technology. For example, he referred to a 'technology of sex' as a type of resource (Foucault 1990: 123), and in relation to discipline he explains that:

> 'Discipline' may be identified neither with an institution nor with an apparatus; it is a type of power, a modality for its exercise, comprising a whole set of instruments, techniques, procedures, levels of application, targets; it is a 'physics' or an 'anatomy' of power, a technology.
>
> (Foucault 1991: 215)

Rooney (1997: 402) defines technologies of power in relation to consumer administration processes as those which 'have the potential to influence the conduct of individuals'. He includes marketing, public relations and advertising, which aim to control consumers' behaviour in how, what and where they buy goods and services. Cheong and Miller (2000: 374–375) add that power relationships occur in the 'seemingly nonpolitical business and banter of tourists and guides, in the operation of codes of ethics, in the design and use of guidebooks, and so on'. Hollinshead (1996: 324–325) also describes marketing communication channels as networks of power-relations through which images of Indigenous Australians are conveyed. The technology of tourism helps to determine the way in which particular histories of place and people are depicted, and the types of attractions that are created, developed and promoted.

Tourism can also be seen as a particular type of travel that is characterised by its close links with consumerism (Watson and Kopachevsky 1994). This difference is clearer when tourism is contrasted with non-consumptive forms of (voluntary) travel. For instance, Bender (1993) writes that while tourists are accepted at Stonehenge, other visitors who do not pay for their experiences are not. Watson and Kopachevsky (1994: 646) argue that the commodification of signs and images of consumption is an issue of power in which commoditisation is 'shaped and honed by specific, influential groups in society utilising a mixture of social, cultural, and political resources'. As Palmer (1999) argues, the problem of commodification in relation to national identity is that only selected images are used. Moreover, a high profile is given to them by the various marketing campaigns which further elevates their importance as the main defining attributes of the nation. For her, 'alternative, or even contested, interpretations are not so forcefully promoted. Furthermore, in utilising aspects of the past choices are being made about what to include and what to leave out' (Palmer 1999: 319). These questions, and the mechanisms involving tourism through which the selective identity of nation is created, are an issue of power-knowledge.

The positive purpose of power: national identity

Foucault (1990) argues that the primary intent of power is positive, and directed towards the satisfaction of specific social goals. Power works *not* because it is repressive, but because it is productive in the creation of knowledge as truth. For

Foucault, the apparently repressive sexual norms of the bourgeoisie were a mechanism which would distinguish them from the elite and the lower classes,

> as to be seen as the self-affirmation of one class rather than the enslavement of another: a defense, a protection, a strengthening, and an exaltation that were eventually extended to others – at the cost of different transformations – as a means of social control and political subjugation.
>
> (Foucault 1990: 123)

In other words, he argues that in an effort to promote their own interests the bourgeoisie practised on themselves, not on those who became subjugated as a result.

One of the benefits of a (singular, predominant) national identity is that it helps societies distinguish themselves and the products they offer for sale in order to compete successfully with other nations. Nations are, though, 'imagined communities' rather than physically identifiable entities (Anderson 1991), and they must, to an extent, be invented. Foster (1991: 252) refers to nations as artefacts which are continually imagined, invented, contested and transformed by individuals, the state and global commodity flows. The formation of nations, which Anderson (1991) and Kapferer (1988) describe, suggests that a national identity is not an inherent feature of society but is an effect of power in which order is achieved by the definition, classification and production of people and their culture according to a specific model (Bhabha 1985: 175). By categorising and ordering society, in such a way that emphasises difference to other societies, culture becomes the prime element in the creation of the nation (Kapferer 1988: 1).

Part of the nation-building process involves establishing links with other nations and between members within the nation (Anderson 1991; Hobsbawm 2000; Triandafyllidou 1998). It involves a process of classifying and distinguishing elements of commonality and difference, or as Triandafyllidou (1998: 594) argues:

> the identity of a nation is defined and/or re-defined through the influence of 'significant others', namely other nations or ethnic groups that are perceived to threaten the nation, its distinctiveness, authenticity and/or independence.

In other words, the process of nation-building is a disciplinary process of its self. It occurs through both inclusion and exclusion of individuals and groups, and as Langton (2000) has noted, the issue of race is a persistent one in contemporary Australia. The power-knowledge of travel creates different types of travellers who are given differential access to spaces in the nation and the opportunity to gaze on, and reflect about, constructs of inclusion and exclusion. First, tourists are constructed and differentiated from others such as refugees, asylum seekers and gypsies. The Irish and European travellers were, for example, defined through bourgeois property rights which were grounded in property and permanency; this contrasted with nomadic people who did not own land. According to MacLaughlin (1996: 47) 'by the end of the century, Irish nationalists had managed to exorcise tribalism and nomadism from nationalist history and from the national psyche'.

A further requirement is that a credible nation must also show temporal continuity from ancient times to the present day (Anderson 1991). It is not necessary, however, that traditions and links with the past are actually ancient, but they can be invented (Hobsbawm 1992; Meethan 2001). In Australia, the myth of Outback helps to create and maintain these links and as McGrath (1991: 122) argues, 'it is intrinsic to our colonial identity, symbolising our history, our soul, our acquisition'. The tourism industry participates in providing visual images in the form of signs and attractions (MacCannell 1999) and touring 'experiences' which reflect the attributes of the created model of the nation. 'Australia' the nation, the individual body of an 'Australian', and the activities one undertakes in being a part of the model, are examples of tourism's participation in creating and reinforcing images of difference.

The structure of power: agents and objects

The key to the efficiency of Foucault's concept of power is that individuals are subject to its effects, yet they participate in its execution. Foucault (1980: 98) argues that power must be analysed as:

> something which circulates, or rather as something which only functions in the form of a chain. It is never localised here or there, never in anybody's hands, never appropriated as a commodity or piece of wealth. Power is employed and exercised through a net-like organisation. And not only do individuals circulate between its threads; they are always in the position of simultaneously undergoing and exercising this power. They are not only its inert or consenting target; they are always also the elements of its articulation. In other words, individuals are the vehicles of power, not its points of application.

Thus, individuals hold dual roles: as agents having access to power, and also as targets which are subject to power. These relationships are formed and maintained not by grand acts or overt physical displays of violence, but through the relationships and activities that are a part of daily life (Foucault 1990). There are differing views about the position of tourists within this system of power-knowledge. Cheong and Miller (2000: 381) see tourists as 'targets' in that they are subject to the actions of tourism's technologies; that is, through power brokers such as government officials, tour guides, hotel staff, academics and market researchers. For example, technology is used to create tour packages as well as through mechanisms like other marketing campaigns, and can direct tourists towards particular destinations and activities. However, tourists can also be seen as agents who are not merely limited to the choreography of technology, but who make their own decisions and interpret their experiences differently to advertising campaigns (see Crouch, Cater – chapters 2 and 3 – in this volume). Crawshaw and Urry's (1997: 178) research to determine whether tourism involved 'systems of panopticon-like surveillance' found that tourists were subjected to photographic images that reflected dominant cultural constructions of nature. They also found that tourists interpreted these images through their own experiences and attitudes. Rojek (1985) points out the relevance of Foucault's theory, and argues that leisure should be conceptualised as both freedom and control.

Tourism plays an important function in creating products for tourists to enjoy and at the same time to create themselves in the images of 'Australians'. Travelling helps form both national and personal identity, especially for long-haul destinations (Desforges 2000; Palmer 1999). Visitors can feel that if they consume the products of a nation that are imbued with the national quality, they will gain these as part of their own personal identity (Foster 1991: 250). Kapferer (1988: 17) suggests that for Australian citizens the relationship between national and self-identity is an effect of the notion of egalitarian individualism in society. Australian national identity is made up of the individual citizens in comparison with other forms of nationalism, where the individual's sense of self-identity stems from the nation. Individuals invest a part of their own selves and their own identity into the nation. National identity can be very real to many Australians because it is so closely bound to a sense of individual self and personal identity (Kapferer 1988: 19). With respect to travel and tourism, given their links with consumption, this means that selected travellers (tourists) have greater access to this process of identity formation than other travellers. Thus, travellers who cannot participate in consumption may be excluded from full participation as citizens of the nation.

The importance of the visual

The visual sense has been accepted in western culture as the main way to establish wisdom, truth and knowledge. Foucault's gaze refers to the way in which the visual operates to instil discipline in a subject and is most clearly captured in his description of Jeremy Bentham's panopticon as the ultimate tool for control of an individual. For Foucault (1991: 170) states that,

> The exercise of discipline presupposes a mechanism that coerces by means of observation; an apparatus in which the techniques that make it possible to see induce effects of power, and in which, conversely, the means of coercion make those on whom they are applied clearly visible.

He comments that 'the perfect disciplinary apparatus would make it possible for a single gaze to see everything constantly' (Foucault 1991: 173). Rather than operating in the dark or operating elusively, power operates most efficiently in the full light and Foucault argued that its very visibility enhances its effects. Tourism also is inherently visual; tourists are known as 'sightseers' taking photographs as visual memories of places visited and seen. Urry's (1990) 'tourist gaze' describes how hosts and tourists are subject to technologies of power that direct and control how and what is seen. Although tourist pursuits are often perceived as trivial, the adoption of a Foucauldian approach can reveal how power operates through these activities (Crawshaw and Urry 1997: 178). According to Hollinshead (1999: 12), Urry does not explicitly explain the nature of Foucault's gaze but rather he presents his work in a Foucauldian style, where description and theory are combined rather than being written in distinct components. Urry (1990) describes the development of themes, such as the romantic gaze and the heritage gaze, within which specific attractions, people and places are situated. Hollinshead (1999: 9) provides a more theoretical interpretation of Foucault's and Urry's 'gaze' as,

> the way its members learn to see and to project preferred versions of reality, and historically, the way that such seeing and projecting privileges certain persons and their inheritances, and subjugates certain others and their inheritances.

He points out that the 'universalising surveillance' described by Urry shows that tourism is 'notably dangerous because it is so under-recognised and thereby so easily tolerated' (Hollinshead 1999: 12). Furthermore, the essence of tourism is the association with attractions which can be argued are signs that represent important social values. In particular, 'the key point about such signs is their ability to convey meaning, to transmit very particular messages about a nation, its culture' (Palmer 1999: 316). Tourism is important in making some narratives dominant while denying, even suppressing, others. Such differentiation involves individual managers, developers and researchers in small and large games of cultural, social, environmental and historic cleansing (Hollinshead 1999: 8).

Thus, some of these attractions and their associated narratives play a particularly important role in acting as sign for the nation. There is an extensive field of research describing the direction of the tourists' gaze towards attractions that are formed from particular views of people, such as Native Americans (Albers and James 1988) and Indigenous Australians (Simondson 1996), who are portrayed to suit tourists' needs rather than their own identities. Landscape imagery has also been used in tourism to create and portray particular 'brand' images of national identity (Duffy 2002). In Australia, use of landscape and Aboriginal art was a dominant theme for tourism promotions in the 1980s (Spearritt 1990) and continues to be so. In this respect, Foucauldian analysis can illuminate knowledge which has been subjugated by acknowledging how 'the gaze' has been directed away from certain sights and images.

The execution of the gaze is, though, linked with the operation of targets and agents so that the idea of the gaze as unidirectional is contested. Hollinshead (1999: 13) argues that even the 'gazer' is subject to the limitations of the gaze he/she is trying to impose; that is, an individual's access to power is illusionary. In relation to the colonial gaze of tourism, Winn (1999: 94) argues that the objects of power can return the gaze and in so doing are able to subvert the order which the gaze aims to impose. MacCannell (2001) describes these effects as a 'second gaze', which can 'see' beyond the way in which an object is presented by power, and determine that which has been hidden or subjugated.

The productive nature of power

In addition to efficiency, Foucauldian power is concerned with productive capacity. Foucault illustrated this feature in his description of the expansion of 'types' or social roles within the broader categories of madness, criminality and sexuality, each with increasingly detailed fields of knowledge. The increase in 'types' encouraged their further investigation and development of a field of knowledge which, in turn, expanded the number of relationships of power between them.

The proliferation of academic study and research of tourist phenomena can also be perceived as producing and reproducing the effects of tourism (Cheong and Miller 2000: 385). In particular, by processes of identification, classification and naming, academics help to bring into existence new typologies of tourists. Where once there were simply 'tourists', there were later 'mass tourists' and 'alternative tourists' (Weaver and Lawton 2002). Now there are profiles for many new segments of the total tourist market (cf. Mowforth and Munt 1998). For example, there are now many types of 'special interest tourism' (Douglas *et al.* 2001), multiple typologies of cultural tourists (McKercher and du Cros 2002) and ecotourists (Weaver 2001). Identification of these typologies then encourages further development of specialised products to suit each of these clienteles.

Participation in the creation of national identity is an area in which tourism is enormously productive. The use of its widely spread linkages, the technology of advertising and communication, the development of typologies, use of visual imagery in producing designs for culture and landscapes all mean tourism is highly effective in the creation of attractions which act as signs for national identity. Virtually any part of a culture can be put up for sale to tourists and in Australia the Ghan Train is part of this process.

Up to this point the chapter has described how some of Foucault's main constructs of power-knowledge relate to tourism and the nation. Specifically, the adoption of a Foucauldian analysis shows that tourism is itself a creation of power-knowledge which now operates as a network propagated by sophisticated technologies of power-knowledge which direct certain travellers to particular sites. Within this network, tourists and citizens of the nation operate both as targets of power and as agents in redistributing the effects of power. The visual is particularly important to both tourism and nation, and it is the basis upon which certain people and objects are classified as belonging within, or excluded from, the national agenda. In the next section a particular attraction, the Ghan Train, is examined. The visual is used to define and distinguish various roles, such as citizen and tourist. Indigenous Australians and Afghan cameleers are used as attractions for tourists and to help validate and reinforce current dominant discourses of the nation.

The Ghan Train and its role in creation of the nation

The Ghan is a train that from 1929 ran the 1,500 kilometre journey from Adelaide to Alice Springs (Fuller 1975). Today, it is an important icon within the Australian tourism industry (Figure 5.1). It can be analysed as one of the themed products

Figure 5.1 The Ghan Train

Source: reprinted courtesy of Great Southern Railway (Adelaide, Australia, www.gsr.com.au).

associated with the branding of Australia and the creation of a national identity using the imagery of the Outback. The Ghan's image has been created through the use of visual imagery and selected aspects from its own history, the desert environment, Afghan cameleers and Indigenous Australians.

The nature of power-knowledge implies that any artefact within a discourse can be read to decipher its underlying meaning (Foucault 1990). Clearly, some objects have potential for greater links with power, and to provide more effective means to promote the nation than others. These are created and promoted as national attractions to which tourists are directed. In Europe, for example, 'landscapes, castles, country houses and their associated paraphernalia of everyday life are presented as embodying the essence of nationhood' (Palmer 1999: 315). In addition to their role in unifying the national identity, these artefacts play a second role such that 'cultural traits, myths, traditions, historical territories form an integral part of the distinction between "us" and "them"' (Triandafyllidou 1998: 597).

One of the benefits of a Foucauldian approach here is that it allows analysis to occur at the micro level of interaction between targets, agents and the apparatus of power (Rojek 1985: 156). Foucault also encourages us to examine the silences of discourses, the subjugated knowledges, the objects excluded from distinction or, as Hollinshead (1999: 15) asks us, to trace what it is in tourism that we do systematically to privilege and similarly to deny and frustrate. For this research, ephemera associated with the Ghan were examined to determine their meanings in relation to power-knowledge. The items examined included some of the advertising materials such as: brochures and leaflets for the period 1998 to 2003; post cards and posters; historical and contemporary accounts of the Ghan, such as history texts, newspaper and travel magazine articles about Outback touring; and tourism reports. Some informal interviews were also conducted with people who had associations with the Ghan, including: travellers of earlier times, staff on the new train and those working at the rail museum in Alice Springs. A number of artefacts such as the bed linen, crockery and the menus in the dining rooms were examined during a return journey to Alice Springs on the train in 1999. In keeping with Foucault's (1991) analysis of the form of the panopticon – on the behavior of its interns – the physical configuration of the train (including its interior decor) was also examined for its potential influences on the behaviour of travellers.

Origins of the Ghan

The relatively late arrival of transport to Australia did not have the huge impact that it did in Europe, and for this reason its historical contribution towards progress and nation-building was less than it could have been (Adam-Smith 1983). Despite this relatively small historical role, the contemporary Ghan is being used to evoke ideas of nation through its connections with the desert environment, Aboriginal Australians and the Afghan cameleers.

Before the Ghan and motorised transport, camels were used extensively in central Australia for communication and transport of goods. Afghan cameleers and their camels were first imported to Australia from Afghanistan in 1867 by Thomas Elder who set up a stud at Beltana (Fuller 1975; Stevens 1989). Camels provided the primary transport and communications link for the Outback from the 1860s to the 1920s and they were important in the construction of the overland telegraph line, as well as the Ghan and Indian Pacific railways (Stevens 1989; Adam-Smith 1983). It is generally agreed that the Ghan was named in honour of these hard-working cameleers (Adam-Smith 1983; Fuller 1975).

The most memorable feature of the Ghan (prior to 1980) was its frequent failure to run on time. It was known as the slowest train in the world, and as Downes and Daum (1996: 7) reflect, 'the timetable was a matter more of hope than of fact'. The Ghan was known to be up to several weeks late and delays of a week or so were fairly common (Fuller 1975). These huge delays were created by the difficulties of the harsh environment, and the fact that, for various reasons, the track had been built through flood-prone areas of the Finke River (Fuller 1975: 8). The effects of lightweight track, and the peculiar circumstances of the route, formed the conditions in which reality came to match reputation and image. Stories about the trials of the passengers and staff under these difficult conditions in the Australian interior formed the basis for the image of the 'Old Ghan'. The train stopped frequently along the way and passengers mixed with each other and the people who lived in the Outback areas. The Ghan was, and still is, held dearly by the people of northern and central Australia and, in its earlier days, upon its arrival in Alice Springs the whole town turned out to welcome it, regardless of the time of day (or night) or whether they were greeting someone or not (Slarks 1998). In 1980 the route was changed to avoid the flood-prone areas and the new train now provides a sleek, modern and reliable service to Alice Springs.

In 1989 the decor of the Ghan was updated to portray a much more 'Australian' image, and to attract a larger and international clientele. These changes incorporated visual imagery based upon Afghan cameleers and Indigenous Australians, the use of a logo, changes to the menu, carriages and the bestowal of a new name – *The Legendary Ghan*. These designs have not changed significantly since then and form the basis of the later discussion. More recently in 1998 the government-owned line was sold to a private consortium (ABC 1997) and major changes to the route including extension of the line to Melbourne, Sydney and Darwin followed.

Given the history of the Ghan, other decor schemes could have been used in 1989 since the Ghan carriages did not have a particular theme in the past. In fact, it was once common practice that carriages from other lines were modified and added to the Ghan as passenger numbers increased. Horne (1978: 120) likened the 'Old Ghan' to the Orient Express and European notions of good taste rather than an Australian design, and he wrote that 'within our compartments, crisp white linen, dark panelling and silvery fittings offer classic symbols of nineteenth century opulence – among which we sleep'.

The Ghan itself – the train

One of the important ways through which the Outback image is created is through the physicality of the train itself. First, it transports tourists to the 'heart' of the nation and as such the Ghan represents one of the important Outback heritage assets that tourists can not only see but also experience for themselves through the journey into the desert (McGrath 1991). The viewing of heritage sights by tourists can form a key aspect of their notion of national identity (Pretes 2003) and in Australia the physical reality of travelling to the Outback and meeting real people reinforces their idea of the nation (McGrath 1991: 114). As Palmer (1999: 315) states, 'the symbols that help to construct and to convey a sense of national identity are imagined to lie at the heart of a nation's soul', and they are especially important in attracting tourists. As advertising materials for the Ghan indicate, the centre of the Australian continent is positioned not just geographically, but as the nation's heart: 'The legendary journey into the heart of Australia . . . it is the ultimate journey to the heart of the continent' (Trainways 2003: 4–5).

Moreover there appears to be an increase in 'Ghan-ness' the more money one pays for the ticket. In third class, the decor of the carriage is much like any other train, and the only example of the logo is found on the paper food containers purchased from the snack bar. In second or holiday class, the lounge features ochre colours of the Outback, and the logo is prominent on the bed linen and other features of the service. In first class, the imagery of the Ghan, the Outback and Indigenous culture are much more intense, and these carriages are most frequently illustrated in the advertising materials. It is here that the Afghan logo, the Outback colours and the designs based on Indigenous art are displayed on the interior surfaces of the train. In addition to these visual representations, the carriages are named after actual places along the route and the menu incorporates ingredients that are native to Australia. In other words, the opportunity for enhancement of one's personality with the identity of 'Australia' the nation by consuming its products is greater in first class than in the other classes – if you can pay for it.

The landscape as outback

A landscape is a particular social view of a natural area, and in Australia the interior regions have been perceived in a number of ways (Heathcote 1987). One of the uses of the arid Australian landscape has been as a mythological and timeless land containing the plant and animal relics that supposedly existed in antiquity (Markus 1990). This then provides the basis for the 'ancient past' that a credible nation requires. In the 1980s, the landscape, together with Aboriginal art, was used extensively in tourism promotions (Spearritt 1990) to promote 'Australia'. The Ghan now provides a symbolic link between the ancient past and the early development of the nation-state that was undertaken by the cameleers. The body of the train also provides an illusory barrier between the travellers and the real environment. This removal of the physicality of the environment helps to facilitate its image as Outback and as part of a distant past. The carriage becomes a vehicle for visual surveillance which is primarily directed by the discourse of the tourism industry. The positioning of the Afghan cameleers and Indigenous Australians within this landscape is also perceived to exist in the past. The comments below reflect an image of the complacent traveller for whom the land is provided as an attraction for his or her entertainment.

> We sipped our drinks and gazed complacently through the double glazing at the sun-battered mulga. . . . It must be hot out there, we thought, adding an ice cube.
>
> (Whitelock 1986: 19)

The advertisements for the Ghan promote lounge chairs that:

> afford splendid views of the Outback through a double glazed panoramic window.
>
> (GSR 2000: 14)

And passengers are encouraged to:

> enjoy the passing parade of Australian landscapes.
>
> (GSR 2000: 12)

The imagery of the landscape is further enhanced by the interior design of the carriages. The decor reflects the colours of a desert landscape with the effect that the two worlds, inside and outside, appear to merge. The passenger sits surrounded by the colours of the Outback depicted in the furnishings and walls of the train. In this sense the Ghan capsule really is a physical representation of the environmental bubble described by Cohen (1972: 166). The Ghan shields the tourist from the outside world and promotes the notion that the landscape can be known and controlled from inside the train. This knowledge and control is an illusion that bears little resemblance to the world outside the train. It denies the physicality of the land and the experiences of the people who live in it. For example, an important part of Indigenous Australians' and Afghans' ability to survive in this harsh environment results from their extensive knowledge, their capacity for endurance and their close links with the land itself. Once a barrier is inserted between humans and the land, an illusion is created that the importance and meaning of such peoples and their cultures is irrelevant. For example, the 1998 brochure descriptions for the dining room focus on colour and style rather than the harsher realities of the environment:

> In the Stuart Restaurant the theme is Art Deco, with acid etched dividers and a colour scheme inspired by the lilacs of a twilight sky blended with sand dune terracottas.
>
> The colour palette takes inspiration from the rich ochre of Ayers Rock [*sic*] and green spinifex grass. . . . Hand painted fabrics capture a stylised impression of Dreamtime art forms, landscapes and reptile wildlife.
>
> (GSR 1998: 3)

The imagery of people and land as Outback is not just to subjugate objects however, but to promote other aims positively. Moran (2002) argues that the European settlers wanted to establish their own connection with, and ownership of, the land and to do this they needed to replace the indigenous people. In this view, this has shaped the type of nationalism in Australia and encouraged the silence about Indigenous Australians. In Australia, the manipulation of social histories to favour a European perspective rather than an Aboriginal viewpoint is well documented (Attwood 1989; Markus 1990; Willis 1993). The analysis of photographs created and used by the tourism industry supports the view that the basis of national identity has been selected from the people of European culture and not from people of Aboriginal, southern European or Asian descent (Willis 1993). It is apparent that people of Middle Eastern origin are also excluded. Thus, the positioning of Indigenous Australians in the past, in the central desert and Outback helps emphasise the presence of European-Australians and their more recent and contemporary connections.

The Afghan cameleers

Thus, to bolster the status of white Australians' nation-building, the real achievements of the Afghan cameleers are subjugated. This is not to say the use of the Ghan is to repress the Afghan people. A Foucauldian perspective argues that the Ghan acts to promote the interests of the ancestors of contemporary white Australians, and that the misrepresentation of the Afghan people is a 'side-effect' of the operation of power-knowledge. For instance, in a recent advertising brochure, credit is given to a young South Australian for the idea to import camels along with their handlers (Trainways 2003: 4). It also notes how early explorers and later telegraph linesmen used camels, but does not state the amount of work actually done by the Afghan people. It does not credit the combination of their special skills in handling camels, their ability to work in the desert environment and their capacity to work long days, all of which were instrumental in their success. Instead, most credit is given to the early explorers and telegraph linesmen (presumably white, not Afghans) to 'blaze a permanent trail into Central Australia' (Trainways 2003: 4). In Outback imagery this is not unusual. As McGrath (1991: 122) points out, the Stockmen's Hall of Fame in Longreach pays only scant attention to the contribution of Indigenous Australians, even though they played a critical role in the pastoral industry.

The historic relationship between Afghan cameleers, Aboriginal people and white Australians was violent and racist. This view is generally hidden in contemporary stories of the Ghan including those of the tourism industry. Stevens (1989: vii) has uncovered this other history, and she states that Afghan people were subject to a life of alienation, racial and religious intolerance and often economic exploitation. The cameleers were perceived as an exotic Other and were excluded from society by Indigenous Australians as well as Europeans. They lived a life of hardship, taking on long hours of heavy work for which white people were less capable and willing. By the time the Ghan line was completed in 1929, motorised transport was making an inroad to transport in the Outback, signalling the end of the cameleers' livelihood. The end of the era, as described by Stevens, was violent towards both the Afghans and their camels, and many Afghans returned to their homelands leaving their camels behind.

Although the real importance of the Afghan cameleers in the development of the nation is de-emphasised, a highly stylised version of them is extensively used in promotion of the Ghan Train. The Afghan cameleers have been commodified as the logo which in turn features on almost all Ghan artefacts; it is etched on glass window panes, printed on the crockery, the bed linen, on the outside of the carriages and on advertising materials. There are no references to contemporary Afghan people in the discourse of the Ghan. Thus, a single representation of the Afghan people is highly visible within the image created by tourism, but this very visibly traps them into a particular role and to a place only in the past from which no link to their contemporary lives is made. A closer examination of the logo reveals that it is more an image of a camel (with a man sitting on top). As Stevens (1989) notes ironically, the subject of tourists' interest is the camels – around Alice Springs in recent times there has been a growth in camel farms and tours – and not so much in the Afghan cameleers. With the recent extension of the rail to Darwin in the northern part of Australia, some of the promotional photographs showed an Afghan man with a European man. Both were dressed in period settler costume, more like the props described by Dann (1996) than playing a contemporary role.

Not all travellers are welcome or eligible to participate in the nation however, and in this repect Afghans have further suffered from exclusion. Some, such as refugees and asylum seekers, are perceived as a threat to the nation (Triandafyllidou 1998; Perera 2002). The arrival of the *Tampa* in 2001 with its cargo of around 400 refugees, many of whom were from Afghanistan, was turned away from Australia. This was accompanied a few days after the *Tampa*'s arrival by the idea that asylum seekers from countries like Afghanistan were potential 'terrorist sleepers' (Perera 2002: 16). Ironically, one of the detention centres for the boat people in Australia was in the central desert regions of Australia at Woomera. Perera (2002: 13) commented that 'the media spectacle of the *Tampa* re-enacts the colonial adventure classic as an occasion for national self-affirmation'. While Afghan people have been celebrated in the past, they are clearly unwelcome in the present.

Indigenous Australians

What involvement of Indigenous Australians there has been with the Ghan appears largely a part of the 1989 'Australianisation' of the train. The culture of Indigenous Australians is depicted inside the Ghan in a similar way to the Afghans' – stylised with little relevance to their historic or contemporary lives. Stevens (1989) noted that there were no indigenous cameleers and that Indigenous Australians did not get on with the Afghan people. Although they may have used and continue to use the Ghan for their own travel, Indigenous Australians' historic connections with its operations seems to have been limited.

There has been considerable academic comment on the notion that the culture and image of Indigenous Australians have been selectively appropriated and used by the (now) dominant white Australia (Hamilton 1990). One use has been to enlighten the supposedly spiritually deficient population through a connection with the land (Lattas 1992). Lattas (1990: 67) argues that this is 'produced within a field of power relations that have an interest in discovering and monopolising the authenticity which we are ascribed as lacking'. Marcus (1988: 254) argues that Indigenous Australians have been used as curios to form a distinctive image for 'Australia'.

One important role for which Aboriginal culture is used is to provide a sense of an ancient past for the nation through links with the desert landscape. Lattas (1992: 51) argues that the alienation of modern life is linked with 'an aesthetic appreciation of the Australian landscape as a site of authenticity'. As if to evoke this sense of spirituality with the landscape inside the train, characteristics of cultural expression are featured on the furnishings in the first-class compartments and the design features the Dreamtime art forms, which is a stylised Aboriginal dot painting. The first-class lounge which features these designs is prominently featured in many of the advertising materials.

In 1997, Cathy Freeman, a famous Indigenous woman athlete launched an Australian national locomotive, painted all over with an Aboriginal Dreaming story, a Warmi dot painting of a snake, bush tucker and women's footprints originally created by Aboriginal artist Bessy Liddle (Jopson 1997). The locomotive featured on the first of Great Southern Rail's brochures to introduce the extension of the Ghan to Melbourne. The exercise is similar to the 1995 launch by Qantas of 'Wunala Dreaming', a 747 Boeing 'Jumbo' painted in a bright red Aboriginal design designed for the Japanese air route. Like the interior of the Ghan, the colours are significant replicas of the land; red for Uluru, green for Kakadu, blue and purple for the Flinders Ranges (Qantas Airways 1995). These displays are made particularly for the international market.

The 1998 advertisements for the Ghan showed no contemporary or traditional Indigenous Australians. Thus, like the Afghan people, their absence signalled that they lived only in the past, and their image and aspects of culture were used to invoke their memory in particular ways to enhance the Ghan. Change is, however, taking place in the depiction of Indigenous Australians. From a complete absence in the 1998 materials, the 2000 brochure showed some pictures of Aboriginal men as porters. This is not unlike advertising in other destinations, and it reflects Dann's (1996) observation that most pictures of indigenous people depicted them in service roles for tourists, such as hotel staff, vendors or entertainers. Like the Afghan cameleers, Indigenous Australians on the Ghan are visible, but trapped within a singular, straitjacketed view of their lives in the past.

Conclusion

The adoption of a Foucauldian approach can help to illustrate the highly productive system of networks and the effects of a 'technology' of power-knowledge that operate through tourism. The phenomenon of tourism is an effect of power-knowledge which distinguishes tourism from other travel such as that of migrants, asylum seekers and refugees. Tourism is itself a system of power-knowledge which exerts an influence through its own networks and operations.

From a cultural inspection of the Ghan Train, it becomes evident that the interaction of two systems of power-knowledge (tourism and the nation) can help increase their respective field of operations and extend the capacity of power-knowledge to direct and control the individuals within them. That is, tourism can extend the interests of an individual nation to the global stage and, in turn, the nation provides an attraction to which tourists both domestic and international will be drawn. The very physicality of travelling helps to make real the concepts promoted by the nation, thus providing a link for the material and conceptual spaces of the nation.

As Foucault (1990) noted, the increasing proliferation of types or roles within a system of power-knowledge allows for more intensive examination and direction of individuals, and for differential distribution of benefits. Thus, within tourism and the nation a number of roles are created, in particular those of tourists and non-tourists (refugees, asylum seekers), citizens and non-citizens. One prominent means of classifying travellers is based on their capacity to consume travel products. This then provides them access to the various attractions and assets which represent the nation, such as the Ghan Train and certain visions of the nation's people.

Tourism relies upon the visual which creates particular representations of places and people that can be used to achieve specific social objectives which, in this case, are those of the nation-state. As Urry (1990) demonstrates, a Foucauldian approach illuminates the visual aspect of tourism as more than mere sightseeing but as a 'tourist gaze', which is an effect of power-knowledge. The notion of race is a second method of classification based on the visual which classifies both travellers and citizens and then provides differential access to the benefits of the nation. This interaction of tourism with the nation allows people and places to be branded and identified as part of the nation (or not). Indigenous Australians and the Afghan cameleers may be citizens but they are placed in the space of the past and are not granted access to the present. The concept of the Outback is a visual artefact of power-knowledge which creates a space of the central Australian desert where the operations of tourism and national identity can be played out. Although Afghan and Indigenous Australians are highly visible within the Outback, they are trapped by a gaze which places them in the past and limits their existence in the present. In the space of the Outback then, the aim of power-knowedge is to benefit society overall, but the roles into which individuals are classified determines whether or not they will receive positive or negative outcomes.

The operation of power is not just conceptual but has a tangible effect on peoples' lives, their travel and their acceptance as citizens. The Ghan presents a useful object for the dispersal of a particular version of national identity through the technology of tourism. It provides a symbolic pathway to the ancient past through landscape and Indigenous culture, and links with the more recent past of white settlement using selected aspects from the history of the Afghan people. The train offers a physical journey to the spiritual heart of the nation which is thought to enhance the reality of the experience for travellers. The contribution of Indigenous and Afghan people towards the nation has been subjugated, but to create a more positive outcome that favours the stories and efforts of white Australians. It also provides an opportunity for international visitors to refine their identity by selecting some of the characteristics of Australia, and for Australian citizens to experience (some of) their own nation.

Indigenous Australians and Afghan people have been severely short-changed in the stories developed about the Outback. They and their culture are portrayed and described in tourist advertising in ways that trivialise and merely commodify. These techniques have had the effect of placing some Australians in the background and in the past without giving them the due recognition they deserve. To ignore their historical reality is to ignore their place in contemporary Australia and to decrease the richness and quality of the current national identity. This practice also ignores a large part of white history. Because their identity is determined by specifications set up by power-knowledge, both the gazers (tourists and citizens) and those gazed upon (Indigenous Australians and Afghans) are targets of power-knowledge. These two groups are also agents in defining the identity of the other. The effects of power are such that the social position and benefits received by each group are significantly different. As Langton (2000: 36) warns:

> History should not be a selective grab bag from which are drawn only those events deemed to support national pride. Australia cannot use the highlights of its history as a backdrop for contemporary Australia, on one hand, and also, on the other, ignore as irrelevant the darker side of its past. If the past is irrelevant then it is irrelevant in its entirety.

References

ABC News (1997) Federal government announces sale of national railways. 28 August, Online document. Available from: http://www.abc.net.au/rural/news_states/nrn/nrn-28aug1997-6.htm [last accessed May 2005].

Adam-Smith, P. (1983) *When We Rode the Rails*. Sydney: Lansdowne.

Albers, P. and James, W. (1988) 'Travel photography: a methodological approach', *Annals of Tourism Research*, 15: 134–158.

Anderson, B. (1991) *Imagined Communities*. London: Verso.

Attwood, B. (1989) *The Making of the Aborigines*. Sydney: Allen & Unwin.

Bender, B. (1993) 'Stonehenge – contested landscapes (medieval to present-day)', in B. Bender (ed.) *Landscape: Politics and Perspectives*. Oxford: Berg.

Bhabha, H. (1985) 'Signs taken for wonders: questions of ambivalence and authority under a tree outside Delhi, May 1817', in H.L. Gates (ed.) *'Race', Writing and Difference*. Chicago, IL: University of Chicago Press.

Cheong, S. and Miller, L. (2000) 'Power and tourism: a Foucauldian observation', *Annals of Tourism Research*, 27: 371–390.

Cohen, E. (1972) 'Toward a sociology of international tourism', *Social Research*, 39: 164–182.

Coles, T.E., Duval, D.T. and Hall, C.M. (2004) 'Tourism, mobility and global communities: new approaches to theorising tourism and tourist spaces', in W. Theobald (ed.) *Global Tourism*. Oxford: Butterworth-Heinemann.

Crawshaw, C. and Urry, J. (1997) 'Tourism and the photographic eye', in C. Rojek and J. Urry (eds) *Touring Cultures: Transformations of Travel and Theory*. London: Routledge.

Dann, G. (1996) 'The people of tourist brochures', T. Selwyn (ed.) *The Tourist Image: Myths and Myth Making in Tourism*. Chichester: Wiley.

De Cillia, R., Reisigl, M. and Wodak, R. (1999) 'The discursive construction of national identities', *Discourse and Society*, 10: 149–173.

Desforges, L. (2000) 'Travelling the world: identity and travel biography', *Annals of Tourism Research*, 27: 926–945.

Douglas, N., Douglas, N. and Derrett, R. (2001) *Special Interest Tourism: Context and Cases*. Milton, NSW: Wiley.

Downes, J. and Daum, B. (1996) *The Ghan – From Adelaide to Alice*. Cromer, Victoria: lichtbild.

Duffy, R. (2002) *A Trip Too Far: Ecotourism, Politics and Exploitation*. London: Earthscan.

Foster, R. (1991) 'Making national cultures in the global ecumene', *Annual Review of Anthropology*, 20: 235–260.

Foucault, M. (1980) *Power/Knowledge*. New York: Pantheon.

—— (1990) *The History of Sexuality, Vol. 1*. Harmondsworth: Penguin.

—— (1991) *Discipline and Punish*. Harmondsworth: Penguin.

—— (1994) *The Order of Things*. London: Routledge.

Fuller, B. (1975) *The Ghan: The Story of the Alice Springs Railway*. Adelaide: Rigby.

GSR (Great Southern Railway) (1998) *The Train Journey to the Heart of Australia*, advertising brochure. Adelaide: Australian National (AN).

—— (2000) *Australia's Great Train Journeys 2000–2001*, advertising brochure no. 68. Marleston, SA: GSR.

Hall, C.M. (2004) *Tourism: Rethinking the Social Science of Mobility*. Harlow: Pearson.

Hamilton, A. (1990) 'Fear and desire: Aborigines, Asians and the national imagery', *Australian Cultural History*, 9: 14–35.

Heathcote, R.L. (1987) 'Images of a desert? Perceptions of arid Australia', *Australian Geographical Studies*, 25: 3–25.

Hobsbawm, E. (1992) *Nations and Nationalism since 1780: Programme, Myth, Reality*, second edition. Cambridge: Cambridge University Press.

—— (2000) 'Introduction: Inventing traditions', in E. Hobsbawm and T. Ranger (eds) *The Invention of Tradition*. Cambridge: Cambridge University Press.

Hollinshead, K. (1996) 'Marketing and metaphysical realism: the disidentification of Aboriginal life and traditions through tourism', in R. Butler and T. Hinch (eds) *Tourism and Indigenous Peoples*. London: Thomson Business Press.

—— (1999) 'Surveillance of the worlds of tourism: Foucault and the eye-of-power', *Tourism Management*, 20: 7–23.

Horne, D. (1978) *Right Way – Don't Go Back*. Melbourne: Sun.

Jopson, D. (1997) 'Loco becomes a moving Aboriginal picture show', *Sydney Morning Herald*, 14 November: 8.

Kapferer, B. (1988) *Legends of People, Myths of State*. Washington, DC: Smithsonian Institute.

Langton, M. (2000) 'Why "race" is a central idea in Australia's construction of the idea of a nation', *Australian Cultural History*, 18: 23–37.

Lattas, A. (1990) 'Aborigines and contemporary Australian nationalism: primordiality and the cultural politics of otherness', in J. Marcus (ed.) *Writing Australian Culture: Text, Society and National Identity*. Special issue of *Social Analysis*, Adelaide: University of Adelaide.

—— (1992) 'Primitivism, nationalism and individualism in Australian popular Culture', in B. Attwood and J. Arnold, *Power, Knowledge and Aborigines*. Special edition of *Journal of Australian Studies*.

MacCannell, D. (1999) *The Tourist: A New Theory of the Leisure Class*. Berkeley, CA: University of California Press.

—— (2001) 'Tourist agency', *Tourist Studies*, 1: 23–37.

McGrath, A. (1991) 'Travels to a distant past: the mythology of the Outback', *Australian Cultural History*, 10: 113–124.

McKercher, B. (1993) 'Some fundamental truths about tourism: understanding tourism's social and environmental impacts', *Journal of Sustainable Tourism*, 1: 6–16.

McKercher, B. and du Cros, H. (2002) *Cultural Tourism: The Partnership Between Tourism and Cultural Heritage Management*. New York: Howarth.

MacLaughlin, J. (1996) 'The evolution of anti-traveller racism in Ireland', *Race and Class*, 37: 47–63.

Marcus, J. (1988) 'The journey out to the Centre: the cultural appropriation of Ayers Rock', in A. Rutherford (ed.) *Aboriginal Culture Today*. Sydney: Dangeroo Press.

Markus, A. (1990) *Governing Savages*. Sydney: Allen & Unwin.

Meethan, K. (2001) *Tourism in Global Society: Place, Culture Consumption*. New York: Palgrave.

Moran, A. (2002) 'As Australia decolonizes: indigenizing settler nationalism and the challenges of settler/indigenous relations', *Ethnic and Racial Studies*, 25: 1013–1042.

Mowforth, M. and Munt, I. (1998) *Tourism and Sustainability. New Tourism in the Third World*. London: Routledge.

Palmer, C. (1999) 'Tourism and the symbols of identity', *Tourism Management*, 20: 313–321.

Perera, S. (2002) 'A line in the sea', *Cultural Studies*, 8: 11–27.

Pratt, M. L. (1992) *Imperial Eyes: Travel Writing and Transculturation*. London: Routledge.

Pretes, M. (2003) 'Tourism and nationalism', *Annals of Tourism Research*, 30: 125–142.

Qantas Airways (1995) *The Journey of the Kangaroo*, brochure.

Rojek, C. (1985) *Capitalism and Leisure Theory*. London: Tavistock.

Rooney, D. (1997) 'A contextualising, socio-technical definition of technology: learning from ancient Greece and Foucault', *Prometheus*, 15: 399–407.

Sheller, M. and Urry, J. (eds) (2004) *Tourism Mobilities. Places to Play, Places in Play*. London: Routledge.

Simondson, C. (1996) 'Primitivism, discourse and national identity: analysis of advertising literature of the Australian tourism industry', *Olive Pink Society Bulletin*, 7(1/2): 22–27.

Slarks, C. (1998) 'Do you remember?' *The Advertiser*, 10 January: 38.

Spearritt, P. (1990) 'How we've sold images of Australia', *Australian Society*, November: 15–19.

Stevens, C. (1989) *Tin Mosques and Ghantowns*. Melbourne: Oxford University Press.

Trainways (2003) *Australia's Great Train Holidays 2003–2004*. Marleston, South Australia: Trainways Australia, Great Southern Railway Travel Pty Ltd.

Triandafyllidou, A. (1998) 'National identity and the "other"', *Ethnic and Racial Studies*, 21: 593–612.

Urry, J. (1990) *The Tourist Gaze*. London: Sage.

Watson, G.L. and Kopachevsky, J.P. (1994) 'Interpretations of tourism as commodity', *Annals of Tourism Research*, 21: 643–660.

Weaver, D. (2001) *Ecotourism*. Milton, Australia: Wiley.

Weaver, D. and Lawton, L. (2002) *Tourism Management*. Milton, NSW: Wiley.

Weaver, D. and Opperman, M. (2000) *Tourism Management*. Milton, NSW: Wiley.

Whitelock, D. (1986) *Gone on the Ghan*. Adelaide: Savvas.

Willis, A. (1993) *Illusions of Identity*. Sydney: Hale & Iremonger.

Winn, P. (1999) 'Tourism, gazing, and cultural authority', *Tourism, Culture and Communication*, 2: 85–98.

Part II

Power, property and resources

6 The politics of bed units

Growth control in the resort of Whistler, British Columbia

Alison M. Gill

Globalization and the continued growth of tourism have resulted in changing power relationships within the resort sector. On the one hand fiscal responsibility has frequently been downloaded to the local level of government which, while resulting in more autonomous decision-making, has demanded greater emphasis on private–public partnerships (Barlow and Wastl-Walter 2004). At the corporate level, consolidations have characterized major players in the tourism industry such as hotels, airlines and resort operators as they strive to remain competitive in the global marketplace (Archambault 2002; Crotts *et al*. 2000; Keller 2000; Porter 1985). Debbage (1990) points to the emergence in the United States of what he calls 'tourism oligopoly' whereby the tourism supply (e.g. hotels, airlines) is increasingly concentrated in the hands of large transnational corporations (TNCs). Concurrently, the concept of sustainability has been introduced not only into the public policy arena but also into the corporate lexicon. In the latter, concern over social and environmental responsibility (Berry and Rondinelli 1998; Flagestad and Hope 2001) is increasingly common in the corporate environment. This approach to achieving sustainability has resulted in some resort locations, but by no means all, in greater public involvement in decision-making by local residents and organized groups (NGOs), and an expansion by corporations to include stakeholders as well as shareholders in their decision frame. Thus, the nature of a resort will reflect the varying coalitions, partnerships and discourses that emerge from the relative power of all these actors within the dominant political regime.

In this chapter, I examine politics and power in the case of Whistler, a year-round mountain resort located 120 km north of Vancouver in the Coast Mountains of British Columbia, Canada (Figure 6.1). Whistler is distinctive from many resorts in that it is a comprehensively planned new resort that became fully operational in 1980 as a major ski resort and subsequently a year-round destination. This study demonstrates how changing local planning and development priorities are reflected in evolving discourses between key stakeholders. In particular, this chapter explores how one planning tool, 'bed units', designed to monitor and manage growth, has acquired political valence in the power relationship between the resort community and Intrawest, the corporation that owns the mountain operations and has significant real estate investments in the resort. I begin with an examination of the literature on power relations relevant to resort settings as a context to the specific theoretical

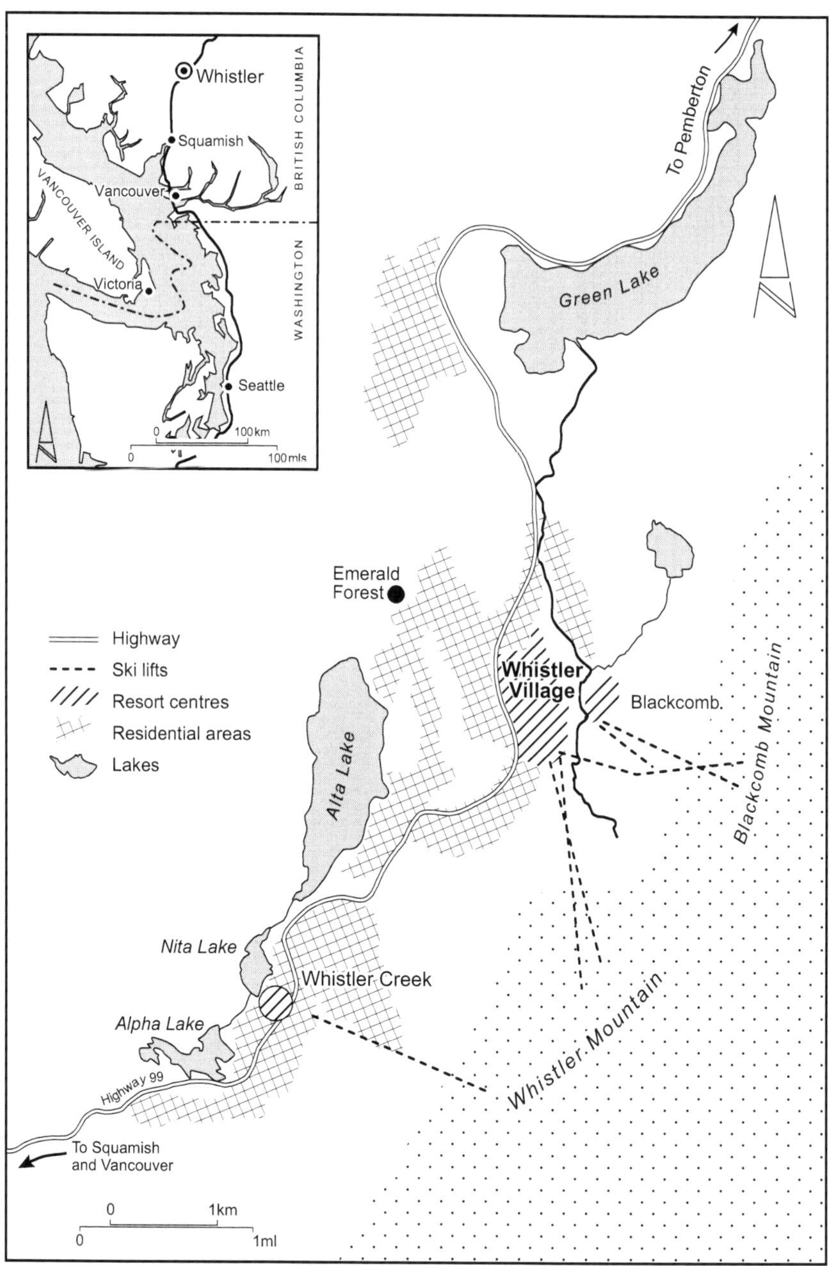

Figure 6.1 Whistler: the case study location

Source: author.

approach adopted in this chapter. This is followed by an overview of the regulatory framework at Whistler and the way in which growth management issues, and more specifically the notion of bed units and a bed unit cap, have evolved into policy. I then explore the discourses over development issues involving bed units by examining four key events that have created much public debate within the resort community.

The empirical data are derived from ongoing field research in Whistler over the past decade. This has employed multi-method approaches that have included analysis of official community documents and reports; newspaper reports; key informant interviews (both formal and informal); and participant observation at community meetings. The analysis of power relations in the resort development process is grounded in an urban regime approach which views power as a function of collaborative arrangements whereby local governments and local actors build coalitions across institutional boundaries (Stone 1989). I also draw on stakeholder theory (Freeman 1984) to enhance an understanding of the drivers behind corporate strategies regarding collaborative arrangements.

Power and politics in resort settings

Resorts range from comprehensively planned enterprises with one or more major developers (Stanton and Aislabie 1992) to catalytic developments that result from a concentration of smaller tourism enterprises that evolve over time (Pearce 1987). The degree to which local institutions and residents exercise control over development will vary depending on the institutional arrangements of the resort (Gill 1997). As Fainstein *et al.* (2003: 7) observe, 'tourism operates within a more tangled web of regulations than other industries'. This may explain in part why Butler (1999) concludes that most tourism researchers have ignored issues of the control of tourism and thereby the politics of tourism.

In Canada, the setting of the present study, regulatory parallels can be drawn with that found in Mullins's (2003) study of Australia's Gold Coast where a 'natural resources mode of regulation' was institutionalized (cf. Church and Ravenscroft, Chapter 8 in this volume). As in Australia, the dominant modes of regulation in Canada were established to support the exploitation of natural resources (forests, minerals, agriculture). With globalization and restructuring, this mode of regulation has been contested as new post-productive activities including tourism and recreation are introduced (Reed and Gill 1997). The result has been an institutional remapping of the resource peripheries (Hayter 2003). In all parts of North America, national and regional governments not only engage in environmental regulation but they also play a major role in the administration of public lands upon which much recreational use is dependent. Resort developers who are dependent on access to these lands generally must negotiate lease or purchase agreements with the government agencies. In recent years growing environmental concerns over the impacts of recreational developments have often resulted in prolonged negotiation and stalled development proposals. Recent examples include Lake Catamount, Colorado; Mt Tremblant, Quebec; and Mt McKenzie, British Columbia (Loverseed 2005).

A common theme in discussions of power relations in tourism is the portrayal of tourism destinations as powerless in the face of external investors who commodify and overwhelm local communities (Rothman 1998; Clifford 2002). This perspective, which often equates power with domination (Allen 2003), is evident in recent commentaries on North American ski resorts (Rothman 1998; Clifford 2002). In response to global competition, a common strategy for mountain resort corporations during the past fifteen years has been the consolidation of properties. This is seen as a vehicle for leveraging the capability and resources of individual mountain operations to achieve corporate goals. In addition, by consolidating several individual businesses under one corporate structure, the ability of the individual mountain resort operations to address risks can be reduced, in particular those associated with: seasonality; real estate development; competition; unfavourable weather conditions; economic downturns; capital leverage and financial commitments; and environmental impacts (Poon 1993; Bullock 1998; Buhalis 2000).

In some developing countries new resort developments have sometimes been typified by events that suggest that the power of large corporations does allow them to wield a dominant influence. For Rothman (1998), the consolidation processes in Canadian ski resorts involve the 'corporatization of place'. The in-depth case study presented here, however, suggests that local power relations are not simply typified by corporate domination and instead, in keeping with some of Allen's (2003) more general arguments, power is more nuanced and displays a combination of modes dependent on space and time.

In unravelling the complexities of power at a community level, Hall (2003) has noted the difficulty of separating social, economic and political processes within a community from the conflict that occurs between stakeholders. He suggests that this accounts for increased interest in the literature on collaborative and cooperative approaches (e.g. Jamal and Getz 1995; Timothy 1998; Bramwell and Sharman 1999). In turn, this raises issues surrounding the distribution of power amongst different stakeholder groups (Reed 1997). Jamal and Getz (1995) view power as an instrument to be managed and balanced. They suggest applying organizational theories to improve mechanisms for collaborative approaches and that it is possible to address the issue of uneven power and authority in the planning process by the inclusion of legitimate stakeholders and identification of a suitable convener (often the local authority) at an early stage in the collaborative planning process (Jamal and Getz 1995). Reed (1997), however, has conceptualized power relations in community-based tourism planning as falling within three policy arenas: developmental, allocational and organizational. For her, accounts of collaborative tourism planning still rely on rather weak theories of power relations within community settings. Reed (1997: 598) contends that,

> theories of collaboration must incorporate power relations as an explanatory variable that explains why collaborative efforts succeed or fail, rather than as an instrumental variable that suggest how power can be balanced and convened.

This line of argument suggests that attempts to understand collaborative partnerships in tourism must go beyond considering which stakeholders are included or excluded and must be based on theoretical approaches which enable a detailed examination of how power relations underpin the changing nature of collaboration and alliances. The remainder of this chapter explores the potential of theoretical perspectives developed to examine urban politics for understanding the power and politics associated with resort development.

Theoretical approach

The empirical study of Whistler is conceptually informed by urban regime analysis (Stone 1989) – an approach that has been widely applied to understand urban politics and power (Stoker 1996). In a retrospective examination of his original concept, Stone (2005: 328) observes, 'at its core, urban regime analysis centers on how local communities are governed'. In a complex society, this calls for public bodies to form coalitions with private interests in the making of policy decisions. In most localities, elected officials and business interests are the key participants but other community interests may be represented (Stone and Sanders 1987). As described by Haughton and While (1999: 5):

> coalitions . . . reflect the leadership skills, vision, commitment, and resources that various partners can contribute by working together, building trust, sometimes combining resources, and seeking to resolve differences in ways that bring both collective benefits for the city and individual benefits for those involved.

Urban regime analysis is more of a concept or model than an actual theory (Mossberger and Stoker 2001), although theoretical constructs have emerged from numerous case studies employing the approach, especially in the US, but increasingly elsewhere (e.g. Dowding *et al.* 1999). Case studies are used to historically examine such key factors as how agendas have been framed, what factors brought coalition partners together and how coalition partners have worked collaboratively (Stone 2005). On this basis, the degree of stability exhibited by regimes can be assessed, although the model is not necessarily explanatory nor predictive (Stone 2005).

Core properties of regimes include the partners (both business and other) drawn from government and non-governmental sources; collaboration based on social production (rather than social control); identifiable policy agendas; and a long-standing pattern of cooperation (Mossberger and Stoker 2001; Stone 2005). Criticism of regime analysis has focused on issues of localism and empiricism (Haughton and While 1999; Stoker 1996) and the fact that it does not travel well beyond its original North American context (Pierre 2005). However, the approach has been widely applied in North America and elsewhere and, as Mossberger and Stoker (2001) contend, the concept has been instrumental in helping analysts develop a new understanding of how power works in urban settings. The internal dynamics of coalition building represent a model of the social production of power whereby the various actors, in response to an agenda, collaborate to constitute the 'power to' (as opposed to domination or 'power over') accomplish their aims (Mossberger and Stoker 2001; Stone 2005).

Stone (1989) identifies four types of urban regime: a maintenance regime; a development regime; a middle-class progressive regime; and a lower class opportunity expansion regime. In the study of Whistler only the development (growth) regime and the progressive regime are of relevance. In the former, the policy focus is on changing land use to promote development, whereas the progressive regime includes more emphasis on issues such as environmental protection, growth control and affordable housing (Mossberger and Stoker 2001). A maintenance regime is not relevant to a resort that has only developed since 1980 and as will be shown the nature of the local community that emerged in the resort has militated against a regime based on lower class opportunity expansion. Identifying regime change is difficult and not well theorized (Mossberger and Stoker 2001), and it is not necessarily linked to political or administrative change.

Regimes do not operate on the basis of formal hierarchy but operate as networks (Stoker 1996). Stoker (1996: 272–273) observes that 'under the network approach organizations learn to cooperate by recognizing their mutual dependency'. This same perspective underpins other network theories such as stakeholder theory (Freeman 1984). In the study of Whistler I draw upon stakeholder theory to provide some further insight from a corporate perspective on motives for coalition building within the urban regime.

In the context of corporate stakeholder relationships, the association between a company and its stakeholders is increasingly being conceived as one of corporate social responsibility (CSR); also termed 'corporate citizenship' and 'corporate community involvement' (van den Berg *et al.* 2004). Socially responsible corporations seek to balance their obligations to shareholders with obligations to other stakeholders (Cragg 1996; Banerjee 1998). As Ryan (2002) notes, there are two reciprocal aspects of power in this relationship: 'power over' and 'responsibility for'. A prerequisite for this reciprocity is that mutual benefits exist. For instance, emerging CSR practices offer new solutions to environmental and social problems through collaborative and resilient stakeholder relationships that benefit both businesses and communities (Marcoux 2004). There are varying perspectives concerning the benefits of stakeholder engagement that include: the view that social responsibility is only pursued to the extent that it furthers business objectives; the belief that it secures the social viability of the firm; an argument that the firm has responsibilities to society as a whole; and, a belief that CSR is a form of investment that produces competitve advantages (van den Berg *et al.* 2004).

The difficulty of stakeholder identification is often cited as a limitation of stakeholder theory (Robson and Robson 1996), especially in the corporate strategic management literature where it has received extensive critical consideration (cf. Ambler and Wilson 1995; Sternberg 1997; Banerjee 2001). A further problem is that of unequal power amongst stakeholders. Even if all stakeholders are identified, there is often an assumption that the act of collaborating will overcome power imbalances (Reed 1997). While being cognizant of the constraints of stakeholder engagement from a corporate point of view, several researchers have employed stakeholder analysis in the study of tourism development (Bramwell and Sharman 1999; Marwick 2000; Ryan 2002). In this chapter, however, the examination of coalition building by corporate and other stakeholders draws on some of the specific conceptual developments in actor-network theory (ANT).

While empirically the study in Whistler does not fully employ ANT, the approach does provide a more explicit perspective on power relations (Davies 2002) than stakeholder analysis, and some specific ANT constructs were incorporated in the analytical approach. In ANT all influencing factors linked to an act, including the role of both human and non-human actors, are considered as part of the network (Murdoch 1995). That an understanding of power is derived from an examination of the intricacies of associations is underscored by Murdoch and Marsden (1995: 372) who, drawing from the ideas of Latour (1986), observe that 'the analysis of power becomes the study of *associations*'. The approach has been widely applied in the areas of environmental planning and management, and it has played a significant role in policy analysis for more than a decade (Murdoch and Marsden 1995; Selman 2000). ANT seems appealing to planners because it is process oriented, defining various stages in the 'sociology of translation' (Selman 2000). It thus embodies a similar historical perspective to that of urban regime analysis in understanding how networks (collaborations and coalitions) evolve and change.

A specific concept drawn from ANT which is used here is that of 'intermediaries' which is the term used to conceptualize objects such as plans, maps or codes of practice, as actors that assist in the transfer of information and power within the network (Selman 2000; Selman and Wragg 1999), The notion of 'intermediaries' is pertinent to the present study which focuses on the evolving role, meaning and application of 'bed units' as a planning tool. The empirical study, therefore, focuses on how the regime in Whistler has evolved and the specific role played by the debate over bed nights in the construction of power relations and alliances especially those involving corporate stakeholders.

Power politics and the regulatory framework in Whistler, British Columbia

Whistler was officially designated as a 'Resort Municipality' in 1975. As such, the Resort Municipality of Whistler (RMOW) was granted powers – for example the ability to borrow money without referendum – that were distinct from other municipalities, and that were aimed at supporting and encouraging tourism growth as well as controlling and managing tourism development. The provincial government was also a major influence as it had provided public (Crown) lands and financing, and it had established the regulatory system. While at the local governance level there was an elected mayor and council, in the early stages the provincial government retained some control through an appointed member on the council. Importantly, the development in the resort has been regulated through an official community plan (OCP) that contains detailed policies that make up the legal framework for regulating land use development, servicing and protection of the natural environment. Further regulation of the built environment is imposed through by-laws that establish zoning regulations and design guidelines (Dorward 1990). By July 2004 the permanent full-time resident population had grown to 9,480, and there were a further 4,000 seasonal employee residents (RMOW 2004a). There are also a large number of second-home owners who account for about 72 percent of property ownership in Whistler, although a considerable proportion of these properties are part of the rental pool for tourist accommodation for at least part of the year (RMOW 2004a). There are approximately 2 million visitors to Whistler each year, 46 per cent in winter months and 54 per cent in the summer (RMOW 2004b). Under the Resort Municipality Act, all commercial enterprises are required to become members of Tourism Whistler (formerly Whistler Resort Association). This organization engages in collaborative marketing for the resort, although this does not preclude independent marketing by individual members.

Power relationships in Whistler have shifted as the resort community has grown and evolved. In the initial stages of development, beginning in 1975, the key actors were the municipal government, the provincial government and the developers. This has been characterized as the 'growth machine' phase (cf. Logan and Molotch 1987) when – as the resort was new and untested and the investment risk high – considerable power lay with the developers (Gill 2000). Although development was comprehensively planned and regulated, all efforts focused on attracting investment and development. While the two companies that owned the ski operations were the largest developers, there were many other developers involved in building hotels, tourism businesses and residential properties. In some instances properties were acquired although not immediately developed, but held for speculative purposes. In 1981 just after the village and ski operations became fully operational, recession resulted in a major financial bail-out by the provincial government and a restructuring of the agency responsible for development. This company was a subsidiary of the RMOW and all members of council were automatically directors, with the mayor as chairman. Thus, the role of local government was integrally bound to land development with several elected members actively involved in development projects (Gill and Reed 1997).

The resident population was initially only a few hundred in 1980 but grew rapidly to around 4,000 by the end of the decade (Gill 2000). However, during the 1980s the 'community' was in a nascent stage with little community organization or 'voice'. The main private sector stakeholders in the resort were the two companies that owned the ski lifts and mountain operations on the two adjacent mountains, Whistler and Blackcomb. Although there were some corporate changes over the first decade of the resort's existence, a major driving force of growth in the resort was the competition between the two ski company operators, Intrawest Corporation and Whistler Mountain Skiing Company. Both operators aggressively matched each other in developing ski lift infrastructure, snow-making equipment and on-hill facilities. This led to accelerated tourism growth that stimulated the development of hospitality and tourist services in the village. The RMOW encouraged this in order to reach the planned critical mass of development necessary to establish the economic viability of the resort. The RMOW was also clearly building alliances with corporate stakeholders but such collaborations also involved encouraging competition between the business bodies (Gill and Williams 1994).

Following a period of rapid development in the late 1980s and increasing numbers of permanent residents with a demographic profile that now included young families with children, the community showed signs of organization. One prominent group to emerge was Whistler Area Residents for the Environment (AWARE), a local environmental NGO. Other resident groups were concerned about the lack of community facilities. The community had no high school, recreation or community centre, daycare facilities or library. In the 1990 election the voice of the community was effectively heard for the first time and the local council elected represented a more balanced perspective between the needs of the resort and the community (Gill 2000). While still pro-development, community needs were given prominence. It was this change that heralded the era of 'growth management'. This approach calls for a high degree of resident involvement with growth decisions. As Gill (2000: 1099) observed, 'growth management is a compromise that appeals to both growth and no-growth forces – with business interests focusing on growth and antigrowth interests emphasizing management'. The introduction of a resort and community monitoring programme that tracked indicators of environmental, social and economic change provided a basis for annual town hall meetings with residents to discuss the resort community's growth. Throughout the 1990s there were numerous opportunities for residents to be involved in development issues with public consultations and hearings on issues ranging from affordable housing and transportation to environmental management (Williams and Todd 1997).

The relationship between the community and the corporate interests changed in 1997 when Intrawest Corporation and Whistler Mountain Skiing Company merged under the publicly traded 'Intrawest' banner. Although the ski companies were always dominant players in the power structure of the resort, they were required under the Resort Municipality Act to coordinate their marketing strategies with all other commercial enterprises in the RMOW through the Whistler Resort Association (Williams *et al*. 2004), thus engaging in a degree of collaboration that is relatively rare in tourism destinations.

The emergence of a single dominant corporation within the community was cause for some concern amongst residents who envisaged problems of a 'single-industry town'. However, while Intrawest does operate all the skiing facilities and has developed a large amount of both commercial and private real estate as well as other business enterprises, Whistler as a resort still presents an image that is not dominated by Intrawest's corporate presence. There are several large hotel chains operating in the resort and numerous small retail businesses, restaurants and tourist services. Intrawest and its on-mountain operations, Whistler-Blackcomb, have a more placeless brand image than that of the RMOW (Williams *et al*. 2004).

In the past few years an ongoing priority for the RMOW, residents and also Whistler-Blackcomb has been maintaining environmental quality. For the RMOW, this has evolved from environmental monitoring to the development of a more comprehensive environmental management system (Waldron 2000). Since 2000, and in the face of reaching the agreed upon maximum level of development in terms of bed units, the focus has shifted to the broader issue of 'sustainability' including not just environmental but also social and economic concerns. Intrawest, at least in its on-mountain operations, has also shown concern for environmental management. Whistler-Blackcomb has developed a highly acclaimed environmental management system that also includes a number of community outreach programs such as bear management and fish habitat restoration (Whistler-Blackcomb 2005). It has also entered into a collaborative sustainability programme called 'The Natural Step' (TNS) with RMOW and several other Whistler businesses. TNS is an international programme that seeks to advance sustainable practices (The Natural Step 2005). Concurrent with Intrawest's evolving interest in corporate environmentalism, the RMOW has adopted a strategic business-oriented approach to its operations (Flagestad and Hope 2001). Thus, the resort and the community are increasingly 'speaking the same language'.

Development in Whistler has always been carefully managed through a detailed and increasingly sophisticated planning process whereby there have been structural constraints on the nature and type of development. As growth has been a common goal of both the RMOW and the private sector there has been close collaboration between public and private interests. Although many developers have been involved in development projects, Intrawest has played the largest role in land development, although regulatory power to allocate and control development has resided with the RMOW. One of the planning tools that the RMOW has used to assist in achieving its development goals has been 'bed units'. The following section examines how, in the context of discourses over growth issues, the concept of bed units and of a bed unit cap has evolved to be a key 'intermediary' device in the development of power relations and alliances. Examining a series of key events involving discussions over bed units provides insights into the nature of coalitions and regime stability.

Key events in the discourse over growth

Planning policy: bed units and the bed unit cap

A bed unit is a measure of the number of beds in the resort; elsewhere (for example Aspen, Colorado) it is referred to as a 'pillow count'. Initially the allocation of bed units was linked to estimates of sewage capacity and later also to water supply capacity. In development planning in Whistler estimates of bed unit capacity are based on the 'highest and best accommodation use under zoning regulations' (RMOW 1993a: 3). Examples of bed unit allocation are shown in Table 6.1. Detached dwellings, for example, are allocated 6 bed units regardless of size whereas the allocation for a hotel room is based on the number of square metres. Table 6.2 presents statements taken from official planning documents that offer a definition of bed units. The first four statements reflect the 'official' view of bed units as simply planning tools that assist in 'tracking and managing growth' The final statement hints at a different perception of bed units but makes it clear that, from an official planning perspective, bed units do not have value with respect to development rights or property status.

Table 6.1 Examples of bed unit allocation in Whistler

Type of unit	*Sq. metres*	*Bed units*
Multiple residential/commercial	0–50	2
	100 +	4
Bed and breakfast	N/A	6
Campsite	N/A	1
Detached dwelling	N/A	6
Duplex	N/A	12
Dormitory bed	N/A	0.5

Source: RMOW (2003a).

Table 6.2 Statements in Whistler planning documents defining bed units

- Measures of the amount of residential development
- Includes all types of accommodation
- A bed unit reflects the servicing and facility requirements for one person
- Bed units are simply a measure of development that assists the RMOW in tracking and managing growth
- Bed units do not equate to development rights nor do they have property status.

Sources: RMOW (2002, 2003a).

Initially the development cap was set at 40,000 bed units but in the face of rapid development it was raised in 1989 to 52,500. This figure is still officially the current limit (RMOW 2003a) and one that 'should not be exceeded except under extraordinary circumstances' (RMOW 1993a: 3). Notwithstanding, as shown in Table 6.3, the total number of bed units built and committed in 2003 totalled 55,175. Although a minor point, it is worth noting that this total number was not included in the RMOW publication from which the figures for the different types of accommodation were drawn (RMOW 2003b). One might speculate that this omission was intentional in order not to draw attention to the fact that the 52,500 cap had been exceeded.

Table 6.3 Allocations of bed units in Whistler, 2003

Tourist accommodation bed units	13,500
Market units in dwellings	30,650
Restricted bed units	3,525
New committed bed units added to current OCP	7,500
Total (built and committed)	55,175

Source: RMOW (2003b).

The debate over bed units will now be used to explore the evolving discourse over development. Three key events surrounding development (at Emerald Forest; Whistler Creek and Nita Lake) are identified as illustrative of the power relations between the main actors, including: the local government (RMOW); Intrawest; other developers; residents (who can be further disaggregated into permanent residents, second-home owners and seasonal workers); and local NGOs. In addition to these three key events, the ongoing debate concerning the relationship between resident-restricted housing and the development cap is also examined as it plays a critical role in understanding how the 'intermediary' device of the bed unit limit is restructured to address political goals.

The Emerald Forest deal

In 1999 a debate over Emerald Forest arose when the owner of a 56 ha (139 acre) plot of land in the valley bottom announced that he was planning to build six detached houses on large lots (Barnett 1999a). The bed unit allocation on this property was 36 units. However, since the owner had acquired the land in the earlier days of development, the RMOW had subsequently established an increasingly rigorous environmental management policy and was attempting to preserve the valley wetlands as an ecological corridor.

The owner of the land was willing to sell the property to the RMOW for CA$10 million but the municipality could not afford this amount. In searching for a solution, the RMOW approached Intrawest to see if it was interested in negotiating a development trade-off. This was agreed, with Intrawest buying the Emerald Forest property for an undisclosed sum and selling it to the RMOW for CA$1 million as well as the rights to develop 476 bed units as a hotel on a parking lot that Intrawest owned near the base of the ski lifts (*Pique Newsmagazine*, 20 August 1999).

While this would appear to be a 'win-win' situation for both the municipality and Intrawest as well as evidence of their close alliance, it evoked a major debate within the resort. The concern surrounded the fact that the RMOW had already committed all 52,500 bed units of the agreed cap; thus, it did not have the additional 476 units to allocate. Many residents believed that the trade-off was in the best interests of the community and the local environmental NGO – the Association of Whistler Area Residents for the Environment (AWARE) – was supportive. By contrast, some residents viewed exceeding the development cap as the 'thin end of the wedge' (Barnett 1999a: 2).

Certainly the decision undermined the perceived close relationship between bed unit limits and the environmental constraints (sewage and water) that had been fostered over the previous 15 years or more. As one local newspaper remarked, 'bed units are not an absolute measurement nor are they a limit to growth' (Barnett 1999b: 2). It was also evident that bed units could 'float' and were not as rigidly fixed in place as zoning regulations had suggested. Most importantly, the Emerald Forest case demonstrated that bed units were, in fact, *currency* (Barnett 1999c).

Further, this case brought to the public's attention that the bed unit cap had been reached and indeed exceeded. Given that it represented such an entrenched concept in the principles guiding development of the resort community, some residents believed that there should have been greater public input into the decision. However, because of impending local elections, the decision process was accelerated as the municipality's priority was to preserve the valley wetlands and it did not want to risk losing the opportunity as the result of election politics.

Whistler Creek

In 1997 Intrawest acquired 2,500 unused bed units when it merged with Whistler Mountain. These bed units were what remained from Whistler Mountain's allocation under the so-called 'lands for lifts' policy. This policy was introduced during the early development stages to encourage the ski companies to invest heavily in on-mountain lift infrastructure. To offset the high cost of capital investment on infrastructure that would take a long time to recoup from users, the companies were granted bed units; in essence, development rights (despite the RMOW definition shown in Table 6.2).

Intrawest, which has a strong real estate division, had previously developed most of its existing allocation. With the new acquisition it had the opportunity to engage in a major development. The corporation produced its Comprehensive Development Strategy that represented the largest single development proposal in Whistler's history (Barnett 2003). The plan was to combine all its remaining undeveloped bed units in a major village development at Whistler Creek, a few kilometres away from the existing village centre, at the original area of ski development at the base of Whistler Mountain. The village would incorporate the design principles and coordinated animation that Intrawest has created as a corporate identity in other resort developments (e.g. Mt Tremblant, Quebec; Blue Mountain, Ontario; Les Arcs, France). In addition, the Comprehensive Development Strategy included a proposal to develop two other residential neighbourhoods: one with mixed residential uses, and another exclusive neighbourhood with large 'trophy' houses.

The negotiation between Intrawest and the RMOW over the proposal was played out over a 2-year period and involved considerable public consultation. Both parties wanted concessions (Barnett 2003). The RMOW held considerable power in the negotiation because Intrawest needed certain variances to the planning regulations including 'floating' bed units from various other locations in the resort. The RMOW realized that there would be few other opportunities to extract community amenities from developers and so negotiated for a new elementary school, a daycare centre, parks and recreational trails.

The main issue of debate, however, surrounded the proposal for an exclusive residential subdivision. From Intrawest's perspective this would provide a lucrative return on investment while consuming few bed units and was considered essential to the financial viability of the proposal. Both residents and council members were divided in their views on this development, with some arguing that such a development would bring 'cold' beds to the community as many of the homes would be purchased by second-home owners who were only in the resort for a short period of time. Others argued that it was contrary to the desire for more affordable development that many felt the community should embrace, whereas some residents were supportive because of the community amenities that Intrawest was offering. Meanwhile, Intrawest was holding out for acceptance of the whole package of development, which compromised the ability of the politicians to reject the high-end neighbourhood. The proposal was finally approved with the tie-breaking vote of the mayor who supported it and who had negotiated the significant level of community amenities (Barnett 2003).

Nita Lake

The most recent controversy in the resort has surrounded the 2002 proposal for a small boutique hotel with 77 suites and 14 additional housing units on the shores of Nita Lake in the valley bottom (Taylor 2003). Unlike the previous two examples, this project did not involve Intrawest but a private developer who had owned the land for some time but not developed it. As it was one of the last remaining developments permitted under the existing OCP, the RMOW was able to negotiate hard with the developer who wished to contravene size and height restrictions. In return for being permitted to develop the property as he desired, the developer offered a number of amenities. These included seven times the required number of employee housing units, CA$1million for the community's health care system and a new railway station (Barnett 2003).

By contributing to two areas of community concern – employee housing and health care – the developer was able to garner (some argued 'buy') support from the RMOW and residents. After protracted public debate, while the RMOW voted in favour of the development, one wealthy second-home owner whose house looked out on the proposed new development argued to prevent the development on the basis that it detracted from his view and hence the value of his property. He eventually took legal action. The courts decided in his favour and the development was disallowed; this was much to the anger of many residents who were supportive of the excellent package of community benefits that were negotiated. However, a few months after the legal ruling, the provincial government intervened and overruled the legal decision through a special government procedure on the basis of community and regional interests (Taylor 2004). This was a very surprising action but greatly welcomed by the community. While not clearly evident, it would appear to be related to the province's interest in seeing the railway station built to support both the Winter Olympic (2010) endeavour and the burgeoning cruise ship industry that planned a high-end train trip as part of the land-based offerings of the cruise industry in Vancouver.

Housing the labour force

The issue of providing employee housing in successful tourism destinations is a widespread problem to which there appears to be no easy solution (Goodno 2004). In Whistler in the early stages of development there was no policy regarding this, in part because RMOW argued that to impose extra costs on developers before the resort had reached a successful critical mass would have deterred investors. It was not until 1997 that the Whistler Housing Authority (WHA) was developed, by which stage house prices were beyond the reach of many employees and many had already moved to the adjacent communities of Pemberton and Squamish.

The WHA is supported by money from developers who are required to either construct the necessary housing for employees required by the development or pay a cash equivalent to the WHA. However, housing studies revealed that there were inadequate bed units remaining under the 52,500 bed unit cap to meet the desired level of housing provision (WHA 2000). The RMOW then decided that resident-restricted housing (referred to elsewhere as employee housing or affordable housing) would not count towards the cap. By 2004 the 52,500 cap was already exceeded as a result of the construction of additional resident-restricted housing (WHA 2004). As shown in Table 6.4 there is clear support by residents for exceeding the cap in order to provide resident-restricted accommodation.

Table 6.4 Residents' attitudes to changes in the bed unit cap, 2003

Statement	*Attitudinal response*
Whistler should not increase the existing bed cap for any reason	2.49
Whistler should increase the bed cap to provide resident-restricted housing	4.00
Whistler should increase the bed cap for market housing units in exchange for the developer building resident-restricted housing for employees	2.78
Whistler should accommodate at least 75% of employees within municipal boundaries	4.09

Source: RMOW 2003c.

Notes: Response scale: 1 = Strongly disagree, 2 = Disagree, 3 = Neutral, 4 = Agree and 5 = Strongly agree.

Discussion

The evolving power of bed units

In the various key events examined, the political nature of 'bed units', especially the establishment of the 52,500 bed unit development cap, is clearly evident. Bed units have been used as a political tool in varying ways to exercise power in negotiating development agreements. However as one candidate for council noted, 'the issue of bed units has become a piecemeal affair where council makes decisions from development to development with no set guide lines to follow' (Lamont 2001: 30).

In the earlier years of development the existence of a bed unit cap, apparently based on scientific facts, gave residents a sense of control and a belief that the

RMOW was engaging in sound environmental stewardship. Also, it was an easy tool to use as it provided an absolute value that was readily understood, and as with the use of many standards in planning, it became entrenched as a central concept in dealing with growth issues. As growth management practices became more developed in the 1990s with the introduction of a resort and community monitoring programme and annual public review meetings, the tracking of bed unit development relative to the cap became a significant indicator. A sense of development control by residents was especially important in the early 1990s when the community experienced a period of very rapid growth. In 1996 the election platforms of most of the candidates for mayor and councillors stated that the 52,500 bed unit cap should not be changed. None of the four candidates that opposed this view was elected. In 1999 a resident was quoted in the local newspaper as saying ‘I have three beliefs in life: death, taxes and the 52,500 bed unit cap’ (Barnett 1999b).

Despite the planning definitions of bed units that conceptualize them as simply measurement tools, it is evident from the Emerald Forest deal that bed units represent currency to individuals who possess them. As Horner (2000: 98) observes, ‘in a place where the supply of real estate is fixed and demand is skyrocketing, bed units are a license to print money’. Bed units also represent political currency and came to have the actor-like features of an ‘intermediary’ as described by ANT. As seen in both the Whistler Creek negotiation and the Nita Lake debate, the greater the scarcity of remaining bed units the greater the negotiating power of the RMOW. Also there seems little doubt that, contrary to RMOW (1993b) claims, the allocation of bed units does carry entitlement to development rights. Nevertheless, the regulatory authority held by the RMOW does strongly direct the type of land use, density and design. Only it can grant variances and these have come at an ever increasing price to the developer and with increasing benefits to the community.

The discourse surrounding bed units has evolved over time in concert with the stages of resort development. As I have noted elsewhere (Gill 2000), Whistler’s development has been driven by changing imperatives. The first of these was an economic imperative that drove the early ‘growth machine’ phase in its attempt to achieve a critical mass. It was too early to think about limits and ‘build out’ seemed far away. The notion of controlling growth through the adoption of growth management practices and the establishment of an upper limit on bed unit development ironically paved the way for an unprecedented period of growth while at the same time giving residents a sense of control, especially with respect to ensuring environmental quality.

This was followed in the 1990s by the introduction of social imperatives in response to the increasing political power of residents who demanded that their needs be met. Although always an important consideration, environmental imperatives dominated the third phase as growth pressures placed stress on the natural environment (Gill 2000). During this period the resort was increasingly portrayed as a ‘resort community’ reflecting the notion of collaboration and partnership between the corporation and the community. Also colouring the environmental agenda were external influences as emergent western societal trends placed greater value on environmental quality and these were reflected within the community with the increasing prominence of AWARE, the local environmental NGO that acted as a watchdog on development along with the development of various environmental management strategies not only by the municipality but also by Intrawest. The concept of the 'bed unit cap' was employed as a signifier of environmental limits in much discourse surrounding growth and environmental quality.

The widespread acceptance by the community of a bed unit cap set at 52,500 – despite the fact that growth management principles state that with community consensus growth limits can be changed – was established in Whistler for about 15 years and, as noted above, it provided an increasingly valuable negotiating tool as the established limits were reached. The RMOW had, however, painted itself into a difficult corner. A 'no-growth' scenario was not attractive, especially in the face of the prospect of competing to host (with Vancouver) the 2010 Winter Olympics. Furthermore, the bed unit cap had created an escalation in housing prices that threatened the ability of the resort to house an acceptable proportion of its labour force.

The solution lay in changing the discourse regarding growth from one that focused on the bed unit limit to one that diffused the tight relationship between a bed unit limit and environmental quality by introducing the concept of 'sustainability' into the policy arena. After an extensive consultation process from 2002 to 2004 on the idea of sustainability with a wide range of stakeholders – including residents, Intrawest and resort businesses – the priority issue that emerged was the need for resident-restricted affordable housing. This allowed the RMOW to move beyond the established cap and develop further bed units that were restricted to resident use. While not abandoning the environmental imperative, the sustainability focus has paved the way for not only social but also economic development. Indeed, the focus on affordable housing serves the resort well in discussions on the needs of the forthcoming Winter Olympics.

Evolving partnerships and alliances

Relationships between the key actors in Whistler have also evolved over time and are event-dependent. While most events include the main actors – the politicians (RMOW), the developers (especially Intrawest) and the residents – their relationships and alliances are complex and nuanced. For example, neither residents nor politicians speak with one voice and even within Intrawest differences in values are evident between, for example, the mountain operations division and the resort development (real estate) division (Marcoux 2004). So alliances have ebbed and flowed over time dependent on the key issues facing those concerned with resort development.

Municipal politics in Whistler is predominantly non-partisan with the mayor and councillors elected on the basis of their independent platforms. What distinguished members of council is their attitudes towards issues such as the rate of growth, the growth cap and related concerns such as environmental quality, community service provision and affordable housing (Gill 2000). In the 'growth machine' years pro-growth advocates dominated council but by 1996 the council had swung to one that supported pro-community/anti-growth approaches (Gill 2000). In the 'growth machine years' the developers and the politicians were closely aligned; but with the growth of a resident population and the emergence of 'community', the politicians have also had to serve community needs sometimes at the expense of the developers. For example, the requirement to provide employee housing has placed an extra financial burden on developers. The Emerald Forest debate was one of the first events that publicly highlighted the changing nature of alliances within the resort community. In this case the environment became the catalyst. The RMOW was firmly on-side to support preservation of the wetland area. This gained them the support of AWARE members who at times had previously opposed the RMOW on development decisions. For example, in the early 1990s they opposed the design of a new golf course (Nicklaus North) that encroached on sensitive wetlands in the valley bottom, and this has resulted in design changes (Xu 2004). Residents were split in their views over the Emerald Forest deal with some suspicious of Intrawest motives and concerned over the precedent of exceeding the development cap, whereas others supported the preservation initiative. Second-home owner 'residents' with property near the proposed Intrawest development also entered this debate, although this in turn caused some residents to more forcefully support the development as a sign of their opposition to the influence of non-permanent residents on local decisions (Barnett 1999b). While some people were suspicious of the close alliance between Intrawest and the RMOW in the resolution of the Emerald Forest debate, Horner (2000) concluded that, although it did benefit from the deal, Intrawest stepped in to resolve the problem as a favour to the RMOW which is indicative of the collaborative nature of their relationship.

There is other evidence that the relationship between Intrawest and the RMOW is one of strong collaboration. This is demonstrated by the sustainability initiatives where various corporate–community partnerships have evolved, most notably by partnering with Whistler-Blackcomb (Intrawest's mountain operation arm) and other local businesses in 'The Natural Step' (TNS, a sustainability programme) (Nattrass and Altomare 1999). Through its environmental management strategy, Whistler-Blackcomb has also engaged in a number of community outreach programmes that have sought to build closer alliances and trust between the corporation and the community. This is especially evident in the case of the relationship between Whistler-Blackcomb and AWARE (Xu 2004). The corporation and the environmental NGO have moved from an earlier position of confrontation to one of collaboration in large part as a result of the trust developed when a senior Whistler-Blackcomb manager became president of AWARE (Xu 2004). While one might argue that this suggests a process of clientization on the part of Intrawest, there is little evidence that it represents anything more than a strong commitment to promoting environmentally responsible behaviour on the part of the individual who is both a resident and an Intrawest manager (Xu 2004). However, while the on-mountain division of Intrawest has built strong alliances, the resort development division responsible for the development of Whistler Creek appears less community oriented. Despite extensive community consultation over the development, the controversial aspect of that development centred around differing values, with Intrawest arguments very much focused on the bottom line.

Other agencies external to the resort have recently entered the power arena in Whistler. In the case of the provincial government, its role was significant in the early stages of development when it held considerable power. However, it has phased out this control and handed responsibility to the RMOW. Its intervention in the Nita Lake development decision, while applauded by many residents, nevertheless highlighted the powers that the provincial government holds especially in relation to issues of provincial or, as in the case of the Olympic Games, national and international significance. Clearly development issues over the next few years leading up to the 2010 Olympics will be influenced by the power of non-local intervention. In particular, the Vancouver Olympic Organizing Committee (VANOC) will play a significant role in negotiating the nature of Olympic-related development in Whistler. While there is no locally organized opposition to the games, concerns have been expressed that local priorities such as the hosting of a 'green' Olympics may be overridden by financial priorities (Sills 2005). Although the Olympic alliance is focused on a specific event, the Comprehensive Sustainability Policy (CSP) for Whistler's overall strategy for sustainable development highlights the importance of partnerships at varying levels of government and recognizes the need to be responsive to external influences beyond muncipal boundaries (RMOW 2004a). As Stoker (1996: 278) observes '[t]he crucial challenge is to connect local and non-local sources of policy change'. Whistler's recently developed CSP suggests that the RMOW recognizes this necessity (RMOW 2004a).

Conclusions

In an attempt to disentangle the complexities of power and politics of tourism development in the resort of Whistler, this chapter has focused on the events and actors associated with the 'intermediary' device of major significance which is the allocation of bed units. By viewing these, it is possible to gain a better understanding of how patterns of power within the community have been negotiated through both the traditional mobilization of resources and the more complex construction of networks of alliances.

In Whistler bed units and the bed unit cap have been used as planning tools in a growth management approach to control the type, degree and spatial patterns of development within the resort. However, they have also acquired value as both political currency in negotiating amenities for the community, and as symbolic currency representing environmental values and community control in the face of development pressures. While clearly defined as a planning tool, the case study demonstrates that bed units have become a political tool that can be used to manipulate development to achieve a desired outcome. The degree to which this allows bed units to be seen as an 'intermediary' actor as defined by actor-network theory is clearly debatable but bed units have certainly acquired a local political significance in the discourse over growth and development that is well above that for other planning tools and guidelines.

The examination of key events stimulating debate within the resort over issues relating to bed units reveals an evolving discourse between the RMOW, residents and developers that reflects a range of factors, from the global to the local, that converge to influence local political decisions. In attempting to understand the power relationships embedded in these events, an urban regime perspective assists in providing an overarching assessment of how evolving coalitions reflect and shape the nature of the regime. Constructs from stakeholder theory help elucidate the forces and devices behind collaborative actions.

From a regime perspective, there is no question that Whistler has predominantly displayed characteristics of a development/growth regime (Stone 2005). This has remained stable through two major phases of growth (the initial growth phase and a period of rapid growth from the early 1990s until recently) and also through periods of contestation when, at the beginning of the 1990s, residents demanded a voice in the decision process and more recently when the growth cap was exceeded. Bed units have played a significant role in supporting the growth regime. For example, following the period of contestation around 1990, the establishment of a bed unit cap at 52,500, which seemed far beyond the existing level of buildout, gave residents a sense of control and security that the community's environmental quality (and related overall quality of life) was not threatened. Meanwhile, the RMOW was able to negotiate significant community benefits from developers thus appeasing opposition from (most) residents.

Some aspects of the local regime resemble those of Stone's (1989) 'middle-class progressive regime', notably the debates over environmental quality and conservation and affordable housing. The provision of these amenities, however, is tied to growth and in a tourism context also serves the purposes of the resort industry. To achieve this, local actors have sought to construct a growth discourse which presents a balance between resort business and community interests. This is evidenced by the use of the now established term 'resort community' to describe Whistler which is articulated by RMOW as follows:

> but we agree on one thing. The health of our community depends upon the economic viability of the resort. Likewise, the continuing success of the resort is only possible through the support of this strong, flourishing community. In Whistler, the whole is greater than the sum of its parts. Understanding this interdependence of the community and resort business is essential. One complements the other.
>
> (RMOW 2002: 9)

In Whistler, the emergence of such a regime is partly linked to the fact that local government possesses significant regulatory power, especially over land use. The long-term strategic objectives of both local and regional governments has been economic growth through planned and coordinated development and this has provided security for development interests. In the more recent period of growth, as the resort community has approached planned buildout and the resident community has matured and become more organized, there has been a shift in power relations which are now more typified by negotiation and collaboration. These changes have also involved Intrawest in negotiating the fine line between shareholders' and other stakeholders' interests.

Given the nature of tourism as a service industry, building good relationships and trust between the resort company and the communities in which they are located is a critical element in establishing a 'licence to operate' (Robson and Robson 1996; Williams *et al.* 2004). As Minger (1991) concludes, good resorts are good communities. While Intrawest as a publicly traded company is in the first instance accountable to shareholders, there is evidence that local alliances with the community have been strengthening especially at the mountain operations level (Whistler-Blackcomb) where good environmental management practices are dependent on adopting a collaborative approach with the RMOW. At the same time a strong degree of trust has been built up between senior management on the mountain and the RMOW as well as with AWARE (Marcoux 2004). In the latter case, this trust represents an evolution from a confrontational relationship to one of collaboration (Xu 2004). Although this study is confined to one resort, evidence from other Intrawest resorts in the USA suggests that corporate–community relations elsewhere are more confrontational (Clifford 2002). The situation in Whistler may partly reflect the very specific circumstances which mean that this is Intrawest's flagship resort and enjoys the advantage of being home to the CEO and other senior corporate managers. Over a period of many years, trust between the corporation and the community has grown and is especially evident in recent years as thinking surrounding sustainability has brought a convergence of corporate and community values.

Despite the fact that Whistler is a highly planned and regulated environment, its distinctive character reflects not only the unique environmental setting but also the overlay of local political decisions. As Horner (2000: 13) states,

> resorts will evolve at the intersection of capital that is simultaneously local and global, public and private; consequently their form, function and image as marketed will be the outcome of the relative power of the actors representing these sources of capital.

The situation that emerged in Whistler reflected the fact that it is located in a relatively wealthy capitalist country with a federal system of government. This should not detract from the fact that the example of Whistler produces quite a positive and socially progressive picture of power relations in a tourism resort. Although Whistler is dominated by one large corporation, its presence in the resort has been moderated by local political actors who have skilfully sought a balance between resort and community needs. Furthermore, a highly consultative approach to local governance has emerged allowing a range of voices to be heard in the negotiation of power. Bed units have played a significant role in how this balance of power has been brokered.

References

Allen, J. (2003) *Lost Geographies of Power*. Oxford: Blackwell.

Ambler, T. and Wilson, A. (1995) 'Problems of stakeholder theory', *Business Ethics, A European Review*, 4: 30–35.

Archambault, M. (2002) *Tourism in the Age of Alliances, Mergers and Acquisitions*. Madrid: World Tourism Organization.

Banerjee, S.B. (1998) 'Corporate environmentalism', *Journal of Management Learning*, 29: 147–164.

—— (2001) 'Managerial perceptions of corporate environmentalism: interpretations from industry and strategic implications for organization', *Journal of Management Studies*, 38: 489–513.

Barlow, M. and Wastl-Walter, D. (eds) (2004) *New Challenges in Local and Regional Administration*. Aldershot: Ashgate.

Barnett, B. (1999a) 'Limiting growth the goal; bed units not the tool', *Pique Newsmagazine*, 8 October: 2.

—— (1999b) 'Emerald Forest deal, panned and praised', *Pique Newsmagazine*, 27 August: 8.

—— (1999c) 'RMOW "buys" Emerald Forest for $1 million and 500 bed units', *Pique Newsmagazine*, 13 August: 5.

—— (2003) 'Creekside developments show need for CSP', *Pique Newsmagazine*, 23 May: 4.

Berry, M.A. and Rondinelli, D.A. (1998) 'Proactive corporate environmental management: a new industrial revolution', *Academy of Management Executive*, 12(2): 38–50.

Bramwell, B. and Sharman, A. (1999) 'Collaboration in local tourism policymaking', *Annals of Tourism Research*, 26(2): 392–415.

Buhalis, D. (2000) 'Marketing the competitive destination of the future', *Tourism Management*, 21: 97–116.

Bullock, G.K. (1998) 'The UK package holiday industry', *Management Accounting*, 76(4): 36–38.

Butler, R.W. (1999) 'Sustainable tourism: a state-of-the-art review', *Tourism Geographies*, 1(1): 7–25.

Clifford, H. (2002) *Downhill Slide: Why the Corporate Ski Industry is Bad for Skiing, Ski Towns and the Environment*. San Francisco, CA: Sierra Club Books.

Cragg, W. (1996) 'Shareholders, stakeholders and the modern corporation', *Policy Options*, 17: 15–19.

Crotts, J.C., Buhalis, D. and March, R. (2000) 'Global alliances in tourism and hospitality management', *International Journal of Hospitality and Tourism Management*, 1: 1–10.

Davies, A.R. (2002) 'Power, politics and networks: shaping partnerships for sustainable communities', *Area*, 34(2): 190–203.

Debbage, K.G. (1990) 'Oligopoly and the resort cycle in the Bahamas', *Annals of Tourism Research*, 17: 513–527.

Dorward, S. (1990) *Design for Mountain Communities: A Landscape and Architectural Guide*. New York: Van Nostrand Reinhold.

Dowding, K., Dunleavy, P., King, D., Margetts, H. and Rydin, Y. (1999) 'Regime poltics in London local government', *Urban Affairs Review*, 34(4): 515–545.

Fainstein, S.S., Hoffman, L.M. and Judd, D.R. (2003) 'Introduction', in L.M. Hoffman, S.S. Fainstein and D.R. Judd (eds) *Cities and Visitors: Regulating People, Markets and City Space*. Oxford: Blackwell.

Flagestad, A. and Hope, C.A. (2001) 'Strategic success in winter sports destinations; a sustainable value creation perspective', *Tourism Management*, 22: 445–461.

Freeman, R.E.(1984) *Strategic Management: A Stakeholder Approach*. Boston, MA: Pitman.

Gill, A.M. (1997) 'Local and resort development', in R.W. Butler, C.M. Hall and J. Jenkins (eds) *Tourism and Recreation in Rural Areas*. Chichester: Wiley.

—— (2000) 'From growth machine to growth management: the dynamics of resort development in Whistler, British Columbia', *Environment and Planning A*, 32: 1083–1100.

Gill, A.M. and Reed, M.G. (1997) 'The reimaging of a Canadian resource town: postproduction in a North American context', *Applied Geographic Studies*, 1(2): 129–147.

Gill, A.M. and Williams, P.W. (1994) 'Managing growth in mountain tourism communities', *Tourism Management*, 15(3): 212–220.

Goodno, J. (2004) Living with tourism, *Planning*. Online document. Available from: www.planning.org/affordablereader/planning/tourism0604.htm [last accessed 4 December 2004].

Hall, C.M. (2003) 'Politics and place: an analysis of power in tourism communities', in S. Singh, D.J. Timothy and R.K. Dowling (eds) *Tourism in Destination Communities* Wallingford: CAB International.

Hart, S. (1995) 'A natural-resource-based view of the firm', *Academy of Management Review*, 20(4): 986–1014.

Haughton, G. and While, A. (1999) 'From corporate city to citizens' city? Urban leadership *after* local entrepreneurialism in the United Kingdom', *Urban Affairs Review*, 35(1): 3–23.

Hayter, R. (2003) '"The war in the woods": Post Fordist restructuring, globalization, and the contested remapping of British Columbia's forest economy', *Annals of the Association of American Geographers*, 93(3): 706–729.

Horner, G. (2000) 'Mountains of money: the corporate production of Whistler resort', unpublished MA thesis, University of British Columbia.

Jamal, T.B. and Getz, D. (1995) 'Collaboration theory and community tourism planning', *Annals of Tourism Research*, 22: 186–204.

Keller, P. (2000) 'Globalization and tourism', in W.E. Gartner and D.W. Lime (eds) *Trends in Outdoor Recreation*. Wallingford: CAB International.

Lamont, C. (2001) 'A time to plan', *Pique Newsmagazine*, March: 30–33.

Latour, B. (1986) 'The power of association', in J. Law (ed.) *Power, Action Belief: A New Sociology of Knowledge?* London: Routledge and Kegan Paul.

Logan, J. and Molotch, H. (1987) *Urban Fortunes. The Political Economy of Space*. Berkeley, CA: University of California Press.

Loverseed, H. (2005) 'Ski resort developers face hill of hassles', *Globe and Mail*, 19 April: B13.

Marcoux, J. (2004) 'Corporate environmentalism: stakeholder influence on corporate environmental strategies in mountain resort destinations', unpublished Master of Resource Management research report, Simon Fraser University, Burnaby, BC.

Marwick, M.C. (2000) 'Golf tourism development, stakeholders, differing discourses and alternative agendas: the case of Malta', *Tourism Management*, 21: 515–524.

Minger, T. (1991) 'The green resort: environmental stewardship and the resort community', in A. Gill and R. Hartmann (eds) *Mountain Resort Development: Proceedings of the Vail Conference*. Burnaby, BC: Centre for Tourism Policy and Research, Simon Fraser University.

Mossberger, K. and Stoker, G. (2001) 'The evolution of urban regime theory: the challenge of conceptualization', *Urban Affairs Review*, 36(6): 810–835.

Mullins, P. (2003) 'The evolution of Australian tourism urbanization', in L.M. Hoffman, S.S. Fainstein and D.R. Judd (eds) *Cities and Visitors: Regulating People, Markets and City Space*. Oxford: Blackwell.

Murdoch, J. (1995) 'Actor-networks and the evolution of economic forms: combining description and expanation in theories of regulation, flexible specialization and networks', *Environment and Planning A*, 27: 731–757.

Murdoch, J. and Marsden, T. (1995) 'The spatialization of politics; local and national actor-spaces in environmental conflict', *Transactions of the Institute of British Geographers New Series*, 20(3): 368–380.

Nattrass, B. and Altomare, M. (1999) *The Natural Step for Business*. Gabriola Island, BC: New Society Publishers.

Pearce, D. (1987) *Tourism Today: A Geographical Analysis*. New York: Wiley.

Pierre, J. (2005) 'Comparative urban governance: uncovering complex causalities', *Urban Affairs Review*, 40(4): 446–462.

Pique Newsmagazine (1999) 'Last chance for input on Emerald Forest deal', 20 August: 7.

Poon, A. (1993) *Tourism, Technology and Competitive Strategies*. Wallingford: CAB International.

Porter, M.E. (1985) *Competitive Advantage*. New York: The Free Press.

Reed, M.G. (1997) 'Power relations and community-based tourism planning', *Annals of Tourism Research*, 24: 566–591.

Reed, M. and Gill, A. (1997) 'Tourism, recreational and amenity values in land allocation: an analysis of institutional arrangements in the post-productivist era', *Environment and Planning A*, 29: 2019–2040.

Resort Municipality of Whistler (RMOW) (1993a) *Official Community Plan, Schedule V, South Whistler*, Resort Municipality of Whistler, Whistler, British Columbia.

—— (1993b) *Comprehensive Development Plan*, Resort Municipality of Whistler, Whistler, British Columbia.

—— (2002) *Whistler 2002: Charting a Course for the Future*. Resort Municipality of Whistler, Whistler, British Columbia.

—— (2003a) *Official Community Plan (OCP) 4*. Resort Municipality of Whistler, Whistler, British Columbia.

—— (2003b) *Create Whistler's Future: Comprehensive Backgrounder*. Resort Municipality of Whistler, Whistler, British Columbia.

—— (2003c) *Whistler: It's Our Future*. Resort Municipality of Whistler, Whistler, British Columbia

—— (2004a) *Whistler 2020; Moving Towards a Sustainable Future, Comprehensive Sustainability Plan (CSP)*. Resort Municipality of Whistler, Whistler, British Columbia.

—— (2004b) *Resort Community Monitoring Report, 2003/04*. Resort Municipality of Whistler, Whistler, British Columbia.

Robson, J. and Robson, I. (1996) 'From shareholders to stakeholders: critical issues for tourism marketers', *Tourism Management*, 17: 553–540.

Rothman, H. (1998) *Devil's Bargain: Tourism in the Twentieth-century American West*. Lawrence, KS: University of Kansas Press.

Ryan, C. (2002) 'Equity, management, power sharing and sustainability – issues of the "new tourism"', *Tourism Management*, 23: 17–26.

Selman, P. (2000) 'Networks of knowledge and influence: connecting "the planners" and "the planned"', *Town Planning Review*, 71: 109–121.

Selman, P. and Wragg, A. (1999) 'Local sustainability planning: from interest-driven networks to vision-driven super-networks', *Planning Practice and Research*, 14(3): 329–340.

Sills, B. (2005) 'Olympic metamorphosis. Letters to the Editor', *Pique Newsmagazine*, 28 April: 7.

Stanton, J. and Aislabie, C. (1992) 'Up-market integrated resorts in Australia', *Annals of Tourism Research*, 19: 435–449.

Sternberg, E. (1997) 'The defects of stakeholder theory', *Corporate Governance: An International Review*, 5: 3–10.

Stoker, G. (1996) 'Regime theory and urban politics', in R.T. LeGales and F. Stout (eds) *The City Reader*. New York: Routledge.

Stone, C. (1989) *Regime Politics; Governing Atlanta 1946–1988*. Lawrence, KS: University Press of Kansas.

—— (2005) 'Looking back to look forward: reflections on urban regime analysis', *Urban Affairs Review*, 40(3): 309–341.

Stone, C. and Sanders, H. (1987) *The Politics of Urban Development*. Lawrence, KS: University of Kansas Press.

Taylor, A. (2003) 'Nita Lake Lodge to go to public hearing', *Pique Newsmagazine*, 11 April: 11.

—— (2004) 'Province trumps B.C. Supreme Court over Nita Lake development', *Pique Newsmagazine*, 20 May: 10.

The Natural Step (2005) Services. Online document. Available from: www.naturalstep.org/services/index.php [last accessed 1 May 2005].

Timothy, D.J. (1998) 'Co-operative tourism planning in a developing destination', *Journal of Sustainable Tourism*, 6(1): 52–68.

van den Berg, L., Braun, E. and Otgaar, A. (2004) 'Corporate community involvement in European and US cities', *Environment and Planning C*, 22: 475–494.

Waldron, D. (2000) 'An environmental sustainability strategy for tourism communities: the case of Whistler, BC, unpublished Master of Resource Management research report, Simon Fraser University, Burnaby, BC.

Whistler-Blackcomb (2005) Improving environmental practices. Online document. Available from: www.whistlerblackcomb.com/about/enviro/local.htm [last accessed 1 May 2005].

Whistler Housing Authority (2000) WHA building our resort community Whistler, British Columbia. Online document. Available from: www.whistlerhousing.ca/references.asp [last accessed 4 May 2005].

—— (2004) Whistler Housing Authority – Housing Needs Assessment Study 2004, Whistler, British Columbia. Online document. Available from: www.whistlerhousing.ca/references.asp [Last accessed 4 May 2005].

Williams, P.W. and Todd, S. (1997) 'Towards an environmental management system for ski areas', *Mountain Research and Development*, 17: 75–90.

Williams, P.W., Gill, A.M. and Chura, N. (2004) 'Branding mountain destinations: the battle for "placefulness"', *Tourism Review*, 59: 6–15.

Xu, A. (2004) 'The perceptions and attitudes of environmental non-governmental organizations towards corporate environmental strategies at a mountain destination', unpublished Master of Resources Management research report, Simon Fraser University, Burnaby, BC.

7 Pedestrian shopping streets and urban tourism in the restructuring of the Chinese city

Alan A. Lew

Transforming the Chinese city

Xiangxiajiu Road was a fairly typical busy retail street in the older quarter of the bustling southern city of Guangzhou. On a hot and humid summer day in 2001, the shophouse-covered sidewalks were crowded, while cars filled its narrow two lanes. The bright and fashionable interiors of the shops stood in contrast to the older upper floors of the shophouses. This was a fairly common scene in contemporary urban China. What was different here, though, was that the building façades were less soot-laden than elsewhere in Guangzhou (Figure 7.1). New paint and brighter colours were evident. The stained glass windows were fully intact and not boarded over in parts. These clues indicated that this was probably one of the more popular shopping districts, although still in a traditional neighbourhood and a far cry from the modern shopping malls that are mostly situated toward the suburban edges of the city.

As dusk approached, however, a transformation took place. Barriers came up to keep out automobile traffic and a half-mile long pedestrian-only thoroughfare was created. Some restaurants extended their reach into the street as they set up tables and chairs for after-work drinks and meals (Figure 7.2). A few temporary booths also went up in the middle of the street. These were selling health drinks, toys and household goods. This was all rather different to the infamous tourist attraction Temple Street Night Market in Hong Kong, which consists of densely crowded shop stalls with only a narrow path for pedestrians between them.

As night fell, the whole ambience of Xiangxiajiu Road changed. The darkening sky stood in increasing contrast to the brightly lit buildings, the colours of which now shone brightly down on the streetscape below. Wall-mounted spotlights aimed up at each building. Fluorescent lighting brightened the sidewalks and shops below. The end of the workday brought larger numbers of people onto the street, especially young adults, teens and young families, strolling the street and enjoying the warm subtropical evening. Shop workers called out sales to encourage passers-by to come into their shops, competing with a cacophony of music blasting from music stores. The street was filled with a party atmosphere that extended late into the evening.

Today, virtually every city in China has one or more pedestrian shopping street

Figure 7.1 The location of places mentioned

Source: author.

(*buxingjie*). They exist in a wide range of forms and styles, and they enjoy varying degrees of success. Some are permanently closed to vehicles, while others are only closed at night or on weekends. They are part of the transformation of many Chinese cities from socialist cities of production to cities of play and tertiary employment where tourism for both domestic and international markets is part of a new consumption-oriented economy (Ma 2004; Yang 2004; Wu 2004; Short and Kim 1999). More specifically, they are also representative of the decentralization of power from the central government to local municipalities, and of the development of a more market-oriented land development system. These changes provide valuable insights into how alterations in power relations especially in relation to land development have shaped the evolution of urban tourism in China.

Since the 1980s, and especially in the 1990s, China's larger cities have been at the forefront of experiments with economic reforms, expanding entrepreneurship. They have led the transformation of urban landscapes through large-scale

Figure 7.2 Early evening on Guangzhou's Xiangxiajiu Road

Source: Alan A. Lew.

inner-city urban renewal and expansive suburban new town development. The creation of public shopping spaces, such as the pedestrian shopping street, is one manifestation of this transformation. While much of China's urban transformation is a reflection of the intentional decentralization of government authority from Beijing to provinces and municipalities, the growing economic power of cities has created a state of tension between local and central government control. The growing concentration of wealth in China's larger cities has helped to shift considerable economic decision-making power away from the centre. In the 1980s, creative local expressions that pressed the boundaries of central policies were often met with police crackdowns and visits by leaders from Beijing (Lew 1987). By the end of the 1990s China's larger cities had become free-wheeling marketplaces in which entrepreneurship and economic experimentation were encouraged (though not political experimentation) with varied and sometimes problematic outcomes.

While its economy has largely adopted liberal economic policies, China remains a socialist country in which most property is owned by various government bodies and leased to private businesses and other governmental entities (Lew 2001). There are some private entrepreneurs (more so among smaller businesses), but most large-scale 'entrepreneurial activities' (e.g. factories, shopping malls and hotels) are owned and operated by governmental entities for whom the income generated is a significant part of their overall budget. In addition, the highest authority in local government decision-making is not the city's mayor, but the local chairperson of the Chinese Communist Party. Thus, the division between government, politics and business is more blurred in China than in many Western market economies creating a particular type of power relations often based on elite alliances or business 'networks'. Cartier (2001) notes that these are highly differentiated by family, class, citizenship, ethnicity and dialect.

The result of these different influences and interests is the creation of a wide range of business and entrepreneurial landscapes that reflect centralized, authoritarian control in some instances, private and semi-private entrepreneurial dominance in others, and a mix of the two in most situations. As distinct as China might seem (with its single-party, socialist system), this situation is not so different from what is found elsewhere in the world. Business competition, politics and the potential for political corruption, land markets and values, and the leisure and lifestyle economy are all considerations that shape the Chinese city, as they do everywhere. The Chinese experience presents a different perspective to these issues, which can shed light on urban tourism elsewhere. The range of pedestrian shopping streets that emerges from these dynamics is discussed below to illustrate the changing nature of urban tourism and leisure landscapes. But first it is important to review some of the historical background that influences the development of urban tourism and the pedestrian shopping street in China today.

Modernizing China

Mainland China had a lot of catching up to do when Deng Xiaoping first announced the Four Modernizations campaign in 1978. The country was amongst the poorest in the world, with a built infrastructure that had changed little since the Chinese Communist Party came to power in 1949. The socialist approach to city planning was to focus on industrial development – to create cities of production under the socialist planning principle of 'production first, living conditions later' (*xian shengchan, hou shenghuo*) (Ma 2004). Industrial soot and meagre housing conditions were the norm for city life throughout China. Deng's Four Modernizations (industry, agriculture, defence, and science and technology) led to a refocusing of industrial production, moving away from an emphasis on heavy machinery and toward an emphasis on consumer products in the 1980s. In China's cities, this also resulted in the movement of heavy industry from the city core to the periphery, and the creation of new opportunities to reshape the urban landscape. Tourism landscapes, primarily in the form of large hotels, were the first significant changes to appear in the Chinese urban landscape in the 1970s (Lew 1987). These early icons to Western consumerism were built and/or managed by foreign investors through 'joint-venture' agreements with Chinese government partners (foreign ownership was non-existent in China). They acted as a form of management and technology transfer within a highly controlled space that was off-limits to most Chinese, as were the Friendship Stores where foreigners could buy products without the limitations of ration coupons. These tourist enclaves were tentative steps toward economic modernization in a country that was still under pervasive central government control.

The 1980s was a period of gradually increasing decentralization of authority to the local level and a slow loosening of restrictions on land development and domestic travel. Through the 1980s, China's government slowly expanded the number of cities and destinations where foreign visitors could go without special permits (Zhang 2003). As each new place opened to international tourism, new hotels were built that stood out from the grey, old and dirty buildings of the past. These newer hotels often contained their own Friendship Stores and shopping centres, thereby putting pressure on older Friendship Stores to modernize their facilities and expand their merchandise offerings.

The new hotel, retail and tourism landscapes that emerged in the 1980s were helped by two significant land reform initiatives. The first was the housing reforms that started in 1982, allowing the development of a housing market that was not directly associated with the workplace. Both domestic (government entities) and foreign joint ventures were permitted to participate in diversifying housing opportunities in urban areas of China. These policies were the first steps toward a market-driven differentiation in land values in China (Logan *et al.* 1999; Ma 2004). The second major reform was the introduction of land use rights transfer leases, which were first used in 1987 (Cartier 2001). The first land use rights transfer was from a municipality to a non-governmental company in the Shenzhen Special Economic Zone (SEZ), adjacent to Hong Kong. The policy was formally approved for use in other SEZs by the central government later that same year. These reforms introduced the ability to lease property use rights from government entities through negotiations and bidding. Land values were, in theory, tied to the highest and best uses of the land. This resulted in a form of land market and the emergence of a land rent gradient and real estate valuations somewhat similar to that found in Western cities. A variety of inefficiencies, however, made this land market far from predictable.

Both the housing reforms and the land leases provided major new revenue sources and development tools for local governments and government entities. These revenues have been used to further reshape the urban landscape of China's major cities by constructing multi-level expressways and subway systems, levelling older neighbourhoods to create open space and monumental buildings, and expanding education, recreation and leisure places and spaces. In the process, the absolute control of central government that had once extended well into the neighbourhood level of China's cities now resided increasingly in municipal and provincial governments, at least in terms of economic and development policies. Central government still uses its considerable taxation and fiscal powers to redirect large-scale development initiatives to economic sectors and regions, but with mixed success. It is primarily in the political realm that the centre has most strived to maintain strong control, though even here its power is nowhere near as pervasive as it used to be.

The land reform experiments that started in the Shenzhen SEZ in the late 1980s resulted in a feverish real estate and land speculation boom that spread throughout coastal China through the proliferation of special economic policy zones associated with every major city and county. By 1995 the central government had recognized 422 such zones, up from only four in 1979 (Guo and Li 1995). Informal special economic zones, established by local and provincial authorities without central government approval, were even more common, driven by the huge profits that land development and speculation had engendered. One estimate found 1,874 informal zones by 1990 (Yeh and Wu 1996).

In theory, under the land leasing policy, local government entities (mostly municipalities) have an incentive to invest in developing and marketing lands that they control so as to best position them in the market (Wu 2004). This involves a combination of both place promotion and real estate knowledge. It also generates a realization that the value of any single plot of land is dependent on a host of larger infrastructure, location and amenity factors. The City Planning Law was adopted by central government in 1989 and gave local authorities the power to create comprehensive and local area plans for the first time (Yeh and Wu 1998). The law, however, had two major shortcomings. Local authorities had considerable administrative discretion to overrule adopted plans and public participation in the planning process was non-existent. Planners were, therefore, easily subject to more powerful political and economic pressures and lacked authority to implement plans for the good of the community (Ma 2004).

Another challenge was that China lacked many of the requirements for a Western-style land market, including an adequate level of infrastructure development, an open decision-making process and an established contract process and legal system. Most of all, China has had an inherent contradiction in that the local authorities both own and regulate land that they hope to develop. Thus, although a quasi-land market system has been established, in socialist China it comes in the form of a complex mix of market and non-market forces, and a complex blurring of distinctions between market-power and state-power. This has resulted in a haphazard development process that is open to particularisms and outright corruption (Lin and Ho 2005).

The decentralization of budget and decision-making, along with the establishment of a more open land market gave local authorities new powers that were often abused. Land use rights were transferred through informal negotiations at rates that benefited the local authorities, but were generally lower than true market rates (Cartier 2001). The decision to approve a long-term lease agreement in China can be based more on *guanxi* business relationships than formal business plans (Haley *et al.* 1998). *Guanxi* is a form of social capital that involves reciprocal responsibilities based on familial and clan relationships (Lew and Wong 2004). The higher the local authority stood in the regional administrative structure (e.g. villages, counties, provinces), the larger the tract of land that could generally be negotiated (Keng 1996). Rampant government corruption was one of the major complaints of the 1989 Tiananmen Square protesters, and weak land use planning regulations had contributed to corruption in many locations.

By the 1990s, and after the Tiananmen Square protests were crushed, China began its unmatched sprint toward economic modernization. The number of private service sector businesses in China grew rapidly in the 1990s, especially starting in 1992 when many government-owned enterprises were privatized (mostly smaller enterprises), downsized, or completely closed in the name of efficiency. In response to the large-scale layoffs that accompanied this process, the central government adopted policies to promote tertiary business development (Putterman and Dong 2000).

Factories that were not closed were moved to suburban locations to create less polluted urban centres. The wholesale razing and relocation of large factories and other enterprises freed up valuable land for more profitable uses (Zhou and Ma 2000). Large amounts of foreign investment capital poured into China. In China's larger cities, approximately half of this capital went to manufacturing and half to service industries (Yang 2004). The tertiarization of China's urban centres led to competition, innovation and job growth. Despite government regulations that often made rural–urban migration illegal, many rural residents moved to the cities in search of employment opportunities, and urban residents became increasingly middle class, with rising consumer choices and expectations (Cartier 2001). As a result, many of China's large urban centres were transformed from industrial cities of production to cities of tertiary services and consumption (Glaeser *et al.* 2001).

In the tourism and retail sectors, traditional government-owned department stores either modernized or closed. The modern hotels that stood out so starkly in the 1980s urban landscape, were no longer so distinct as urban renewal replaced old buildings with new department stores, malls, office towers and apartment buildings. Further out of town, suburban shopping malls, with multistoried anchor department stores, were being constructed. In the early 2000s, Walmart Supercentres and other warehouse-size department stores were also appearing on the expanding urban skyline.

The uncontrolled real estate development of the early 1990s was somewhat slowed by the Asian economic crisis (1997–1998), which burst China's urban real estate bubble and led to the passage of the 'revised Land Administration Law' in 1998. The revised law placed a moratorium on large-scale land developments, stopped unapproved special economic zones, and sought to bring order to the disorganized land rights transfer and development process (Cartier 2001). The law required an assessment of all existing land rights transfer agreements, integrated planning for all new land transfer agreements, and provincial or central government approval of all new agreements. Strict enforcement of these regulations reigned in the proliferation of independent and uncoordinated local-level decision-making on converting land to private uses, though it has not completely stopped the corruption associated with local land development agreements. The revised Land Administration Law was considered a hallmark in democratic legislation in China because it was the first to be based on a significant level of public opinion, often opposed to the dislocation caused by large-scale urban renewal and the loss of agricultural land. The revised law sought to limit local excesses in a more open economic system. This was not a full retreat to the era of micromanagement by central authorities (Zhang 2003); instead, it was an effort to move the economy and the land market toward the rule of law, after a period of local experimentation and debate over just what those laws should be (Zhu 2005).

Meanwhile, through the 1980s and 1990s China's tertiary sector grew more rapidly than the secondary sector and came to comprise over 50 per cent of GDP in Shanghai, Beijing and Guangzhou by 2001 (Yang 2004). China has become the principle consumer-goods producing workshop of the world. And the clothing, toys and electronic goods that are now being produced in China were not only going overseas, but were also flooding the Chinese marketplace. By the early 2000s, China's urban consumers (as opposed to its rural populations) were increasingly dreaming and living an international lifestyle defined by a wide variety of choices, global brand names, ubiquitous cell phones, Internet access, fast foods and convenience foods, and increasingly postmodern public architecture and open spaces. Related to this, domestic tourism in China expanded year-by-year through the 1990s, reaching 740 million in 2000, while outbound tourism reached 10.5 million in that same year (CNTA 2004a, 2004b).

In less that two decades, certain groups of Chinese consumers, but by no means all, had evolved from pre-consumer (in the Mao Zedong era) through a consumer-based society to a post/hypermodern consumer. Whereas capitalist modernity was first seen in China's urban landscape in the international hotels of the early 1980s, the post/hypermodern consumer society of today's China is best seen in the country's new, and renewed, retail landscapes. The transformation of the urban built environment in China also reflects the place promotion competition amongst cities to image themselves as modern and attractive environments for economic investment (Morgan 2004; Wu 2004). In China, urban quality of life has become an economic development tool and a real estate marketing tool. Moving factories out of the city, improving the transportation infrastructure, fostering a greater range of housing availability, encouraging foreign investment in office space and retail sectors, and creating international-class 'shoppertainment' and 'eatertainment' venues (Lafferty and van Fossen 2001), are all part of the place promotion efforts of China's cities. This has been especially true of cities, large and small, on China's eastern seaboard. But global consumption habits are quickly expanding to central and western China as well.

Tourism and pedestrian shopping streets

Almost every city of any significant size in China has a pedestrian shopping area which is a retail shopping street or district that is closed to most vehicular traffic either all of the time or part of the time. Even small towns have their evening 'night market' streets. The range of forms and approaches that have been adopted in creating these streets in contemporary China are quite remarkable and indicate the importance of taking account of changing power relations in order to understand geographical variations in the nature of urban tourism spaces. Commercial streets crowded with pedestrians have long been a common site in China, even in the difficult days of the Great Leap Forward (1950s) and Cultural Revolution (1970s and early 1980s) when there was little to buy. Strolling, window shopping and people watching are universal, low-cost leisure activities that are especially focused in pedestrian shopping precincts. This is no different in China, where home living space and personal privacy is often limited. In China, such pedestrian streets serve as regional recreation destinations and are promoted as significant tourist destinations. Indeed, in some of China's best known cities they are central elements in the tourism product, sometimes being part of designated tourism zones, and they are often considered places where the life of the city is on its best public display.

The first contemporary pedestrian shopping streets in China emerged with the land reforms of the late 1980s. Their most recent manifestations have come about following enactment of the revised Land Administration Law in 1998. The case studies presented below reflect the differences between these two periods, and the changes in decision-making and power relations that they represent. The initial period of expansion of pedestrian shopping streets occurred in China's largest cities during late 1980s and early 1990s which were the heyday of unregulated and haphazard land speculation and development. During this period, local government authorities were not well positioned to plan comprehensively and to develop supporting infrastructure. Instead, they relied on the retail building owners, who in turn relied on their tenants to fund site and area improvements. This resulted in minimal coordination of building designs and the organization of public space. Only basic improvements were provided by the municipalities, ranging from the construction of road blocks to stop traffic to more elaborate resurfacing of former vehicular streets. The streets of this period, exemplified by Beijing Road in Guangzhou and Nanjing Road in Shanghai, lacked the coherency and dynamism of the more thematic streets that have evolved in the more orderly land development era of the late 1990s and early 2000s. In this recent period more cities have developed pedestrian shopping streets and the design approaches have been thematically structured, reflecting a stronger role for local planning authorities in the development process. The 1998 Land Adminstration Law gave local municipalities much greater influence on the land development process so that they are in a position to increase the market value of the shopping district overall, and to demand land right lease agreements that are closer to actual market values.

A review of the different approaches adopted to these pedestrian streets illustrates how the changing power relationships between localities, developers, businesses and central authorities are expressed in the tourism and recreation landscapes of contemporary urban China. The case studies are organized into two thematic approaches to pedestrian retail street development. The first approach uses historic architecture as a thematic tool, while the second approach uses modern and hypermodern architectural themes. In the next two sections examples of both design approaches are provided that span the 1990s and early 2000s, as well as retail destinations that differ in terms of their economic success. The final section of the chapter draws out the implications arising from these examples for understanding power relations and the development of urban tourism in China.

Historic architecture and themed streets

Historic preservation streets are based on an effort to maintain the architectural 'authenticity' of the retail district. By authenticity, I am referring to the construction of a contemporary architectural style that is associated with an identified period in the past and which is perceived to be evident in the exterior design of existing buildings in the district. In some places these streets have a distinctive architectural character throughout, while in others distinctive historic buildings are intermixed with a range of other buildings. The economic success of these efforts has been mixed. In part this may be due to a greater emphasis on preservation over the economics of retailing, which has been an issue in some other countries (Lew 1988). Western, mostly European, approaches to historic preservation and pedestrian shopping streets also appear to be a common thread in these efforts. Pedestrian shopping streets that use the historic architecture of the street as a theme come in several forms in China, depending on the degree to which the historic architecture maintains its original design motifs, and the continuity of the historic theme.

Macau may be the site of the earliest attempt at the historic preservation of a pedestrian shopping street in China. Macau was granted under a leasehold agreement to the Portuguese by the Emperor of China in 1557. It was returned to Chinese rule in 1999. The older core area of Macau is famed for its 'European charm' and in the mid-1980s several streets leading from the central plaza of the Largo do Senado to the St Paul's Basilica (early 1600s) were closed to create a pedestrian-only area in which historic preservation efforts were concentrated (Duncan 1986). Many of the most important historic colonial buildings of Macau (dating back to the mid-1700s) face the Largo do Senado, and today this area is the heart of tourist visitation to Macau. The plaza and nearby streets received a major facelift in 1994, with façade improvements, street resurfacing, and the addition of public furniture and special event spaces (Fiala 2004).

Only 40 miles to the north of Macau is the city of Zhongshan, the capital of the county in which Dr Sun Yatsen (the founder of modern China) was born. Because of its proximity to Macau and Hong Kong, Zhongshan County is one of the wealthiest rural areas in China. Taking a cue from its neighbour, the city of Zhongshan replicated central Macau in its old downtown area by creating the Sun Wen Xi Road Tourism Zone (Figure 7.3). The approach was almost identical to that used in Macau, with considerable attention given to preserving the historic character of the buildings (mostly dating to the nineteenth century), although there is more neon in the commercial signage of Zhongshan (Ming 2003). This project, which dates to the early 1990s, was one of the first of its kind in China and has been popular and successful. Large numbers of shoppers are drawn to Zhongshan on a daily basis, especially in the evenings after work.

A somewhat different approach to historic preservation is seen in the Dashilan (also known as Danshilanr and Dazhalan) district of Beijing. This is the old downtown of Beijing, located to the south-east of Tiananmen Square and the Gate of Heavenly Peace. Dashilan has been a retail district since at least 1420 (Ming Dynasty), and some of the retail shops and restaurants on this street, along with numerous small alleyways adjacent to it, have been in the same location for over a century (CIIC 2004). At the same time, this district has gone through continuous change over time, and many buildings are of mixed historical heritage – including recently built department stores that are not historic at all. Because of this, it does not have the same thematic feel of the more homogenous examples from Macau and Zhongshan. The longevity of many businesses in Dashilan is due to the reputations that they have built over time and, despite the dilapidated look of much

Figure 7.3 Zhongshan's Sun Wen Xi Road Tourism Zone. The bronze statues on the left depict a rickshaw worker picking up some clients

Source: Alan A. Lew.

of the area, there is a continuous crowd of shoppers on its pedestrian-only streets who are drawn to its famous restaurants, silk shops, herbal medicine dealers and *hutong* (back alley) shops to explore for unusual products.

The example of Xiangxiajiu Road in Guangzhou, presented at the start of this chapter, is again somewhat different from the other historic preservation examples because of the degree to which its older architecture is embellished in a 'Disneyesque' manner. Its buildings, which are pre-Second World War, but mostly without special historic significance, are brightly painted in a variety of colours and decorated with cornices and other ornaments in ways that do not adhere to the original building architecture. On the surface it appears to replicate the examples of Macau and Zhongshan, and it clearly offers the same pedestrian experience. However, with careful attention to details one can see how the buildings have been embellished in comparison to those on neighbouring streets. Xiangxiajiu Road had some claims to fame as an eating and retail district, but was not nearly so central to the city of Guangzhou as were the preserved streets in Macau and Zhongshan. One end of Xiangxiajiu Road is anchored by a large, modern, multistoried shopping centre, with residential apartment towers above. In 2001 Xiangxiajiu Road was the first major pedestrian shopping street project in Guangzhou, but not the first in China. Within a couple of years, Xiangxiajiu Road was permanently closed

Figure 7.4 Shanghai's Xin Tian Di. Architecture students on the right examine details of the project

Source: Alan A. Lew.

to vehicle traffic, not just in the evening, repaved for pedestrians, and sprinkled with public art, as the street gained in fame and popularity. However, the buildings were not quite as brightly coloured as before and numerous new neon signs had been added to the building façades, making it more reminiscent of San Francisco's Chinatown, or some areas of Hong Kong, and so demonstrating the dynamic nature of place in contemporary China.

One of China's newest historic preservation pedestrian shopping districts is Shanghai's Xin Tian Di, which opened in 2003 (Figure 7.4). In this example historical preservation was a small component of a larger project involving the large-scale clearance of an old residential area. This project took two blocks of an old *shikumen* district (a type of *hutong* tenement found only in Shanghai) and converted them into boutique shops, stylish night clubs, upmarket restaurants and a department store (Gluckman 2003). The site has some historic significance in that Mao Zedong met with others in one of the buildings here to form the Chinese Communist Party. Most of the other *shikumen* buildings in the adjoining area, however, were demolished and the residential uses of Xin Tian Di have essentially vanished. Today the area contains a form of festival retailing, along the lines of Boston's Faneuil Hall and San Francisco's Ghiradelli Square. Indeed, the American architect Benjamin Wood was heavily involved in the development and was originally hired by the Hong Kong-based Shui On development company to demolish the entire site and build a replica *shikumen* shopping and condominium centre (Chu and Lee 2004). Although some original buildings were saved, Shui On still needed to fund the relocation of some 8,000 residents (2,300 families) from the site. Hangzhou is another city building its own festival marketplace, and other major cities will follow suit in coming years as a way of diversifying their urban recreation options by targeting elite and expatriate shoppers and tourists.

One other historic themed pedestrian shopping street model that has been tried in China is the creation of a historic-themed shopping district completely from scratch. An example of this 'theme park' approach is Kunming's JinBi GuangChang shopping centre, which replicates a large, traditional Chinese village. An urban renewal project, the narrow streets in this 'village' were never designed to serve anyone but pedestrians. The architecture is clearly Chinese-themed, but taller and denser than any real village would be. A large and costly complex, it is not well linked to the surrounding city and although it houses some of the more Western-oriented night clubs and restaurants, it has only achieved mixed success for the businesses that have located there.

Modern and hypermodern shopping streets

Elsewhere in China large cities have attempted to adopt modernist architectural elements in their most important shopping streets in the late 1980s and early 1990s by first cutting off the flow of vehicular traffic, and later redesigning the pedestrian pathways. Beijing Road in Guangzhou (Figure 7.5) and Nanjing Road in Shanghai are similar examples of this. Nanjing Road is much longer and busier than Beijing Road, but both suffer from older buildings, mostly covered by aluminium façades and signage. Because both of these streets have long been the downtown centre for their large cities, the mix of retail shops is considerable, with some having a long history and reputation and others being very new. Some of the first contemporary shopping malls in China are found in these districts, and most shops are far more contemporary on their interior than their outside modernist façades would indicate. Because Guangzhou's Beijing Road is smaller, it has had a more difficult time competing with the city's new suburban shopping malls. Shanghai's Nanjing Road is more crowded, but still shows signs of failed businesses, and is less frequented by younger shoppers.

Other pedestrian retail streets have undergone a more significant modernization that might be better called 'hypermodernity'. By hypermodernity, I am referring to postmodern architectural designs that incorporate technology, minimalism and futurist elements (Lowenthal 1985). One example is Wangfujing Street in Beijing, which has been the traditional downtown retail area of Beijing. Wangfujing Street had undergone a similar transformation to Shanghai's Nanjing Road in the 1980s. However, in starting in the early 1990s and finishing in 1999, it underwent a major urban renewal and redefinition. Today only a few of the older retail buildings exist, having been replaced by newer department stores and malls, both large and small,

Figure 7.5 Guangzhou's Beijing Road on a warm weekday afternoon. Barriers to keep vehicles off the street can be seen in the lower right

Source: Alan A. Lew.

interspersed with considerable public open spaces. The environment is completely different from the downtowns of Shanghai and Guangzhou. It is the new modernity (postmodern/hypermodern), as opposed to the old, worn modernity; it speaks more to technology and youth culture, and in that sense is able to more effectively compete with the emerging suburban shopping districts.

Other cities also hypermodernized their downtown retail districts. Guilin has placed an underground shopping mall below a massive central plaza with two large video screens for viewing advertisements, as well as national events, and the mall is overlooked by a 13-storey hotel the width of which is covered by cascading water from its roof, flowing to the sound of music each evening. Some local people proudly compare it to the themed hotel-casinos on the Strip in Las Vegas, Nevada. Smaller pedestrian streets lead to the city's lakes and river, which are now lined with thematic parks and bridges that replicate famous sites throughout the world. Some examples include an Arc de Triomphe bridge, a scaled version of the Golden Gate Bridge, a Korean garden and a large Chinese garden that was built on the site of a razed factory.

In Kunming, a portion of the old downtown has been completely razed and replaced by a series of large, hypermodern shopping experiences. There is a large Parkson department store in a complex (Brilliant Plaza) with a four-storey entrance

Figure 7.6 Kunming's hypermodern shopping street

Source: Alan A. Lew.

plaza overlooked by a massive video screen and topped by an apartment complex (Figure 7.6). Also on the plaza is a multistoried warehouse department store, new office towers, and a shopping centre and apartment complex that sells itself with large images of the Eiffel Tower, Venus de Milo, the Statue of Liberty, Mount Fuji and a Dutch windmill. Here in the far southwest of China, the global consumer identity is as strong an attraction for the up and coming urban elite as it is in Shanghai and Guangzhou. These symbols also position this development for a particular urban middle class market that sees itself as cosmopolitan in taste and globally connected economically.

Discussion and conclusions

Despite the growth and popularity of suburban shopping centres, the pedestrian shopping street remains a vibrant urban form in China. Street markets have a long tradition throughout Asia, but other countries have not sought to turn this enthusiasm into the postmodern shopping phenomenon that is occurring in China. Undoubtedly the popularity of the pedestrian shopping street has something to do with China's large population, dense urban settlement and limited private automobile ownership, but it is also linked to the distinct historical situation that China finds itself in at the turn of the twenty-first century. China experienced an extended period during which consumer and mercantile instincts were suppressed under extreme forms of centralized Communist Party control. Because of the long period of war, followed by the economic and political fluctuations of the Maoist period, the landscape and people of China were ready and anxious for new directions when Deng Xiaoping announced his Four Modernizations policy in 1978 (Meisner 1999). Mao Zedong's communist industrialization had been successful, but at the cost of political suppression, social engineering and micromanagement, environmental degradation and the rationing of consumer goods. Deng's radical approach was to introduce socialist market reforms, which resulted in the gradual relaxation of central government management of most aspects of everyday life. The limits of personal freedom were defined by the 1989 Tiananmen Square incident: political opposition to Chinese Communist Party rule would not be tolerated, but economic experimentation would. Then, through the 1990s, China underwent a rapid economic growth and social transformation. New China caught up with, and in some ways surpassed, New Asia (Chang and Huang 2005; Sheridan 1999). Through the Asian economic crisis of the late 1990s, China served as a steady and stable source of support for the entire region of East and Southeast Asia (Lew 1999).

The foundation for much of the transformation of the Chinese urban landscape and tourism spaces was established in the land reform and administrative decentralization policies of the 1980s. China, however, remains a socialist country in which land ownership is concentrated in the state. Through the land lease mechanism, private development corporations and enterprises can own use rights to land for their projects. Sometimes these development firms work in partnership with local governments, and share their profits with them (Ma 2004). In this case, different branches of local, provincial and state government play three roles in relation to land development. First, they are the landowner, receiving rent on the leased land. Second, they act, often in partnership, as the developer, paying for use of the land and receiving profits from its use. Third, state authorities are the regulator, ensuring proper planning and development for the larger public good. Private, non-governmental entities can also participate in this process, though for most of the past two decades this has required some partnership with government bodies. This situation gave tremendous discretionary power to the local government authorities especially in the late 1980s and early 1990s, and that discretionary power has been a major reason for the rapid, and at times monumental, transformation of the Chinese landscape since the 1990s. The Chinese central government clamped down on the excesses of this discretionary power in the revised Land Administration Law in 1998, instilling some element of order to the chaos of local government land dealings.

Local decision-making power is most readily seen in the transformation of China's urban built environment, much of which was so worn and dilapidated by the 1980s that it offered significant opportunities for creative urban renewal, conservation and preservation, and reuse. Urban renewal has been the predominant approach, with narrow roads being transformed into wide boulevards by razing buildings in their path, and presenting opportunities for new, high-rise commercial and residential buildings. In other instances large areas of older factories and other buildings have been razed to enable new districts of expansive roads, parks and modern architecture. In nearly every city, though, there are areas that have used their traditional architectural and retailing core to create vibrant new retail and tourism districts. Some older residential areas have also been preserved, but the record on this has been less successful (*Economist* 2000; Wang 2003).

The decentralization of economic decision-making in China has also contributed to the diversity of different approaches to pedestrian shopping streets described above. Since the late 1980s, local officials have been given increasing freedom to directly negotiate land development agreements with foreign investors. This occurred in a poorly formed regulatory environment in which local elites could exert direction, control and approval to a much greater degree than is typical in more developed economies. At the same time, because widespread international travel by mainland Chinese citizens was forbidden from the early 1950s to the early 2000s, most local officials in China had limited knowledge of global development experiences upon which to judge investment proposals. As a result, new and different conceptual approaches to large-scale development projects were often received with less scepticism in China than they might be elsewhere. This situation was further combined with a willingness to facilitate dramatic change on the urban fabric of most cities and a growing competitiveness among Chinese cities for both foreign investment and symbols of economic and social dynamism.

As alluded to above, there are no requirements for general public input into development decisions, and government land ownership allows little recourse for previous users to stop the condemnation and transformation of the sites on which their homes and places of work are situated (Wang 2003). This has allowed large-scale urban renewal to take place with little effective opposition, resulting in the loss of many older, well-established neighbourhoods, the dislocation of residents and businesses, and the loss of arable land. It has also created economic inefficiencies as poorly conceived development projects, including some major tourism initiatives, have been constructed and gone bankrupt in short course. This has been especially true in the area of theme parks, with six of the seven large theme parks that were built in Shanghai in the early to mid-1990s being closed by 1998 (Ap 2003). Most of these were financed by overseas investors and suffered from poor siting, poor construction and design, no understanding of the local market, and poor management and services. Ap (2003) identified political interference as an underlying reason for these failures, with local officials concerned only with attracting foreign investment and not with financial feasibility. Similar problems occurred in large residential and resort development projects in the 1990s.

The 1998 revisions to the Land Administration Law brought more order to the land development process in China. The new regulatory environment reflects a growing sophistication in the Chinese economy that comes with experience, experimentation, failure and success. Local authorities have lost some freedom in their decision-making in that they must now gain approvals from provisional, and sometimes central, government authorities. While this added layer of oversight can itself be incompetent and inefficient, there does appear to be a growing degree of quality and success in the more recent urban tourism and retail development initiatives in China.

The conservation of older urban landscapes has been an area of intense criticism of the Chinese path to urban modernization (Wang 2003). Some efforts have been made to museumize selected residential areas, and various forms of historicity were significant in the retail and tourism case studies described above. However, only 25 of some 6,000 of the original *hutongs* in central Beijing remain today (Luard 2004). Despite its long history and reverence for the past, China does not seem to know how to incorporate the past into the present, except in a museumized, commodified and commercialized way. On the other hand, the role of historic preservation in a modern and post/hypermodern world has come under scrutiny by social theorists (Goss 1988). As Lowenthal (1985) has argued, we cannot know the past except through the values of the present. It is the values of the present that define what is preserved and how it is preserved. For the time being, these values are primarily monetary in the new China.

A comparable development environment existed in the USA in the first half of the 1900s, but came to a close in the early 1960s following large-scale public protests over the negative impacts of urban renewal (Ladner 2001). In New York, for example, this was the period in which was built Shea Stadium, the Triborough Bridge, the Lincoln Center for the Performing Arts, the Long Island Expressway, the Northern State Parkway, Jones Beach, the World's Fair Grounds, the United Nations building and many other large public works (Callahan and Ikeda 2004). It was also the era of the Tammany Hall political machine, which was beholden to special interest groups and monopolies, and exerted its political power to gain its ends through any means possible. The era came to a close in the early 1960s under criticisms that the urban renewal and displacement of people had destroyed vibrant neighbourhoods and 'created new slums faster than it was clearing old ones' (Caro 1975). It was about this same time that historic preservation became a major alternative movement in North America (Lew 1989).

There have been some signs that similar protests may be on the rise in China in the early 2000s. Forced demolition of homes and neighbourhoods have been increasing in China as urban land increases in value. The more egregious losses are increasingly being met with violent protest, with 85 arrested protesters and one attempted suicide in Beijing on China's National Day (1 October) in 2004 (Ricci 2004). As China's urban population becomes more sophisticated and cosmopolitan, nostalgia for the past and demands for more conservation efforts are likely to increase. In the meantime, largely in response to government efforts to maximize land values, China's urban landscape has been transformed to reflect the interests and demands of modern and postmodern Chinese consumerism and profit-seeking land speculators, with little regard for deeper values of community and sense of place.

Thus, when we look at how the past is conceptualized, presented and even destroyed in China (through the range of pedestrian shopping streets, for example), what we are seeing is the juggernaut of economic liberation and decentralized decision-making following decades of central government suppression. Some of this is seen in an unreflexive hypermodern retail environment – typically created on the tailings of the past. In many other instances, however, there is a reflexive desire to maintain some core elements of past identities. Both are taking place at the same time; and both are contributing to a transformation of the urban tourism landscape and the urban consumer.

Through this urban transformation, power relationships have evolved from a dominance by the central authority in Beijing, to the local level, and then somewhat back to the provincial and central state level. Clearly, the development of pedestrian shopping streets as tourism attractions has been strongly influenced by power transference between central, provincial and local government relations along with changing power relations affecting the ownership and regulation of land. Central authorities, however, are limited in the degree to which they can re-contain the authority that they once held. Chinese society has grown far too complicated and decentralized (with many more competing local power centres) and far too wealthy (with the associated corruption that money can buy) to be easily reigned in. The core power struggles among these government entities are influenced by foreign and, increasingly, domestic private sector capital. The ability to attract capital investment has become the ultimate symbol of political power in urban China. The manifestation of that political power occurs through the physical landscape created by the investment capital. China's market liberalizations have, however, also unleashed a form of entrepreneurial liberation, and its associated creative energies. This, too, has contributed to the diversity of pedestrian shopping streets that have emerged across China. The marriage of this entrepreneurial creativity with government interests, if it can be linked to residents' interests, may be a major lesson from the Chinese experience; however, it is not one without flaws. Indeed the major drawback to the Chinese approach to urban regeneration and related tourism developments is the disempowerment of the people who live in the city. Almost lost in the governmental struggles for political power and money are the people, whose good the government is responsible for nurturing. As consumers and employees, they are sought after. As a political power centre in their own right, they have yet to exert a coordinated influence on the land development process. It is questionable whether they will succeed without a broadening of democratic participation in China. But China is a large and rapidly changing place. We have yet to see the end result of the transformation of post-Mao China that was begun in 1978. There may be many more surprises and lessons to come.

References

Ap, J. (2003) 'Theme park development in China', in A.A. Lew, L. Yu, J. Ap and G. Zhang (eds) *Tourism in China*. Binghamton, NY: Haworth Press.

Callahan, G. and Ikeda, S. (2004) 'The career of Robert Moses: city planning as a microcosm of socialism', *Independent Review*, 9(2): 253–261.

Caro, R.A. (1975) *The Power Broker: Robert Moses and the Fall of New York*. New York: Vintage.

Cartier, C. (2001) *Globalizing South China*. Oxford: Blackwell.

Chang, T.C. and Huang, S. (2005) '"New Asia-Singapore": a concoction of tourism, place and image', in C. Cartier and A.A. Lew (eds) *Seductions of Place*. London: Routledge.

China Internet Information Center (CIIC) (2004) Dazhalan Market: http://www.china.org.cn/english/features/beijing/30846.htm [last accessed 24 October 2004].

China National Tourism Association (CNTA) (2004a) Review of China's Tourism Industry (II) Domestic Tourism: http://www.cnta.com/lyen/2fact/domestic per cent20tourism.htm [last accessed 4 April 2005].

—— (2004b) Review of China's Tourism Industry (III) Outbound Tourism: http://www.cnta.com/lyen/2fact/outbound per cent20tourism-1.htm, [last accessed 4 April 2005].

Chu, C.L. and Lee, J.I. (2004) 'Living cities as cultural heritage sites: the effects of historic preservation and tourism in Havana and Shanghai', unpublished manuscript, Harvard University, Project on the City.

Duncan, C. (1986) 'City Profile: Macau', *Cities*, 3(1): 2–11.

Economist (2000) 'That was Beijing', *The Economist*, 356 (9 September; 8187): 39–40.

Fiala, R.D. (2004) Oriental Architecture: Streets of Macau: http://www.orientalarchitecture.com/macau/macaustreetsindex.htm [last accessed 19 October 2004].

Glaeser, E., Kolko, J. and Saiz, A. (2001) 'Consumer city', *Journal of Economic Geography*, 1(1): 27–50.

Gluckman, R. (2003) Shanghai's stylish Xin Tian Di: http://www.gluckman.com/XinTianDi.html [last accessed 4 April 2004].

Goss, J. (1988) 'The built environment and social theory: towards an architectural geography', *Professional Geographer*, 40(4): 392–403.

Guo, K. and Li, H. (1995) 'A study of regional variation in China's opening to the outside (in Chinese)', *Zhongguo Gongye Jingji* (China's Industrial Economy), 8: 61–68.

Haley, G.T., Tan, C.T. and Haley, U.C.V. (1998) *New Asian Emperors: The Overseas Chinese, their Strategies and Competitive Advantages*. Oxford: Butterworth-Heinemann.

Keng, C.W.K. (1996) 'China's land disposition system', *Journal of Contemporary China*, 5(13): 325–345.

Ladner, J.A. (2001) *The New Urban Leaders*. Washington, DC: Brookings Insitution Press.

Lafferty, G. and van Fossen, A. (2001) 'Integrating the tourism industry: problems and strategies', *Tourism Management*, 22 (1): 11–19.

Lew, A.A. (1987) 'The history, policies and social impact of international tourism in the People's Republic of China', *Asian Profile*, 15(2): 117–128.

—— (1988) 'Tourism and place studies: an example of Oregon's older retail districts', *Journal of Geography*, 87(4): 122–126.

—— (1989) 'Authenticity and sense of place in the tourism development experience of older retail districts', *Journal of Travel Research*, 27(4): 15–22.

—— (1999) 'Tourism and the Southeast Asian crises of 1997 and 1998: a view from Singapore', *Current Issues in Tourism*, 2(4): 304–315.

—— (2001) 'Tourism development in China: the dilemma of bureaucratic decentralization and economic liberalization', in D. Harrison (ed.) *Tourism in the Less Developed World*, second edition. Wallingford: CAB International.

Lew, A.A. and Wong, A. (2004) 'Sojourners, Gangxi and clan associations: social capital and overseas Chinese tourism to China', in T.E. Coles and D.J. Timothy (eds) *Tourism, Diasporas and Space*. London: Routledge.

Lin, G.S. and Ho, S.P.S. (2005) 'The state, land system and land development processes in contemporary China', *Annals of the Association of American Geographers*, 95(2): 411–436.

Logan, J.R., Bian, Y. and Bian, F. (1999) 'Housing inequality in urban China in the 1990s', *International Journal of Urban and Regional Research*, 23(1): 7–25.

Lowenthal, D. (1985) *The Past is a Foreign Country*. Cambridge: Cambridge University Press.

Luard, T. (2004) China diary: Vanishing Beijing. *BBC*, http://news.bbc.co.uk/2/hi/asia-pacific/3679170.stm [last accessed 21 October 2004].

Ma, L.J.C. (2004) 'Economic reforms, urban spatial restructuring and planning in China', *Progress in Planning*, 61: 237–260.

Meisner, M. (1999) 'China's communist revolution: a half-century perspective', *Current History*, 98(629): 243–248.

Ming, L. (2003) Top ten sights in Zhongshan: Sun Wen West Road culture: http://www.newsgd.com/travel/toursites/200306030045.htm [last accessed 18 October 2004].

Morgan, N. (2004) 'Problematizing place promotion', in A.A. Lew, C.M. Hall and A.M. Williams (eds) *Companion to Tourism*. Oxford: Blackwell.

Putterman, L. and Dong, X. (2000) 'China's state-owned enterprises: their role, job creation, and efficiency in long-term perspective', *Modern China*, 26(4): 403–447.

Ricci, T. (2004) Forced demolitions: an attempt on man and culture, *AsiaNews.it*, http://www.asianews.it/dos.php?l=en&dos=&art=442

Sheridan, G. (1999) *Asian Values Western Dreams: Understanding the New Asia*. St Leonards, Australia: Allen & Unwin.

Short, J.R. and Kim, Y-H. (1999) *Globalization and the City*. Harlow: Longman.

Wang, N. (2003) 'Chinese vernacular heritage as a tourist attraction', in A.A. Lew, L. Yu, J. Ap and G. Zhang (eds) *Tourism in China*. Binghamton, NY: Haworth Press.

Wu, W.P. (2004) 'Cultural strategies in Shanghai: regenerating cosmopolitanism in an era of globalization', *Progress in Planning*, 61: 159–180.

Yang, F.F. (2004) 'Services and metropolitan development in China: the case of Guangzhou', *Progress in Planning*, 61: 181–209.

Yeh, A.G. and Wu, F. (1996) 'The new land development process and urban development in Chinese cities', *International Journal of Urban and Regional Research*, 20(2): 330–353.

—— (1998) 'The transformation of the urban planning system in China from a centrally-planned to transitional economy', *Progress in Planning*, 51(3): 167–252.

Zhang, G. (2003) 'China's tourism development since 1978: policies, experiences, and lessons learned', in A.A. Lew, L. Yu, J. Ap and G. Zhang (eds) *Tourism in China*, Binghamton, NY: Haworth Press.

Zhou, Y. and Ma, L.J.C. (2000) 'Economic restructuring and suburbanization in China', *Urban Geography*, 21(3): 205–236.

Zhu, J. (2005) 'A transitional institution for the emerging land market in urban China', *Urban Studies*, 42(8): 1369–1385.

8 Power, resource mobilisation and leisure conflict on inland rivers in England

Andrew Church and Neil Ravenscroft

The complexities of power theory present a range of challenges for empirical research into power relations and tourism. The elusiveness of power partly stems from its conceptualisation as relational, originating in social interaction and being empowering, restraining and resistive. Lukes (1982) captured the vagaries of power theory when he noted that a reading of existing writings suggests that:

> power is something which is possessed; it can only be exercised; it is a matter of authority. Power belongs to the individual; it belongs only to collectives; power doesn't belong to anyone, but is a feature of social relations. Power usually involves conflict but it doesn't have to. Power presupposes resistance; power is primarily involved in compliance (to norms); power is both. Power is tied to repression and domination, power is productive and enabling. Power is bad, good, demonic or routine.
>
> (Lukes 1982, quoted in Haugaard 2002: 223)

Not surprisingly, given the diversity of viewpoints summarised here some have even suggested abandoning power as a concept (Latour 1986). One factor that unites many power theorists is the concern to understand how power can become concentrated in social groups, institutions or agents (Sharp *et al.* 2000). Identifying the exercise and concentration of power in relation to tourism is made more challenging by the mobilities inherent in contemporary tourism and the ever-changing nature of the industry. Previous empirical studies have adopted a range of theoretical and conceptual starting points. Actor-oriented approaches focus on how the power imbued in actor strategies and practices in tourism plays a role in the emergence of social orderings (Franklin 2004). Writers concerned with tourism and salient social differences, such as gender and sexuality, often aim to identify how power and power relations emerge from the social-cultural interactions between structural differences, agency and space (Aitchison 2003). A growing number of studies have sought to examine the power concentrated in state institutions with responsibilities for tourism development (Hall 2003). In this chapter we aim to show that the nature and sources of power relating to tourism can be revealed by empirical studies which focus on the concept of resource mobilisation. A number of power theorists stress that resource mobilisation is central to the development of power relations and the exercise of power (Allen 2003; Giddens 1993; Mann 1993). By understanding the role of resources it is possible to examine how concentrations of power develop within the complex mobilities that make up contemporary tourism.

Resources and power

The resources of power and how they are mobilised are viewed in quite contrasting manners by differing theoretical perspectives. For example, and at the risk of being oversimplistic, Foucauldian studies stress that knowledges, truths and technologies, such as the gaze (Urry 2002), are important resources in the dispersed organisation of power. Post-structural writings drawing on the work of Deleuze stress the centrality of texts and discourses to the social orderings (see Coles and Church, Chapter 1 in this volume). In a review of power theories, however, Allen (2003: 63) stresses 'the critical role that resources play in the generation of power as an effect'. Allen (2003) also notes that extreme care is needed when considering resource mobilisation since resources may only imply power and their scale cannot be simply linked to the degree of power. Resources can be unused, misused and simply given away but, despite their uncertain uses, resources still play a key role in the development of power relations.

Certain theorists have sought to outline the details of the resources mobilised in the development of power. For Giddens (1984), power and resources are central to revealing the interactions between structural principles, institutions, practices and human agents in the development of social systems. Resources play a role in co-ordinating social systems and can be separated into authoritative and allocative resources. Giddens (1984) concluded that authoritative resources had been the traditional emphasis of social theory and these included the features of the environment, the means of production and reproduction, and produced goods. Allocative resources stem from the organisation of social time–space, the production/reproduction of the body and the organisation of chances for self-development and expression. Both forms of resources mutually augment each other and are 'stored' in containers such as the media and technology. Critiques of Giddens have tended to question the fixed description of power as being stored (Allen 2003), and they have noted that the processes of resource mobilisation portrayed by Giddens are usually linked to the exercise of power as domination rather than other modalities of power, such as negotiation (Haugaard 1997). Nevertheless, these debates clearly highlight the centrality of a wide range of resources to the mediation of power. Giddens (1993) has noted that allocative resources can also include knowledges, rhetoric and he confirms that the understanding of power requires an analysis of how resources are organised and interact with structural principles, institutions, practices and agency.

The broad historical overview of power provided by Mann (1986, 1993) also highlights the importance of resources in the organisation of social power but stresses how resource mobilisation changes over time and can suddenly alter in unexpected ways. For Mann (1986) there are four historically consistent sources of power: ideological, economic, military and political. Each has a different organisational basis in terms of the constitutive socio-spatial relations. For example, military power is portrayed as being coercive and existing in concentrated networks whereas economic power is based on more extensive circuits of praxis (Mann 1986).

Mann's (1993) writings reinforce the argument that the role of resources in the emergence of concentrations of power can only be understood through an examination of the socio-spatial relations between agents, institutions and structures that co-construct the mobilisation of resources. In analysing power relations, however, power theory is concerned not only with structural relations or salient social differences, such as gender relations, but also the form (or modalities) of power relations, such as domination or authority, that emerge from social interaction and practices. Allen (2003) argues that Foucault and other post-structural writings, while providing important insights on the dispersed nature of social power, have also resulted in a propensity in power studies to specify inadequately the types of power relation that emerge in different social situations. In an attempt to 'name' power more precisely, he identifies eight key theoretical modalities of power each based on specific relations: domination, authority, coercion, inducement, manipulation, persuasion, negotiation and seduction – the latter is presented as almost the opposite of domination (Allen 2003). These modalities are theoretically seductive in themselves as they appear to provide categories for summarising the various power practices, techniques and tactics of agents and social groups. Allen (2003: 101, emphasis original), however, argues strongly that modalities are,

> quite specific ways of exercising power which, I would maintain, entail only *certain* practices and techniques in *particular* modal arrangements . . . modalities of power are not entities that can be understood by breaking them down into an endless stream of practices which are then added together to realize the whole.

This argument reflects Allen's (2003) general desire for more precision in identifying the *organisation* of power and also his claims that power concentrations can often be associational. Modalities such as negotiation offer associational opportunities for autonomy and co-operation as opposed to dependence and exclusion that stem from domination and authority.

In the context of tourism research we would support Allen's (2003) general arguments, in that too often research has focused on the practices of power and has not always made that extra conceptual step to stating what form of power relations emerge in the tourism context. It is, however, a significant challenge for empirical research to distinguish between the general relations or modalities of power as opposed to the more specific tactics, strategies and practices that imbue the negotiated interactions between agents, groups, institutions and structures (cf. also Coles and Scherle, Chapter 10 in this volume). For example, persuasion (as a modality), often based on discourse (as a tactic) may be used by one group interacting with another, but it may not always have an influence and thus power effect on the other group. In a case study of conservation planning and tourism developments in Belize, Few (2002) attempted to overcome this problem by including tactics as a specific subset of more general power relations. While this provides a conceptual neatness, the difficulty of specifying the general nature of power relations and modalities remains a significant issue for empirical research.

The increasingly diverse and generalised theoretical discussions of power appear to leave research into tourism and power chasing after elusive possible starting points for empirical studies. In fact, inspection of the theoretical debates offers some clear pointers as to how we can proceed to reveal the organisation and concentrations of power in tourism. If tourism research is to respond to the long-standing theoretical debates over power, then the basic foundation of an understanding of power in tourism must be to reveal the concentrations of power that stem from the power relations based on the socio-spatial and historical interactions between actors, social groups, institutions and structures. In this chapter we intend to show that a focus on resource mobilisation allows empirical analysis to tease out the general power modalities from the details of practice and tactics.

Of course a number of the writings on tourism mentioned in Chapter 1 have already embraced aspects of this agenda. The notion of the tourist gaze (Urry 2002) considers how the resources of tourist sites, signs and discourses combine with state policies and actor practices to produce a gaze that is both self-disciplining and disciplining of others. Few's (2002) study of tourism in proposed nature conservation areas in Belize provides a rare example of empirical research that explicitly attempts to specify the nature of power relations, resources, motives and tactics. The empirical analysis examines how planning outcomes can be understood as the product of actors, in the form of planners and local stakeholders, pursuing their motives by drawing on tactics, such as manipulation and compromise, and a range of resources including social contacts and knowledges (Few 2002). While Few (2002: 33) argues that his study of actors is mainly concerned with the microfoundations of power relations, he does acknowledge the importance of resources in that 'they refer not only to personal skills and social connections, but also to the structural properties of social systems including discourses'. This statement has a resonance with the arguments of Giddens (1984) outlined above for separating allocative resources, such as the skills mentioned by Few (2002), from authoritative resources embedded in structural principles. The value of focusing on the organisation of resources in the study of power is that it provides insights not only into actor strategies but how these interact with the structural and institutional processes that are embedded in resources. It is important not to essentialise resources since their use can only be understood in the relational contexts of power. This chapter seeks to illustrate that while there are hegemonic possibilities in resource mobilisation, it is also a thoroughly contingent activity involving conflicts and mutual dependencies amongst groups and individuals seeking to mobilise similar types of resource. Furthermore, resources are not static and are organised in often unpredictable ways not just by actors but also by processes linked to structures and institutions. The rest of this chapter provides an empirical case study of the process of power concentration in relation to certain specific tourism and leisure activities. The conceptual focus is on how organisation of resources contributes to specific power modalities based on forms of power relations that are rooted in the socio-spatial practices of different actors interacting with a range of structural and institutional processes.

Power, tourism, property and the law

The case study examines the current struggles and conflicts over the use of inland rivers in England between landowners, anglers (fishermen) and canoeists (a canoe here is a generic description for kayaks and open canoes). These are often intense conflicts for those involved, and they are currently the subject of a range of national political actions. This may seem a rather parochial focus given that the theoretical discussions of power outlined above are often part of broader endeavours to understand systemic societal change. However, the case study aims to focus on the resources of law and property which are issues at the core of discussions of power and also central to tourism as an activity. There is a long-standing recognition amongst power theorists that the law and property are important power resources drawn on not just by state institutions but also by individuals and social groups (Mann 1986). Haugaard (1997: 138) argues that private property plays an important role in maintaining the uneven distribution of power,

> Take the structures and institutions which are central to the constitution of private property as a resource. In principle, within modern capitalism everyone is equally entitled to private property. . . . In many forms of feudalism only certain individuals were entitled to own property. The principle that every person has an equal right to the possession of property is part of the arsenal of the legitimacy of capitalism.

An analysis of the role of law and property in power relations is particularly valuable in the context of tourism as it allows an examination of the interactions between mobile resources and those that are more spatially bounded. Tourism currently is typified by the 'seeming' mobility of resources, relations, actors, structures and institutions and these different mobilities contribute to the exercise of power both through presence and from a distance. Considerable attention has been paid to how mobilities contribute to power relations (Urry 2002, 2003) and is examined further by other chapters in this volume. As Allen (2003: 115) notes, however, 'some resources, such as the infrastructural resources of the state . . . namely taxation, law, property and the like, are more grounded, whereas others, such as finance, information, ideas, people and contacts, appear less fixed by nature'. Tourism writers have often noted that the physical sights and resorts of tourism, unlike the factories and offices of other industries, cannot be readily relocated to take advantage of changing economic circumstances and factor costs (Shaw and Williams 2004). In the study of tourism and power, therefore, a concern with the mobile, fluid and distanciated exercise of power must be accompanied by an understanding of more spatially fixed power resources, such as property and the law. Furthermore in discussions of power and tourism, the resources of law and property have received relatively little attention compared to other resources such as discourses, knowledges, embodied practices and state tourism policies.

An area where leisure and tourism studies have considered legal issues and property rights is in the examination of tourist access to private and public property. McIntyre *et al.* (2001) note that there is a long history of research on access in developed northern-hemisphere countries but there is a lack of studies of access to land in developing countries and southern-hemisphere states such as New Zealand and Australia. Indigenous land rights disputes in many different countries have led to an expansion of research interest into struggles over land for use by tourism (Curry 2001; Higgins-Desboilles 2004). Often such studies seek to identify the broader processes that are shaping the interplay between tourism, access to land and indigenous rights, such as the increasing commodification of nature, changing public and group values and new state policies (Kaltenborn *et al.* 2001). Rarely, however, are property and the law considered as specific resources of power in relation to leisure and tourism. Blomley (1994, 2005) notes that the discipline of geography has also been neglectful of the spatiality of property.

Clearly, the law and property as resources cannot be considered in isolation from the other resources that will be utilised concurrently. The case study in this chapter also examines the discourses, knowledges and social interactions that are involved in the mobilisation of the law and property as resources in the conflicts between canoeists, anglers and landowners. Furthermore, the mobilisation of these other resources ensures that attempts to utilise the law and property in order to concentrate power becomes a contingent process involving complex negotiations amongst different actors and social groups. As Allen (2003: 105) notes, ‘resources are the media through which power is exercised they are not power itself’. Thus, the case study seeks to reveal not simply the role of resources in organising power but also the specific power relations that emerge. The next section starts by summarising the current disputes over the recreational use of inland rivers in England. This is followed by a section that focuses on the mobilisation of property and the law by examining the role of property rights and navigation law in shaping public access to inland rivers. The analysis reveals how the power relations surrounding the conflict over inland rivers are shaped by the historical role of social institutions and structures that prioritise private property rights and market exchange. A further section considers recent attempts to alleviate the conflict through voluntary agreements between users and landowners. This illustrates how the resources of the law and property interact with the practices of actors. The concluding section seeks to identify the power relations and modalities involved in the conflict and the implications for research into power and tourism.

The empirical analysis has developed out of a long-term research project undertaken in England for a variety of central government departments and agencies which afforded the opportunity to study the issues and parties in-depth and over time (University of Brighton 2001, 2004; Countryside Agency 2004). The studies focused on England due in part to the different legal systems for access to water operating in other areas of the United Kingdom (see below for a discussion of

Scotland). The data presented here is drawn from the documents published by organisations representing the leisure and tourism uses of inland water, such as boaters, anglers and canoeists, and also from a series of individual and group interviews undertaken in different parts of England between 2001 and 2004. A total of 16 group interviews were undertaken, 5 with canoeists, 5 with anglers and landowners and 6 with mixed groups of stakeholders. In addition, 58 face-to-face interviews were conducted with the senior staff of national and regional organisations representing water sports and other forms of countryside tourism and leisure. All group and individual interviews were taperecorded and their contents analysed to identify various themes and sub-themes but the advantage of this mixed methods approach was that views could be compared in differing situations to give a sense of how stakeholders responded to the presence of competing interests and user groups.

This type of long-term research inevitably involves the researchers being drawn into continually changing power relations surrounding tourism and leisure. The various government-commissioned reports produced by the authors of this chapter became part of the persuasive discourses and negotiations of actors and organisations, many of whom also sought to influence the research agenda and questions. Often some organisations would be highly critical of our research at one point in time but these criticisms would be less marked at a later date when the politics of access to inland rivers had changed. As researchers, this required us to continually reflect on our changing positionality and to maintain a critical perspective on a process that sought to subsume us into the conflicts and power relations.

Tourism, leisure and conflict on inland rivers in England

In many countries inland rivers passing over or situated on land will be subject to some type of specific property legislation and rights. In some countries, partly because of their historical role in transportation, major rivers and canals are treated as a form of public highway which facilitates the activities of contemporary leisure and tourist users, although riparian property legislation will usually determine access for bank-based activities such as angling. In England the system of rights differs between tidal and inland water which is defined as those spaces above the estuarine tidal limit. Below the tidal limit there is a recognised public right of navigation. On rivers above the tidal limit, navigation is more restricted and much of the river space is effectively privately owned restricting the opportunities for certain leisure and tourism users (Telling and Smith 1985). England is by no means alone in having legal restrictions on the use of inland water and similar situations exist elsewhere (Funck 2002). Also legal rights are at the core of ongoing debates in developed and developing countries concerning the 'privatisation' of beach spaces (Mongeau 2001, 2003).

As freight transport has reduced markedly on inland rivers and canals in England so the main users physically on the water are often those involved in leisure and tourism activities such as motorised boating, sailing, canoeing, water-skiing and rowing. Participation levels in these pursuits are relatively low with each activity in any year involving between 3 and 4 per cent of the adult population (Mintel 1998; University of Brighton 2001). Participation will include regular and infrequent participants. For example, canoeing involves about 100,000 regulars and perhaps 1 million people who take part at some point in a year (University of Brighton 2001). In some rural areas, such as in the Broads National Park in East Anglia, water-based activities contribute significantly to the local tourism sector. Inland water is also an important component in other tourism and leisure activities with higher levels of participation. The UK day visitor survey for the late 1990s suggests that annually 14 per cent of countryside day trips by adults in the United Kingdom involve water usually as a backdrop to activities such as walking and picnicking (University of Brighton 2001). In terms of participation, angling is the most significant activity in the United Kingdom to make physical use of the water with approximately 3–4 million (7 per cent) of people aged over 12 participating in any one year although this represents a fall on participation levels for the 1970s (University of Brighton 2001). The two main types of angling activities are coarse and game fishing, with the latter involving trout and salmon species often generating important sources of revenue for fishing rights owners and local tourism industries. Recent estimates suggest that freshwater angling may be worth £3.5 billion to the British economy (Labour Party 2005) and angling tourism is now a global industry involving trips to locations as varied as Alaska, Siberia and New Zealand.

In England the nature and spaces of inland rivers used for tourism and leisure are markedly influenced by historical law and property rights which partly contribute to the significant conflicts involving not just leisure users but also landowners and the state. Overall, there are about 17,000 km of major rivers and canals in England and a further 43,000 km of minor rivers. The major rivers and canals are the focus of most water-based leisure activities but only 4,300 km are legally available for navigation (University of Brighton 2001). Boaters and canoeists, therefore, are largely confined to canals and major rivers, such as the Thames, where navigation is permitted and also tidal estuaries where there is a public right of navigation. This level of supply means that sailing, motorised boating and rowing have access to the majority of the major river and canal spaces that are suitable for their leisure activities, although there are still vigorous campaigns to reopen derelict canals and navigations (Inland Waterways Amenity Advisory Council 1999).

For canoeists the current situation is far less satisfactory since the legally navigable rivers and canals mainly comprise 'flatwater' as do the lakes and other enclosed waters they are able to use. These are useful for novices, training and specialist activities such as sprinting and marathon racing. Many of the whitewater, slalom and 'play' sites, which are graded according to their degree of technical challenge, are on rivers where navigation is not permitted as a public right. In a few locations canoeists have been able to negotiate access through voluntary agreements with landowners and other rights holders but these rare cases make little difference to the overall level of supply of space for canoeing. Overall, the situation makes it difficult for canoeists to pursue aspects of their sport. It also turns many canoeists into tourists as they have to travel long distances to access legally navigable rivers with suitable water (University of Brighton 2001).

In contrast, the legal system and property rights enable angling in rivers to take place relatively easily. Property law is well established concerning access to watersides for fishing and the exercise of rights to fish. These rights have evolved since the late nineteenth century when angling clubs, often from urban working-class areas, wanted to develop a more formal structure for their activities (Franklin 2002). The legal system is underpinned by a process of payment with anglers paying owners for various rights either directly as individuals or indirectly through membership of a club that rents or owns rights. Anglers must also possess a rod licence which they purchase from the government's Environment Agency which uses the resulting income to maintain fisheries.

Partly as a result of these very different situations regarding access to inland rivers, the last twenty years have seen increasing evidence of conflict between canoeists and anglers, especially when canoeists use inland rivers where anglers and landowners claim their presence is illegal. Our interviewees referred to a rising incidence in verbal abuse, physical threats, vandalism and acts of violence. The following quotes were quite typical and indicate the emotional and practice-based dimensions to the conflict:

> 90 per cent of canoeists don't lift their paddles to avoid disturbance when they pass you.
>
> The attitude of canoeists is 'tough, I'm using the river as well as you'.
>
> I've had everything thrown at me by people fishing . . . bait, bread, stones, hooks.

There have been a number of government initiatives to ameliorate this conflict. The national Environment Agency convenes a national Angling and Canoeing Liaison Group (ACLG 1999) but there have been occasions when this group has failed to meet as certain members refuse to attend due to previous disagreements. More recently, the Countryside Agency (2004) has explored the feasibility of voluntary agreements between canoeing organisations and landowners to increase the supply of river space for canoeing. The profile of the conflict was recently raised at the national level when the British Canoe Union (BCU), the national governing body for the sport, revitalised its river access campaign by presenting a petition to parliament to highlight the lack of rivers available for canoeing and further to press for changes in legislation (BCU 2005).

Within tourism and leisure research there have been a number of different approaches developed to examine conflicts between users of natural resources (see, for example, Jackson and Wong 1982; Jacob and Schreyer 1980; Mikolic *et al.* 1997; Owens 1985; Schneider and Hammitt 1995; Suls *et al.* 1998). These have included examinations of the conflicts between anglers and boaters (Driver and Bassett 1975; O'Riordan and Paget 1978; Gramann and Burdge 1981; Adelman *et al.* 1982; Ruddell and Gramann 1994). Few of these have considered the impact of either property-based rights claims or the type of national level political actions that are connected with the conflicts on inland rivers in England (for an exception see Sports Council 1985).

The different perspectives on user conflicts led Carothers *et al.* (2001) to categorise leisure conflict into two principal classifications: 'interpersonal conflicts' arise from goal interference caused directly by the actions of another user; and 'social value conflicts' result from divergent views about the social acceptability of different behaviours in particular leisure settings (Vaske *et al.* 1995). The principal utility of this distinction is that it makes clear that conflict can arise in a number of ways, without people needing to be in physical proximity to one another. It also implies that recreational conflict over natural resources can assume at least two identifiably different constructs: physical contact and perception. In many cases, however, the distinction between these two constructs is blurred, and can elide in situations to physical conflict (Ravenscroft *et al.* 2002).

Clearly, a social values perspective has some relevance to the conflict over inland rivers in England. For example, anglers and canoeists have developed differing discourses reflecting their concern for the environment and often seek to present themselves as having superior environmental values compared to other users. The British Canoe Union (BCU 2003: 3) argues that 'the canoe causes no pollution, and leaves no trace of their passing. Canoeing is not a threat to environmental conservation'. The director of the Salmon and Trout Association, a national representative body for game angling, was more explicit in his condemnation of the social and environmental values of some canoeists when he claimed that 'canoeists do little to help protect the environment or contribute to the maintenance of the resource' (Knight 2003: 12). These environmental discourses are just one of several discourses developed by anglers and canoeists to justify their approaches to the current conflict and they indicate that perceived differences in social values are imbued in these discourses.

Other discourses show similar social values. For example both anglers and canoeists make a virtue of the fact that they rely largely on volunteers for the development of their sports and voluntarism is stressed as a positive social value (Ravenscroft *et al.* 2005). Indeed, differences in social values between anglers and canoeists cannot be neatly linked just to the (current) conflicts between the two groups of users. The origins of this conflict are complex, long-standing and involve the state, non-users (e.g. landowners), property rights and legal agreements which give rise to a market for access. Consequently, the conflict cannot be readily encapsulated within a goal interference, social psychology or social values conceptual framework but requires an analysis of power resources and relations.

Power, property, legal rights and inland rivers

At the heart of the conflict between canoeists and anglers is a situation in which one leisure user, the anglers, mobilise the law and property rights to claim a superior legal right to the use of a natural resource. The canoeists mobilise property rights rather differently by drawing them into a 'moral' discourse which stresses their 'right' to enjoy their sport and the inequity of private landowners and angling clubs excluding other users from rivers through their ownership of property rights. The laws central to this conflict are, as always, open to interpretation and this section sets out how laws and rights have been interpreted as part of their mobilisation by rights owners and the state. This reveals how laws and property rights play a major role in the changing power relations between anglers, canoeists and landowners.

In English law, 'land' generally includes the surface itself and all that lies above or beneath, but there are many qualifications (Law of Property Act 1925). In the case of rivers, the ownership of the land beneath the water determines the broad uses of the river but this does not imply rights to extract or use the water. In the case where a watercourse marks the boundary to a property, the legal rule is that unless there is evidence to the contrary, the boundary will be the mid-point of the watercourse. The situation for canals is different since the Transport Act of 1947 when, due to declining commercial use, they passed into public ownership and are now mainly controlled by British Waterways (established under the Transport Act 1962).

Generally, in English law a Lockean approach is taken to 'ownership' so that the owner has control over a bundle of rights, as well as the most extensive form of control over the property known to the law (Becker 1977; Denman 1978; Munton 1994; Parker 2002). From these rights an owner can allow other people some rights over, and access to, property by defining certain lesser interests, such as leases and licenses (Law of Property Act 1925). However, the control that can be exercised by owners, even if they retain the entire bundle of rights, is regulated by the state in accordance with the broader public interest. Thus, for example, certain rights of access to land are reserved in law. These include public rights of way, easements for official uses, such as water pipes, electricity cables and the like (including reasonable access to them) and a generalised right for people to enter, and remain on, property for the purpose of conducting legitimate business.

The 'ownership' of water follows similar principles. The owner of the controlling interest in a section of riparian land owns, subject to existing public rights of way, the right to control access to the water but they do not actually 'own' the water itself as it passes over the submerged land which they do own. A series of subsidiary rights can be sold to third parties by an owner which can result in the establishment of separate markets for leisure use and leisure access. For example, an owner may sell a permit to fish which might imply a right to cast a line, hook or other device into the water and to extract the fish caught. However, it may or may not include an implied right to reach the waterside, which could be the subject of a separate negotiation or fee involving access over riparian land. Permits to fish can range from informal 'day tickets' to long-term tenancies, usually granted to angling clubs and syndicates. In some cases angling clubs have used their resources to buy riparian stretches of land giving them access to the water and fishing rights. An owner could, subject to market conditions, sell separate bundles of rights to both anglers and canoeists to use the same piece of river at the same time. In reality, they tend to sell rights to just one type of user, normally anglers. The arrangements any owner makes are underpinned by the common law concerning trespass which covers water in the same way as it does land. Individuals travelling along a privately owned watercourse without permission can be stopped by the owner of the riparian land. The exception is when a public right of navigation exists.

Public rights of navigation, however, are an area of some legal debate and consequently have been a topic of considerable contestation as canoeists have argued for more access to rivers. Foster (1985) notes that both judges and textbooks have often confused the legal and physical descriptions of navigations, and they have sometimes failed to differentiate between the different types of legal navigability. Thus, there is uncertainty constituting even the definition of what constitutes a 'navigable river' and the term 'navigation' (Foster 1985).

Rights of navigation can come about through a variety of means, leading Foster (1985: 449) to propose the following legal understandings in English law of the term 'navigation':

> Legally navigable rivers (navigable here referring to legal rather physical attributes) are either:
>
> – Navigable by statute (tidal or non-tidal)
>
> or
>
> – Navigable at common law:
> (a) tidal
> (b) non-tidal:
>
> (i) navigable through immemorial use
> (ii) navigable through express dedication
> (iii) navigable through implied dedication.

This means that there is a common law public right of navigation on tidal waters which may, in some cases, be confirmed or regulated by statute (Caffyn 2004). Telling and Smith (1985) explain this situation by describing such waters as 'arms of the sea'. On non-tidal waters there is no right of public navigation as the title of the river bed is prima facie vested with the river owner. A public right of navigation can derive either from statute, prescription or dedication. Statute can be used to create new rights of navigation or to confirm existing ones. Most of the major navigable rivers are regulated by statute where navigation authorities often have the right to levy tolls as a means of recouping their investment. The right of navigation, like a right of way over land, is simply a right of passage so that, no matter how often it is used, no rights in property are conferred to users and no additional rights can be inferred such as fishing or embarking and disembarking from the water (Gray 1993). The right of passage, according to Foster (1985), appears to involve commonly accepted movement along the water and the possibility of stopping for short breaks.

While Foster (1985) was seeking to bring some clarity to a complex legal area, considerable dispute remains. The legal basis for establishing a public right of navigation based on immemorial use is the fact that a right has enjoyed long-term use to the extent that a court could infer that there once was an actual grant of the right. Although in legal theory it is possible to establish immemorial use through proof of public use during living memory, such evidence is hard to sustain; this means that in many cases the precise extent of public rights of navigation remain in doubt. Dedication is even harder to identify precisely as it arises when there is evidence, very rarely available, that a landowner expressly or by implication has dedicated the river a highway. In such cases, acceptance by the public is indicated through subsequent use.

Given this legal background it is not surprising that there is still considerable debate over public rights of navigation. Recently, Caffyn (2004), a former voluntary British Canoe Union access officer, published the results of a very detailed analysis of the legal situation pertaining to a number of English rivers. This argues that the Magna Carta along with legal writings from the twelfth, thirteenth, fifteenth and seventeenth centuries can be used to argue that 'in common law there is a public right of navigation on all non-tidal rivers which are naturally physically navigable by small boats and on those rivers which have been made physically navigable at public expense' (Caffyn 2004: 7). Such a claim has yet to be subject to further legal scrutiny and would no doubt lead to considerable debate as to what constituted public expense.

Nevertheless, Caffyn (2004) serves to highlight the disagreements surrounding public rights of navigation. Currently, about 2,000 km of the major rivers in England are generally recognised as having public rights of navigation, which is only 13 per cent of the total resource of major rivers (University of Brighton 2001). Such a situation feeds the canoeists 'moral' claims for increased access to rivers. The disputes over the legal rights to navigation are part of wider attempts by both canoeists and anglers to mobilise the resource of the law in pursuit of their goals.

The British Canoe Union and many canoeists have the goal of increasing access to water through legislative change. The BCU's access policy (BCU 2003: 7) has the overall objective 'to secure legislation to improve the legal rights for canoeing' and in its recent campaigning documentation argues that 'legally protected access would provide clarity and certainty for those visiting our rivers'. One of the senior canoeing interviewees argued that, 'I don't care what anyone says, navigation law isn't clear cut, in fact it's a mess . . . there are rivers where some of us believe history is on our side and we should be allowed to paddle'. As this position would suggest, canoeists and their representative organisations seek to mobilise the law through a discourse stressing uncertainty.

In contrast, national bodies representing angling mobilise property and legal rights rather differently with a discourse stressing the certainty of public rights of navigation. The National Association of Fisheries and Angling Consultative (2003: 17) contended that:

> The law of access to inland waters for the use of navigation or other rights is simple and quite clear. Uncertainty and complexity are created by those who claim that public rights of navigation exist on waters where it is not generally

> accepted that such rights exist. The onus is on the claimant to prove that any other right of navigation exists. It may be hard to establish sufficient evidence to support such a case, but that is a different issue. Confusion has been created by the BCU's reference to 'disputed rights of navigation'; even the Environment Agency used this term for a while until angling interests objected that under law there was no such thing.

For angling bodies, legal rights over property are presented as clearly superior to any uncertainties over navigation. Furthermore, anglers stress that these rights allow a market-based system of payment to operate. The following is from an advert placed in a fishing magazine by the Salmon and Trout Association emphasising their desire to maintain the current system of rights.

> Access is negotiated on a voluntary basis in a system which has evolved over 150 years. It is also a system that works well because it protects all the participants – riparian owners, anglers and the environment. The same negotiating arrangements are available to canoeists but the principal reason why angling has been so successful in obtaining access is that we are prepared to pay.
>
> (*Hooked* 2001: 17)

This extract encapsulates the goal of angling organisations for the maintenance of the status quo regarding access to inland rivers. The law and property rights are mobilised to defend the current situation. An accompanying 'common-sense' discourse is used to bolster these claims by stressing the value of a market-based payment system that accompanies these rights and the social contribution of a voluntaristic organisational culture (Parker and Ravenscroft 2001).

As a result of the current arrangements for access to inland rivers, canoeists are legally excluded by rights holders from many rivers. Consequently, a considerable amount of the canoeing on some English rivers takes place in an informal manner where the boundary between legal permission and trespass is unclear. The outcome is a series of actions by canoeists that it would be tempting to interpret in terms of power as acts of resistance against the hegemony of landowners and anglers. Some canoeists have chosen to operate on rivers without clear permission, a process sometimes described as 'stealth' or 'bandit' canoeing. In addition, in order to highlight the perceived inequity of the current situation, groups of canoeists in the past have organised mass trespasses on rivers where access is restricted. Canoeists, however, have parodied a simplistic interpretation of these events as acts of resistance to hegemony, preferring to emphasise the way the current system of legal rights restricts canoeing activity. As Mark Rainsley (2002: 23) a leading campaigner for improved access wrote in an article in *Canoeist* magazine considering the access situation:

> Oh no: this article is concerned with banditing rivers! Clearly, it is written by some reckless yoof who wants to stir up trouble! Clearly, it is intended to incite

> folk to break the law. . . . Now that's cleared away the old clichés. Now let's think about the way that things really are. . . . Anyone who paddles in our country knows that a huge proportion of boating is done outside access agreements or rights of navigation, illegally, some would claim. Incredible as it seems, much of (most of?) the time when a paddler goes paddling on English rivers he or she is doing so without permission and possibly in contravention of the law.

The rest of the article goes on to provide guidelines for paddling 'without censure or abuse' on rivers where canoeing is not permitted. This could be seen as encouraging resistance but the quote above and the article as a whole stress that it is the legal system and property rights that are shaping the experience and spaces of canoeing in England and constraining resistive opportunities. The resources of the law and property rights are mobilised by anglers and landowners to develop discourses and exclusionary practices which give them a hegemonic control of many inland river spaces. As one of our canoeing interviewees noted:

> At the end of the day the law's the law and the fishing clubs have the upper hand . . . we can go on trying to get new access agreements but they always have their legal rights behind them.

The mobilisation of resources, however, is not some one-sided process in which one group simply commandeers resources for its benefit. Rather, it is subject to the complex interdependencies between the practices of anglers and canoeists. The power stemming from property rights and the law arises not so much from the possession of land but more from the ability to exclude others wishing to access property. If canoeists were not present, there would be limited value in mobilising property rights or the power to exclude. Ironically, anglers, and perhaps all land and resource owners, need the actions of the non-landed to legitimate their power. The demands and practices of the non-landed provide the justification for exclusionary behaviour. Anglers and landowners require an abuse of their property to justify the mobilisation of the defence of their resource. When the ownership of property is questioned then anglers and landowners mobilise the legal system and a series of accompanying discourses to justify their ownership of land and rights.

Indeed, this dependent aspect of the power relations between the landed and non-landed is enshrined in English law. Trespass is a civil offence and only repeated or aggravated trespass enables the mobilisation of court orders by rights owners. This legal situation partly explains the anomalous situation arising on England's rivers in that canoeists seek to claim legal rights through a change in legislation when many canoeists have in the past exercised these rights de facto through discrete use of rivers. In doing this many canoeists are aware that they effectively legitimate the recourse of anglers and landowners to the exclusionary power of property rights. The process of being excluded, however, is not without benefit in that it legitimates the canoeists' own power of 'moral' authority by enabling canoeing organisations to argue that while the law must protect rights it should not be used to exclude others. Furthermore, the conflicts over the issue of exclusion serve to divert attention from the issue of canoeists purchasing rights using a market system. The British Canoe Union's access policy (BCU 2003) argues that, while payment is appropriate for static activities, such as canoeing facilities, it is not relevant for passage along a river. Such an argument is hard to maintain when faced by the system of market payments developed for angling. A conflictual discourse focused on the exclusionary activities of rights owners is more suited to the canoeists' goal of accessing more river space through legislative change. The practices and relational interdependencies between anglers, landowners and canoeists make resource mobilisation a contingent and changeable process.

The state, property and access to inland rivers

The mobilisation of the law and property rights in relation to inland rivers is also contingent on the changing role of the state in regulating legal rights in the public interest. In England the Labour government has historically viewed improving access to open countryside on foot as being in the public interest, in part reflecting Labour Party members' involvement in mass trespasses in the 1930s to gain access to upland areas near urban working-class communities in northern England (Shoard 1999). In 2000 the Labour government introduced the Countryside Rights of Way Act (CROW) 2000, dubbed the 'Right to Roam Act', to extend statutory access on foot to open countryside defined as uplands, moorland, heathland and commons. In the run up to this legislation it was national government who decided that, despite the campaigns of canoeists, extending the act to inland waters was too problematic (Countryside Agency *et al.* 2000). In making this decision, central government chose to accept that private property rights should be maintained because fishing rights and private rights of access to riparian land were historical legal rights. Ironically, this decision was made at the same time as CROW was removing the rights of some landowners. Support for the government position had been expressed in a national government report, *The Salmon and Freshwater Fisheries Review* (MAFF 2000: 81) which concluded that:

> The ownership of fishing rights and private rights of access are long standing rights under English law. Any proposal to amend them would need to be considered in this wider context, not simply as a fisheries issue. Those bodies that wish to make water-based recreation more widely available need to take into account the legal rights of existing users such as anglers. Public funds should not be used to promote access to waterways at the expense of anglers.

More recent statements by political parties on the issue of access to water convey the same argument and do not support any changes to the legal system. The Labour Party (2005: 21) has recently launched a *Charter for Angling* which supports the view that 'managed solutions need to be sought to canoe access issues which do not adversely impact on existing users'. The Conservative Party (2004: 5) in a consultation paper *A Fair Deal for Country Sports* argued for 'ways of allowing greater canoe access to rivers while safeguarding fishing interests'. While such statements seek to maintain the legal status quo they also indicate that the campaigns of canoeists are not without influence since both major political parties have felt the need to present a view on access to inland rivers.

The government decision, however, to prioritise property rights and not extend access to inland rivers maintains the system whereby property rights are well suited to anglers who wish to negotiate access with single landowners to occupy a piece of land for a largely static activity. Some landowners generate only a small income out of this process but on desirable salmon rivers anglers may be willing to pay over £1,000 a week for exclusive use (Bingham 2005). Canoe clubs can use the same legal system to negotiate agreements with owners and pay for access for a short stretch of river next to a sporting facility. The process is highly problematic for those who wish to travel any distance along a river since it is necessary to negotiate with a large number of landowners usually on both banks of a river.

Overall, anglers and landowners have to date been successful in convincing government of the need to maintain the superiority of property rights over the 'moral' claims of canoeists. The mobilisation of property and the law by anglers and landowners to maintain their power and control over inland rivers has required a particular set of institutional arrangements that prioritise their property-based rights and a system of market payments. In capitalist societies private property rights and markets are embedded in structural principles which in this case interact with the institutional features of the state to shape the nature of leisure and tourism activities on inland rivers. Furthermore, as the conflict between users continues, the state has been highly influential in establishing the institutional framework within which action for future change can be considered without fundamentally threatening property rights. By not pursuing legislation to enhance access to inland rivers, national government has chosen to work within the current system of property rights and has sought to promote negotiated voluntary agreements.

'Negotiating' voluntary access to rivers

The concept of voluntary negotiated agreements to facilitate public access to private property was first mooted (in connection with public access to private land) in a series of parliamentary bills in the early twentieth century. These bills, designed to overcome the legal issues of granting access rights to individual people, proposed giving walkers limited rights of access to unenclosed uplands through regulated agreements devised by public authorities. They culminated in the Access to Mountains Act 1939 which, although never fully enacted, paved the way for the formal construction of access agreements, in the National Parks and Access to the Countryside Act (NPACA) 1949. This construction was later modified by the Countryside Act 1968, which now forms the legal basis for all voluntary management agreements relating to public interest in private land. In general terms after the Countryside Act, an access and management agreement is:

> a formal written agreement between a public authority and an owner of an interest in land . . . who thereby undertakes to manage the land in a specified manner in order to satisfy a particular public need, usually in return for some form of consideration.
>
> (Feist 1978: 7)

It is clear from this definition that such agreements were designed to facilitate access of the type that might reasonably have been expected if the land had been in the ownership of the negotiating public authority. Rather than enter into difficult and expensive negotiations to purchase or lease the land, the public authority was granted the power to 'buy' general access on behalf of the public. Also the central construct of the access agreement is less its voluntary nature than its ability to create a legal vehicle for granting access without the establishment of permanent legal rights.

Thus, despite ensuring that individual people cannot gain any permanent rights of access to the land, the intention of the acts was to create long-term binding agreements over the land through registration of a charge on the Land Charges Register to bind successive owners. In practice, most negotiated agreements had a term of 20 years, after which they were subject to renegotiation or removal from the register. Yet, although it was widely assumed that, under the 1968 Act, local councils and national park authorities would negotiate agreements with private landowners to allow the public to walk over fine but previously inaccessible countryside (Blunden and Curry 1990), this was not the case. By 1980, agreements had been negotiated successfully on just 34,000 ha of land, mainly in the Peak District National Park. These were mostly negotiated prior to 1960, since when Blunden and Curry (1990) suggest that local landowner attitudes to access in the Peak District have hardened with many reluctant to discuss access agreements because increased public access is perceived to generate a range of management problems.

Some progress towards gaining voluntary public access over private land was later achieved through various agri-environment schemes (see Ravenscroft 1995), although few of these agreements provided for the certainty and longevity envisaged by the NPACA 1949. The continued failure of the access agreement mechanism to achieve large-scale increases in public access to private land partly lay behind the Labour government's introduction of CROW to provide access to open countryside and registered commons.

Given the limited success of voluntary agreements on land, it is not surprising they have been problematic in relation to inland rivers. The British Canoe Union (2003) has negotiated large-scale voluntary access agreements to allow all 45,000 members the use of rivers where there is also a public right of navigation in England. This means that members do not pay individual dues to certain major navigation authorities, such as British Waterways. Instead a collective charge for use is paid from membership fees.

There are also in the region of 50 smaller agreements on rivers where there is no public right of navigation. These are mainly with single riparian owners and few are for river stretches in excess of 15 km. While most of the smaller agreements are in writing and include a prescribed code of conduct, there is little consistency between them and, in many cases, limited certainty over the conditions under which the agreement operates (Countryside Agency 2004). It is also apparent that few of the agreements are for long terms, with most being annual and involving time constraints and very small fees payable to the riparian owner. Only five agreements allow all-year use and over half are only open in the fishing closed season when anglers would not require access (University of Brighton 2001). Consequently, many of these smaller agreements are essentially informal and local, offering little opportunity for outsiders or those new to the sport to benefit. Typically, this type of access agreement consists of a series of letters between the British Canoe Union, landowners and rights holders agreeing to certain types and times of use. This is then conveyed to members in the form of an annotated map and details of the restrictions and booking requirements. There is often no indication of whether the agreement remains in force, nor for how long the parties have negotiated access. In some cases landowners have withdrawn from the agreement via a solicitor's letter; the British Canoe Union has then been left to contact each affiliated club to inform them (University of Brighton 2001).

In part, this informality reflects the temporary nature of these agreements but it is also connected to the British Canoe Union's lack of resources for tackling negotiation professionally as it is reliant on local volunteers to undertake negotiations on its behalf. This leads to very uneven use of the rivers available to canoeists, with many land and riparian owners prepared only to make agreements with one person or club. Indeed clubs, in a bid to protect such arrangements, are often reluctant to tell other clubs about the access that they have negotiated and effectively exclude canoeists who aren't members of the local club. One canoeing interviewee highlighted such a situation in his own local club stating that,

> over the last thirty years I've gone round talking to farmers, landowners making sure they are happy for us to be on the river . . . they wouldn't be happy with other clubs turning up and I'll tell you they do turn up.

This typifies that dilemma of canoeists seeking to avoid their own version of the 'tragedy of the commons'. If legislative change were to occur, it could make rivers common property which might undermine some of the canoeists' own controlling strategies which use club and British Canoe Union membership to promote certain forms of behaviour and to manage levels of usage. As befits a member organisation such as the British Canoe Union, most current voluntary agreements are limited to its members, are not widely advertised and thus fail to address many of the broader problems related to access to inland rivers for canoeing. This situation highlights the paradox of negotiating voluntary agreements in that they can involve landowners, anglers and canoeists making arrangements that effectively exclude certain other types of canoeists whose behaviour and usage is not easily policed or managed by the informal features of many agreements.

Furthermore, under the current legal system the agreements themselves remain extremely difficult to develop, implement and sustain. So rather than reflect on the technical issues involved in negotiating and managing access agreements, canoeing interests have been strong in their criticism of the very concept of negotiated voluntary agreements; they argue that the problems lie not so much in their implementation, but in the very unevenness of property rights that have caused them to be required at all. One long-standing campaigner for access improvements noted that:

> Access arrangements are negotiated by BCU or WCA access officers. These people are unpaid volunteers who have a thankless and often impossible task, operating against feudal laws, deeply selfish interests and general indifference to achieve access arrangements on our rivers. In most instances, and it is hardly surprising, given the opposition, they fail to achieve worthwhile access to their nominated river.
>
> (Rainsley 2002: 23)

Clearly, anglers and landowners can use voluntary agreements at the local level as a potential way of controlling canoeing activity. For example, angling clubs have informally traded managed canoe access on one river for non-use of another (Fisher 2003). While the context for voluntary access agreements to rivers has been established by the interdependencies between the practices of canoeists and anglers, it is also the case that national government's prioritisation of property rights relating to inland rivers significantly shapes how the agreements develop. The resulting agreements are not the broad legal agreements envisaged in the NPACA (1949) and set out by Feist (1978). Rather, as the House of Commons Select Committee on Environment, Transport and Regional Affairs (2001) has suggested, they are in many ways the diametric opposite, as more informal and much narrower in scope, in terms of their geographical and social dimensions. The institutional arrangements of the state have asserted private property rights in relation to inland rivers. As a result, voluntary agreements concerning inland rivers despite the best attempts of canoeists and their representative bodies have developed in a particular manner not envisaged by legislation promoting such agreements as a way of improving public access to the countryside. The result is that to date voluntary agreements have made little difference to the situation although attempts to develop them further have been progressed by government (Countryside Agency 2004).

For the time being, the current legal and political frameworks also mean that the British Canoe Union, which receives government money to fund elite aspects of the sport, has little choice but to support attempts to develop voluntary agreements as an approach to increasing access. The BCU's current access policy as well as seeking legislative change also states that it 'will continue to implement the course of action, recommended by Governments since 1992, that canoeists should work to achieve this on a local basis through voluntary agreements with riparian landowners and other water users' (BCU 2003: 7). This paradox of supporting large-scale legislative change alongside incremental agreements is reminiscent of the dilemma encountered by campaigners for increased access to the British countryside. Shoard (1999) compares the partialist position which aims to increase access through incremental steps with the universalist approach which argues for a universal right of access. Shoard (1999) notes that universalists will often support incremental improvements if they do not conflict with their ultimate goal of universal access. The situation is made all the more complex in respect of inland rivers in that alongside the expressed universalist and partialist viewpoints many canoeists and clubs simply work within a situation of customary access negotiated informally at the local level. This provides an extra dimension since, as noted above, such informality may provide new river spaces for some canoeists while effectively excluding others.

Discussion and conclusion

The legal and property rights that are at the heart of the conflict between canoeists, anglers and landowners emphasise the need for a greater emphasis on power within tourism and leisure conflict studies. The differing ways anglers, landowners and canoeists mobilise the law and property rights along with their contrasting 'moral' claims clearly indicate that the conflict cannot be readily understood by the more traditional approaches to conflicts which emphasise either social values or a social psychological goal interference perspective. Indeed, some of the actors involved in the conflict on inland rivers clearly perceive their role in terms of changing power relations. The Salmon and Trout Association (2003: 1) in discussing its position noted that:

> we are taking a more positive role in trying to divert potential problems with other river users, rather than insisting on the right of anglers to dictate what happens on our waterways. As much as we might like that to be the case, political reality required us to be more diplomatic.

The empirical analysis reveals that the processes by which power is concentrated are contingent and involve a negotiated choreography of resource mobilisation. Anglers whose activities involve a series of market-related payments are also river access rights holders, and they appear in a superior situation compared to the canoeists for mobilising property rights and the law. Private property rights would constitute a dimension of allocative resources identified by Giddens (1984) whereas market exchange would be part of authoritative resources. This combination of both allocative and authoritative resources *potentially* provides anglers and landowners with considerable power to determine access to inland rivers. This power rests not on the simple ownership of rights but on the power to exclude others, in this case canoeists, who desire access to property. Mobilising the potential to exclude requires the canoeists to abuse, or at least appear to abuse, the property of rights owners. This is broadly in keeping with Allen's (2003) argument that resources do not simply equate with power and they will only be mobilised in specific situations. Anglers and landowners feeling their ownership of property is threatened seek to mobilise the legal system and a series of discourses concerning market payments and voluntary organisation to justify their ownership of land and rights. Canoeists themselves can use the resulting exclusion to mobilise their own power resources based on a 'moral' claim discourse which stresses that the legal system should be used to protect property rights not exclude others. These 'moral' claims have limited influence on access to inland rivers compared to the property rights of anglers and landowners partly because currently the state interprets property law in a manner that prioritises property rights and a market-based system of establishing access to rivers. Thus resource mobilisation is not an asymmetrical assertion of rights but a process that is rooted in the interdependencies between actors as well as the actions of state institutions.

The situation that arises both constrains and enables canoeists in their use of certain other resources, such as their own acts of resistance. Mass trespasses designed to highlight the exclusionary behaviour of anglers have in the past raised the public profile of the arguments of canoeists (Rainsley 2002). The British Canoe Union, however, as a governing body feels it cannot condone illegal actions and has been accused by the canoeing media of being more interested in maintaining membership levels and subscription income than tackling the access problem (Fisher 2003). Angling organisations are always ready to remind canoeists of the potentially negative effects of their actions on rights holders. The director of the Salmon and Trout organisation argued 'canoes have the ability to wreck a day's fishing over many miles of river, if their crews chose to do so' (Knight 2003: 12). Again the interdependencies surrounding resource mobilisation become apparent. The potential resources of canoeists involving practice-based resistance can be manipulated by the anglers to become part of the discourse that anglers develop to strengthen their own position. Such a situation is not uncommon in power relations as a subject's power contains resistive and empowering possibilities but is often a resource that can be mobilised by others to create power over that subject. As Allen (2003: 52) argues 'the "power to" achieve is made to serve the power held "over" others. The circle is closed in an altogether familiar way'. The case study of inland rivers in England indicates that, while researching the degree to which power in the context of tourism takes the form of power over and/or power to, it is also imperative to explore the interconnections between these different forms of power as they co-construct each other.

In their attempt to change the situation on access to inland rivers, the organisations that represent canoeing are faced with a number of dilemmas. They find themselves in an invidious position; whatever action they take appears to confirm the power of the property rights of anglers. For example, support for the government's preferred partialist approach of promoting voluntary agreements may be interpreted as a rejection of the universalists' claim that canoeists should have universal access to rivers. In addition, canoeing clubs that are willing to operate at the local level with informal customary access arrangements can end up working with anglers and landowners to exclude other canoeists whose behaviour does not adhere to an informal, self-regulated system.

The outcome is that the situation regarding access to inland rivers, where there is currently no public right of navigation, can best be described as a gift relationship (Parker and Ravenscroft 2001) in which those holding rights 'gift' access to others while in no way weakening their right. Such a relationship does not constitute a power modality of domination if one accepts Lukes's (2005: 118) argument that domination is when 'the power of some affects the interest of others by restricting their capabilities for truly human functioning'. There are, of course, other ways of defining the modality of domination (Stewart 2001). The gift relationship, however, contributes to a quite specific situation that would best be encapsulated within a power modality of dependent authority. The anglers and landowners significantly constrain canoeists but their authority in part depends on the actions of canoeists in order to legitimise the use of their property rights and the legal system. This socio-spatial process of resource mobilisation interacts with the structural principles of a neo-liberal society and the history of state institutions which prioritise the maintenance of the legal arrangements and market exchanges that have built up around these property rights. As in many areas of social life property rights are centrally implicated in negotiations relating to power.

Mann (1986) reminds us that power relations creating authority can suddenly change. A time may come in England when the differing practices and discourses encourage the state to decide that a weakening of private property rights regarding inland rivers may favour the public interest. In Scotland the river access situation has recently changed significantly compared to England, as under the Land Reform (Scotland) Act 2003 customary freedoms regarding access to the countryside are now recognised in law. In many countries, however, increased tourism and leisure demands are actually seeing renewed attempts to clarify and strengthen different forms of property rights (Kaltenborn *et al.* 2001). The case of inland rivers in England indicates the value for tourism and leisure research of understanding the role of contingent and negotiated resource mobilisation in order to reveal how power relations develop and concentrate power so that certain groups involved in tourism exert influence over others.

References

Adelman, B., Heberlein, T. and Bonnicksen, T. (1982) 'Social psychological explanations for the persistence of conflict between paddling canoeists and motor craft users in the Boundary Water Canoe Area', *Leisure Sciences*, 5: 45–62.

Aitchison, C.C. (2003) *Gender and Leisure: Social and Cultural Perspectives*. London: Routledge.

Allen, J. (2003) *The Lost Geographies of Power*. Oxford: Blackwell.

Angling and Canoeing Liaison Group (ACLG) (1999) *Agreeing on Access*. Bristol: Environment Agency.

Becker, L. (1977) *Property Rights: Philosophic Foundations*. London: Routledge.

Bingham, D. (2005) 'Property watch', *Trout and Salmon*, December: 63–64.

Blomley, N. (1994) *Law, Space and the Geographies of Power*. New York: Guilford Press.

—— (2005) 'Remember property?', *Progress in Human Geography*, 29(2): 126–129.

Blunden, J. and Curry, N. (1990) *A People's Charter?* London: HMSO.

British Canoe Union (BCU) (2003) *Access Strategy: England*. Nottingham: BCU.

—— (2005) *Access to Rivers Campaign*. Available at: www.bcu.org.uk [last accessed 15 February 2006].

Caffyn, D. (2004) *The Right of Navigation on Non-tidal Rivers and the Common Law*. Eastbourne: privately published.

Carothers, P., Vaske, J.J. and Donnelly, M.P. (2001) 'Social values versus interpersonal conflict among hikers and mountain bikers', *Leisure Sciences*, 23: 47–61.

Conservative Party (2004) *A Fair Deal for Country Sports*. London: Conservative Party.

Countryside Agency (2004) *A Feasibility Study on Improving Access for Canoeing by Voluntary Agreement*. Research Notes, CRN 79. Cheltenham: Countryside Agency.

Countryside Agency, Forestry Commission, Environment Agency and English Nature (2000) *Improving Access to Woods, Watersides and the Coast. A Joint Report to Government on the Options for Change*. Publication CA 33. Cheltenham: Countryside Agency.

Curry, N. (2001) 'Rights of access to land for outdoor recreation in New Zealand: dilemmas concerning justice and equity', *Journal of Rural Studies*, 17: 409–419.

Denman, D. (1978) *The Place of Property*. Berkhamstead: Geo Books.

Driver, B.L. and Bassett, J. (1975) 'Defining conflict among river users: a case-study of Michigan's Au Sable River', *Naturalist*, 26: 19–23.

Feist, M. (1978) *A Study of Management Agreements*. Publication CCP114. Cheltenham: Countryside Commission.

Few, R. (2002) 'Researching actor power: analyzing mechanisms of interaction in negotiations over space', *Area*, 34(1): 29–38.

Fisher, S. (2003) 'Editorial', *Canoeist*, August: 5.

Foster, S.E. (1985) 'Inland waterways: legal aspects of public access and enjoyment', *Journal of Planning and Environment Law*, July: 440–460.

Franklin A. (2002) *Nature and Social Theory*. London: Sage.

—— (2004) 'Tourism as an ordering: towards a new ontology of tourism', *Tourist Studies*, 4(3): 277–301.

Funck, C. (2002) 'Contested landscapes of marine sports', *International Institute for Asian Studies Newsletter*, 28: 9–10.

Giddens, A. (1984) *The Constitution of Society*. Cambridge: Polity.

—— (1993) *The Giddens Reader*. London: Macmillan.

Gramann, J.H. and Burdge, R. (1981) 'The effect of recreation goals on conflict perception: the case of water skiers and fishermen', *Journal of Leisure Research*, 13: 15–27.

Gray, K. (1993) *Elements of Land Law*, secnd edition. London: Butterworths.

Hall, C.M. (2003) 'Politics and place: an analysis of power in tourism communities', in S. Singh, D.J. Timothy and R.K. Dowling (eds) *Tourism in Destination Communities*. Wallingford: CAB International.

Haugaard, M. (1997) *The Constitution of Power*. Manchester: Manchester University Press.

—— (ed.) (2002) *Power: A Reader*. Manchester: Manchester University Press.

Higgins-Desboilles, F. (2004) 'Taming tourism: indigenous rights as a check to unbridled tourism', conference paper at Tourism, Politics and Democracy, Centre for Tourism Policy Studies, Third International Symposium, University of Brighton, Eastbourne, 9 September.

Hooked (2001) 'No right to roam', *Hooked Magazine*, May: 17.

House of Commons Select Committee on Environment, Transport and Regional Affairs (2001) *Fourth Report: Inland Waterways*. London: House of Commons Papers.

Inland Waterways Amenity Advisory Council (1999) *Britain's Inland Waterways: An Undervalued Asset*. Rickmansworth: Inland Waterways Amenity Advisory Council.

Jackson, E.L. and Wong, R. (1982) 'Perceived conflict between urban cross-country skiers and snowmobilers in Alberta', *Journal of Leisure Research*, 14: 47–62.

Jacob, G.R. and Schreyer, R. (1980) 'Conflict in outdoor recreation: a theoretical perspective', *Journal of Leisure Research*, 12: 368–380.

Kaltenborn, B.P., Haaland, H. and Sandell, S. (2001) 'The public right to access – some challenges to sustainable tourism development in Scandinavia', *Journal of Sustainable Tourism*, 9(5): 417–433.

Knight, P. (2003) 'A question of access', *Salmon and Trout*, February: 12.

Labour Party (2005) *Labour's Charter for Angling 2005*. London: Labour Party.

Latour, B. (1986) 'The powers of association', in J. Law (ed.) *Power, Action and Belief: A New Sociology of Knowledge*. London: Routledge.

Lukes, S. (1982) 'Panoptikon: Macht and Herrschaft bei Weber, Marx, Foucault', *Kursbuch*, 70: 135–148.

—— (2005) *Power: A Radical View*, second edition. Basingstoke: Palgrave Macmillan.

McIntyre, N., Jenkins, J. and Booth, K. (2001) 'Global influences on access: the changing face of access to public conservation lands in New Zealand', *Journal of Sustainable Tourism*, 9(5): 434–450.

Mann, M. (1986) *The Sources of Social Power. Volume 1. A History of Power from the Beginning to AD1760*. New York: Cambridge University Press.

—— (1993) *The Sources of Social Power. Volume 2. The Rise of Classes and Nation-states, 1760–1914*. New York: Cambridge University Press.

Mikolic, J.M., Parker, J.C. and Pruitt, D.G. (1997) 'Escalation in response to persistent annoyance', *Journal of Personality and Social Psychology*, 72: 151–163.

Ministry of Agriculture, Fisheries and Food (MAFF) (2000) *The Salmon and Freshwater Fisheries Review*. London: MAFF.

Mintel (1998) *Activity Holidays. Mintel Marketing Intelligence*. London: Mintel International.

Mongeau, D. (2001) 'Public access to the beach: a selective bibliography', *Reference Services Review*, 29(1): 81–88.

—— (2003) 'Coastal access and public policy: a selective annotated bibliography', *Reference Services Review*, 31(4): 374–384.

Munton, R. (1994) 'Rural accumulation and property rights: sustaining the means', paper given at the Institute of British Geographers' Conference, University of Nottingham, January.

National Association of Fisheries and Angling Consultatives (2003) *Water-based Sport and Recreation. The Facts: A Response*. Woking: NAFAC.

O'Riordan, T. and Paget, G. (1978) *Sharing Rivers and Canals. A Study of the Views of Coarse Anglers and Boat Users on Selected Waterways*, Study 16. London: Sports Council.

Owens, P.L. (1985) 'Conflict as a social interaction process in environment and behaviour research: the example of leisure and recreation research', *Journal of Environmental Psychology*, 5: 243–259.

Parker, G. (2002) *Citizenships, Contingency and the Countryside*. London: Routledge.

Parker, G. and Ravenscroft, N. (2001) 'Land, rights and the gift: the Countryside and Rights of Way Act 2000 and the negotiation of citizenship', *Sociologia Ruralis*, 41(4): 381–398.

Rainsley, M. (2002) 'The way it is – access to English and Welsh rivers in practice', *Canoeist*, June: 23.

Ravenscroft, N. (1995) 'Recreational access to the countryside of England and Wales: popular leisure as the legitimation of private property', *Journal of Property Research*, 12(1): 63–74.

Ravenscroft, N., Church, A. and Gilchrist, P. (2005) 'The ontology of exclusion: A European perspective on leisure constraints research', in E.L. Jackson (ed.) *Constraints to Leisure*. State College, PA: Venture Publishing.

Ravenscroft, N., Uzzell, D. and Leach, R. (2002) 'Danger ahead? The impact of fear of crime on people's recreational use of nonmotorised shared-use routes', *Environment and Planning C: Government and Policy*, 20(5): 741–756.

Ruddell, E.J. and Gramann, J.H. (1994) 'Goal orientation, norms and noise-induced conflict among recreation area users', *Leisure Sciences*, 16: 93–104.

Salmon and Trout Association (2003) 'Access to river banks'. Archived news March–April 2003. Available at: www.salmon-trout.org/news_archive [last accessed December 2005].

Schneider, I.E. and Hammitt, W.E. (1995) 'Visitor response to outdoor recreation conflict: a conceptual approach', *Leisure Sciences*, 17: 223–234.

Sharp, J.P., Routledge, P., Philo, C. and Padison, R. (2000) 'Entanglements of power: geographies of domination/resistance', in J. Sharp, P. Routledge, C. Philo and R. Padison (eds) *Entanglements of Power: Geographies of Domination/Resistance*. London: Routledge.

Shaw, G. and Williams, A. (2004) *Tourism and Tourism Spaces*. London: Sage.

Shoard, M. (1999) *A Right to Roam*. London: Grafton.

Sports Council (1985) *Time for a Change? Managing the Public Right of Navigation. A Paper for Consultation*. London: Sports Council.

Stewart, A. (2001) *Theories of Power and Domination*. London: Sage.

Suls, J., Martin, R. and David, J.P. (1998) 'Person–environment fit and its limits', *Personality and Social Psychology Bulletin*, 21: 88–98.

Telling, A. and Smith, R. (1985) *The Public Right of Navigation. A Report to the Sports Council and the Water Space Amenity Commission*, Study 27. London: Sports Council.

University of Brighton (2001) *Water-based Sport and Recreation: The Facts*. Brighton: School of the Environment, University of Brighton. (Report to Department for Environment, Food and Rural Affairs.)

—— (2004) *A Feasibility Study on Improving Access for Canoeing by Voluntary Agreement*. Cheltenham: Countryside Agency.

Urry, J. (2002) *The Tourist Gaze*. London: Sage.

—— (2003) *Global Complexity*. Cambridge: Polity.

Vaske, J.J., Donnelly, M.P., Wittman, K. and Laidlaw, S. (1995) 'Interpersonal versus social values conflict', *Leisure Sciences*, 17: 205–222.

Part III

Power, governance and empowerment

9 Empowerment and stakeholder participation in tourism destination communities

Dallen J. Timothy

Beginning in the mid-twentieth century, development and planning specialists began to realize that development efforts, including tourism, rarely had the desired results that programme initiators expected (McIntyre *et al.* 1993; Stiefel and Wolfe 1994). In recognition of the failure of traditional, top-down approaches to economic development, observers of, and participants in, community development efforts have strongly advocated an approach that involves as many stakeholders as possible in the communities where development occurs. The failure of the majority of development projects was attributed to the fact that most potential beneficiaries of these programmes were routinely excluded from the development process (Botchway 2001; Dobbs and Moore 2002; Friedmann 1996; Watt *et al.* 2000). Since the 1970s, debate has intensified with respect to who should participate, how various stakeholders might participate, and to what extent community members ought to be involved in development initiatives. In response, calls have been made throughout the world for a decentralization of decision-making power in all areas of development.

One possible solution to many of the ills associated with tourism's growth, development specialists argued, might be to devolve power to the people locally on the ground, including primarily, but not exclusively, community residents and other destination stakeholders. Such grassroots involvement in, and control of, development initiatives would be better positioned to support the principles and goals of sustainable development in tourism, such as social equity, holistic growth, harmony, balance, and preservation of ecological and cultural integrity (Davidson and Maitland 1999; Gow and Vansant 1983; Hatton 1999; Simmons 1994; Timothy and Tosun 2003). This process of transferring power to communities and community stakeholders has become popularly known as 'empowerment'. This chapter examines the phenomenon of community empowerment in the realm of tourism by highlighting various levels of empowerment and the processes involved, as well as scales and types of empowerment as they pertain to the growth and development of tourism.

Community empowerment and participatory development

Empowerment is both a condition (a capacity) and a process. Being empowered is a condition whereby authority to act, choice of actions, and control over decisions and resources lie in the hands of destination community members rather than central government authorities, and sometimes from multinational companies and external investors. The word 'empowerment' is suggestive of being in (relative) power and it refers to a process whereby a transfer of control (or devolution of power) to individuals and/or communities takes place (Brown 2002; Chambers 1993; Laverack 2001; Laverack and Wallerstein 2001; Lincoln *et al.* 2002; Lyons *et al.* 2001). In effect one individual, group or association has been authorised or enabled to exercise power by the concession of power by another. The concession of power may or may not be willing, and the empowerment of the individual, group or association implies that the capacity of power is applied to a particular end. The most common contexts associated with community empowerment relate to health, gender studies, technology, fisheries, and forest management (Laverack and Wallerstein 2001; Ngwa and Fonjong 2002; Okoji 2001; Pinkett and O'Bryant 2003), although the concept is beginning to be utilized more commonly by tourism scholars to denote various types and scales of community participation in the tourism development process (cf. Scheyvens 2002; Sofield 2003; Telfer 2003; Tosun and Timothy 2003).

At the core of empowerment is a variety of power relations between those with authority and those lacking it. The very process of empowerment can only take place – and hence is observable – through the exercise of power in one form or another, for someone or something must possess enough authority to be able to bestow power to organizations and individuals (Friedmann 1996; Lyons *et al.* 2001; O'Neal and O'Neal 2003). As well, authority comes in many different forms, with knowledge often being the core element in power relations (Allen 2003; Timothy 2000). Thus, empowerment is a relational concept, although many geographers do not use the word empowerment in their discussions of relational authority (cf. Allen 2003).

While the primary purpose of Arnstein's (1969) 'ladder of participation' was to describe different types of public participation in regional planning and development, it also suggests implicitly that empowerment is not a static or fixed condition. Rather, it is a dynamic process that can lead to higher levels or losses of agency (disempowerment). Agency, or authority, increases as central hierarchies deregulate, promote grassroots efforts and take on a laissez-faire attitude toward development. On the other hand, disempowerment can grow when governments, in an effort to lend a helping hand, become overpowering in legislation, policy and planning. Similary, disempowerment may occur at the hands of private, multinational corporations, when local control of resources and services is superseded by the heavy control of external investors. Once this occurs, destination residents have little control over their own tourism resources and industry development.

In addition to the devolution of power and according to popular research and commentary, participatory development and the idea of empowerment go beyond beneficiary participation, to include issues of citizenship, rights and responsibilities (Cornwall 2003). For example, responsibilities increase when stewardship of resources is decentralized to the community level. The community and its members then have the right to preserve, conserve and utilize, but by the same token they are responsible for failures or successes. The growth of this line of thinking goes even beyond recent considerations of empowerment to suggest that people should not only be involved in and own development decision making, but that community stakeholders have obligations as citizens to become more involved and accountable for the successes and failures of growth initiatives (Osborne 1994).

Degrees of empowerment

A number of authors have attempted over the years to develop various typologies of empowerment and delineate various degrees of empowerment. In most cases these range from complete disempowerment at one end of the spectrum to complete empowerment at the opposite end. In between these two extremes, many sub-levels of empowerment have been described (cf. Arnstein 1969; Choguill 1996). While most of the extant typologies attempt to demonstrate various levels of empowerment, they also suggest implicitly that tourist destinations can pass through a dynamic pathway of power transfer. The following four degrees of empowerment – imposed development, tokenistic involvement, meaningful participation and empowerment – reflect the perspective of destination communities primarily in developing countries rather than the perspective of the tourists. They represent a consolidation of the empowerment and development literatures, and they are highly instructive in so far as they can reflect conditions in specific destinations or a process leading to full empowerment.

Imposed development

This form of tourism growth is a uni-directional, top-down approach that is conceived, planned and carried out by central authorities, who suggest that what they are doing is in the best interests of the destination community (cf. Lukes's third dimension of power, see Coles and Church, Coles and Scherle, Hall – chapters 1, 10 and 11 – in this volume). In fact, imposed development would appear not to be a form of empowerment at all, but rather, the starting point in the evolutionary process of power transference, once the problems associated with it are acknowledged. At this level government agencies rarely, if ever, involve community members in decision making and often exclude them from benefiting in many ways from the growth of tourism. Public agencies typically reject any input by community members, especially various marginal peoples (e.g. ethnic and racial minorities, poor people, women), who are sometimes seen as an embarrassment or a hassle to deal with (Choguill 1996; Jenkins 1980; Johnston 2003; Scheyvens 2002; Tosun 1998). This approach to development rarely succeeds according to contemporary definitions of (successful) sustainable tourism.

The best examples of imposed development were found in the former state socialist countries of Eastern Europe, and today exist in socialist states such as China (Lew, Chapter 7 in this volume) and North Korea. While China is rapidly developing socio-economically in a dualist society, its citizens have not yet become accustomed to participatory development in tourism or other sectors, although they have many opportunities now to work in the industry. Although many people are working in large hotels, restaurants and tourist attractions, there are also many people involved in small-scale entrepreneurial ventures and small business ownership in both rural and urban contexts. This form of involvement with the benefits of tourism is perhaps the most significant of the new developments related to reform in China. In North Korea, citizen empowerment – even of this rather restricted nature – is by and large completely absent, and few people have opportunities to seek employment in tourism. Perhaps not to the extent of North Korea, the planning approaches utilized in several less-developed countries of Africa and Asia still fall within this category or first stage of the empowerment process. For instance, as tourism development occurred rapidly in Indonesia during the 1980s and 1990s, several examples of imposed development were evident, such as the Prambanan Temple complex in Central Java of ancient Hindu temples and god statues. The Indonesian government saw the potential of developing the site into a major tourist attraction, although in the centuries following the demise of the temple, a village had been established in and around the ruins. Rather than consulting with villagers about their concerns, fears or desires for tourism development, the national government imposed its wishes for the area to grow into a major destination. In so doing the villagers were forced out of their homes and removed to the east end of the island, far away from their ancestral homelands. They were not well compensated (financially) for their displacement, and this has been a major point of contention in Indonesia since it occured in the 1970s and 1980s (Timothy 1996).

Clearly, one of the principal issues surrounding imposed development is that it often results in negative social, cultural and ecological outcomes and unbalanced economies. All too often, it denies indigenous people access to their own traditional natural resources, as tourist resort managers – with the support of government leaders – forbid resident right of entry to beaches and natural areas (Faust and Smardon 2001).

Tokenistic involvement

Many destination governments have recently started to involve community stakeholders and members at large in planning future trajectories for tourism. Although more inclusive than imposed development, this usually takes on the form of agencies consulting with community members simply in an effort to erect a façade of their empowering the community in order to pay due homage to contemporary global demands to plan at the grassroots level (Castro and Nielsen 2001; Dobbs and Moore 2002).

This, too, is hardly a form of empowerment; rather, it simply resembles a gesture to satisfy external observers' demands for grassroots participation and to a lesser extent minimize dissent among community members. This 'lip service' (Brown 2002) often takes on the form of holding occasional meetings with destination residents to inform them about plans, development progress and the recruitment of people in predetermined ventures. Rarely, though, does it allow them to question the approaches to, and degrees of, development (Arnstein 1969; Botchway 2001; Choguill 1996; Cornwall 2003).

Unfortunately, this is still the most common level of participation and so-called 'empowerment' in the developing world (Tosun 2000; Tosun and Timothy 2003). Examples can be found throughout much of Asia, Africa and Latin America where planning is carried out by government leaders whose role it is to 'inform' the destination communities of how tourism will grow (Timothy 1999, 2002). A particularly interesting example of this is provided by Timothy and Wall (1997) in the context of Yogyakarta (Indonesia) where a top-down approach to tourism planning is still the norm. In their exercises to develop tourism in the city of Yogyakarta during the 1990s, the national and provincial governments were anxiously pushing for development. However, due to pressures from outside observers (i.e. consultants and researchers), one of their goals was to consult with destination residents and service providers in the planning process. In common with so many places in the developing world, this process was undertaken. But it was purely a token gesture to demonstrate an apparent compliance with growing global standards. Government officials interviewed a small portion of street vendors and small guesthouse owners in the city for their opinions, hopes and desires for tourism. According to those whom they interviewed, it was all 'talk, talk, talk', never amounting to any real or substantive changes in policies or practices.

Meaningful participation

As a third stage, meaningful participation begins to appear when governments and other development agents sincerely desire to involve local residents and business owners in decision-making processes as well as the other economic benefits of tourism development. While government control still exists at this level, its hegemony is no longer as absolute. Partnerships between public agencies, community groups, destination residents and other stakeholders are operationalized. Successful partnerships at this level entail community members and external officials sharing decision-making responsibilities (Arnstein 1969; Choguill 1996), and meaningful dialogue is sought to find shared solutions to real-life situations (Campbell and Marshall 2000).

This is perhaps the most common level of empowerment in the developed world, and there are many examples from the developing world as well. For instance, on the Peruvian island of Taquile residents have worked together to share development ideas and they have benefited relatively equally from the growth of tourism. Nonetheless, power struggles still exist between community groups, government leaders and other outside business interests (Mitchell and Reid 2001). A similar example is that of the Inca-controlled tourism development on Lake Titicaca in Peru and Bolivia (Healy and Zorn 1983).

Notwithstanding, the pattern of development and political participation still follows the same linear (top-down) pathway. Governments still formulate the solutions and then seek endorsement by the destination community. This form of participation begins to resemble a degree of empowerment, but it is not enough simply for participants to be heard. Input of opinions alone is not empowerment of itself, nor desirable in this or other contexts. In the words of Lincoln *et al.* (2002: 285),

> Individuals or groups that do not perceive that real power has been delegated are not empowered. They may hear the words, but when they see that the behaviour is not consistent with the words, they rarely believe that empowerment has occurred.

Empowerment

At this level, community members are able to (i.e. authorized) initiate their own development goals and programmes. This does not, of course, preclude assistance, intervention or encouragement from government leaders or NGOs, but any external involvement should be limited to the role of facilitator rather than decision maker (Logan and Moseley 2002: 7). This level of empowerment suggests that, when communities (however they are variously defined) take responsibility for their own tourism industries, they have ownership of both problems and solutions, such that development should become attainable and sustainable (Botchway 2001; Cornwall 2003; Motteux *et al.* 1999; Simpson *et al.* 2003).

Only a few destinations have achieved this degree of empowerment in tourism, and they are typically in the developed world. An important exception to this can be found in Belize, where the Toledo Ecotourism Association (TEA) was conceived, initiated and implemented by residents of indigenous Maya villages in the southernmost region. The TEA, which provides indigenous village and ecotourism experiences, has found considerable success in preventing outside control, and many community services have improved as a result of the economic gains achieved through tourism (Timothy and White 1999). Outside control in this case was avoided by forcing out national government efforts to establish other competitive guesthouses in the villages and to regulate the local indigenous people's efforts at developing tourism products and services. Other important examples have recently been noted among the Kuna Indians of Costa Rica and among the Maasai people of East Africa (Baez 1996; Berger 1996; Thomas 1995).

One final point here is to note that it is crucial to recognize that these four degrees of empowerment may not always be so clear-cut in their operation and their outcomes. There may be examples that cut across these four degrees with development involvement being tokenistic for some groups (e.g. native Africans) but empowering for others (e.g. whites in South Africa) (Goudie *et al.* 1999). Likewise, according to Mowforth and Munt (1998), it may be that the empowerment of tourists sometimes results in disempowered destination communities when the desires and demands of the outsiders take precedence over local needs and aspirations.

Empowerment may function at and inbetween different geographical scales, and it is to this issue that this chapter now turns.

Scales and forms of empowerment

Empowerment has traditionally been scrutinized on three different scales: national, local/community and personal (Lincoln *et al.* 2002). National empowerment occurs when countries gain independence from (erstwhile) colonial powers or when oppressive majority (or minority) rule is overturned or abolished in some other way. These types of events provide opportunities for minority groups or otherwise marginalized people to have a voice in decision making for tourism and in other areas of development. Community-level empowerment refers more specifically to the collective well-being of the local community where tourism takes place (Campbell and Marshall 2000). At the level of the individual, empowerment should reflect an awareness of self-worth and create some kind of inner transformation (Wilson 1996). This perspective focuses on the rights of individual people to be able to express themselves and pursue their own self-interests (Campbell and Marshall 2000: 324), which nearly always forms the foundation of community and national empowerment.

Despite this tripartite traditional view of the scales of empowerment, it is important to note that empowerment does not always function in such a straightforward, clear manner. Tourism empowerment strategies can be easily mediated through multi-scalar networks and across scales (see Coles and Scherle, Hall – chapters 10 and 11 – in this volume). For example, eco-labelling of tourism products seeks to use international certification to empower local communities. This device and several others like it draw strength and credibility from their international scope. Standards which are understood by consumers across the world are used to certify and to legitimate the activities of those at the local level. It is also important to note that measures at any one scale may disempower actors or stakeholders at another scale, such as when national empowerment may counteract local empowerment. In addition to scales of empowerment, several authors have identified at least four forms of empowerment that are crucial to successful community development in general and to tourism in particular (Brown 2002; Friedmann 1996; O'Neal and O'Neal 2003; Scheyvens 1999, 2003). These include political, social, economic and psychological empowerment, and these forms of empowerment function at, and cut across, different scales.

Political empowerment

Political empowerment is achieved, to some extent at least, when entire communities and their individual stakeholders have a voice in policy decision making. This is crucial because destination residents must have a forum through which they can raise questions and articulate concerns; that is, participate in the decisions that affect their community most. Several observers agree that community residents and other stakeholders should be invited by implementing agencies to provide their opinions, suggestions and concerns through various types of public forums (Arnstein 1969; Scheyvens 1999, 2002; Tosun 2000; Timothy 1999), but as the typology presented earlier suggests, this may be only a tokenistic form of community involvement or, at best, meaningful involvement, but not full-fledged empowerment. Full political empowerment may only exist when ownership of development problems and benefits lies in the hands of the destination community. This is more complex than it might seem, however, because in an increasingly globalized world it might be questionable as to whether or not this can truly occur given that the mobilities of tourism imply that destinations are dependent upon the degree to which visitors will go along with empowerment strategies.

Perhaps the greatest need for political empowerment lies with the individuals and communities, which Wall (1995) refers to as the 'people outside the plans'. These include traditionally marginalized people, such as ethnic and racial minorities, women and the poor, who are kept powerless by ruling elites or colonializing outsiders. These people rarely, if ever, have opportunities to decide their own futures, let alone initiate development programmes or participate in a meaningful way in the development of tourism. The San people and other indigenous Africans of South Africa before 1994 are prime examples of this (Goudie *et al.* 1996), as are the Prambanan villagers of Indonesia mentioned earlier. From the community perspective, when people are able to organize and mobilize themselves for the common good, they will become politically more powerful (Laverack and Wallerstein 2001). Communities then can have the authority to instigate tourism development or reject it, although given the multifaceted nature of communities, not all people will have equal opportunity to accept or reject.

Social empowerment

Empowerment in tourism from the social perspective should enhance the destination community's equilibrium and cohesiveness. As individuals and other interested groups work together for the good of the entire community, social cohesiveness is enhanced, and community services improve (e.g. schools, health care, infrastructure) through the vehicle of tourism (Scheyvens 1999).

Indigenous knowledge is now viewed as a vital component of the development process, including policy decision making. Traditional thinking by governments and some development experts was that indigenous people or those living in the tourist destination were a threat to biodiversity and other resources. Today, however, indigenous knowledge is seen as a crucial part of finding solutions to more sustainable uses of resources (Brown 2002; Faust and Smardon 2001; Ironside 2000; Johnston 2003; Motteux *et al.* 1999; Zanetell and Knuth 2002), instead of relying solely on imported notions fabricated by external politicians and consultants in places far from the reality of local conditions (Cornwall 2003; Timothy 1999). In most recorded cases, indigenous people are seen as the natural guardians of the environment, due to their ecology-based knowledges, practices and coping systems, and they typically have a 'better track record than modern, industrial societ[ies] in [their] treatment of natural ecosystems' (Faust and Smardon 2001: 147). In the past in most development projects in the less-developed world, local people have been forced to choose between either tradition or conservation. These are, though, no longer seen as being mutually incompatible or at opposing ends of a development spectrum. According to Zanetell and Knuth (2002: 821), the cultural traditions of indigenous people and their local knowledges ('knowledge partnerships') are as important in management decisions as allegedly scientific 'facts'.

Thus, social empowerment not only allows participation in development; rather, it demands it. Truly empowered societies do not view control and participation as options, but instead as a social obligation for the greater common good. Community social empowerment has the potential to reduce individual, clan and family-based differences as a common (collective) identity begins to emerge (Laverack and Wallerstein 2001). This community capacity building, or social capital, refers to the development of strengths and assets that individuals, as well as entire communities collectively, contribute intrinsically to the improvement of their own quality of life. Such strengths include, among others, leadership, social networks, culture, skills and local knowledge (Lyons *et al.* 2001; Simpson *et al.* 2003: 278). It is import to note, however, that not all development observers see social capital as being so straightforward, which has led to some critical evaluations of its dynamics (e.g. Elmhirst and Silvey 2003).

In the very process of individual empowerment, a sense of solidarity develops. This is based on a person's involvement in a broader community. The self-esteem that develops through individual empowerment usually gives way to an acknowledgement that one is a part of a larger, holistic system (Lyons *et al.* 2001; Wilson 1996). Thus, individual empowerment leads to quicker and more effective social empowerment in the community, and when local people support and take pride in their tourism industry, tourist satisfaction will be higher than in places where tourism growth is imposed (Hall 1998).

Economic empowerment

Economic empowerment in tourism is realized when the industry produces long-term fiscal benefits (Scheyvens 1999). Presently in the developing world context, researchers (Mahony and van Zyl 2002; Scheyvens 2002; Timothy 1999; Timothy and White 1999; Wallace and Pierce 1996) contend that the most sustainable form of economic empowerment takes place as money is earned in the community and shared through cooperative arrangements and community networks. It is particularly important in the realm of empowerment that disadvantaged segments of the population are supported, enabled and encouraged to participate and share in the earnings. This is often referred to as 'pro-poor' tourism, wherein all members of a given society, especially the least affluent cohorts, are afforded opportunities to share the industry's financial rewards through employment, small business ownership, grant programmes and – in common with social empowerment – basic public services.

One area of concern in economic empowerment is how to alleviate poverty by making rural and poor areas more profitable and economically viable (Logan and

Moseley 2002). Commentators have noted that the provision of agricultural products or making handicrafts for tourists and their sale to visitors is a good example of economic empowerment, and among the most common in many parts of the world (Gombe 2002; Koczberski 2002; Wallace and Pierce 1996). In particular for women, who have traditionally depended on their husbands for financial support, these efforts provide their own money to buy items for their families and themselves. Through handicraft production for tourists, women now have more control over their own financial security. Nonetheless, a lack of skills and modern technology commonly function as considerable barriers to this route to economic development.

Psychological empowerment

Outside recognition of the value of indigenous knowledge about natural resources and cultural traditions is often viewed as a significant self-esteem builder among destination community members. This leads to individuals and entire communities taking pride in their traditions and being willing to share their experiences and knowledge with visitors (Scheyvens 1999). In addition to its role in empowering communities economically, as noted above, the development of crafting skills can be an important element in empowering communities psychologically by helping them preserve cultural traditions and increase their respect for their own heritage.

Through psychological empowerment, destination communities gain a sense of ownership (Nel and Binns 2002), which can lead to stewardship over resource protection and the growth of sustainable forms of tourism. Psychological empowerment deals primarily with individuals, but as individual self-esteem and confidence grow, community empowerment is enhanced through collective social identity (Laverack and Wallerstein 2001). At the *Villa Escudero* Plantation in the Philippines, villagers, who were once the labour force for the coconut plantation, take great pride in demonstrating their agricultural skills to visitors and show traditional Filipino village life. This has created an important community spirit and a strong commitment to the heritage environment (Hatton 1999).

Examples of community empowerment

As noted so far, there are many examples of community empowerment throughout the world. Hatton (1999) provides several case studies that highlight many of the empowerment issues described here. The following two brief vignettes are borrowed from his work and reflect varying degrees and scales of empowerment and varying foci of development.

In 1988 the Malaysian Homestay Programme began in Desa Murni, an area of five rural villages. The programme was developed as a way of providing experiences in rural and traditional Malaysian way of life for tourists and providing a way for the local population to participate in, and benefit from, tourism. Under this programme, guests stay in the homes of local villagers and participate in various cultural activities, games and handicraft endeavours. The homestay programme was designed to meet several goals. First was the involvement of local people in the tourism industry. Desa Murni villagers organized and administered the programme and they reaped its financial rewards. Additionally, there were spin-offs: many small businesses in the villages benefited considerably from the growth of tourism. Second, the programme was considered as a major tool for learning, sharing and reinforcing local customs and values. In the words of the programme's founder, 'by showing off our culture to the guests, we reinforce for ourselves and our young people our rich and diverse cultural heritage'. Third, the villages provided opportunities for urban Malaysians to experience their own rural roots. Finally, the project delivered opportunities for rural Malaysians to come into contact with people from other parts of the country and abroad. So far, from the community's perspective, this interaction has been positive and has delivered positive outcomes in terms of better education, health care, hygiene and standards of living (Hatton 1999: 38–41).

A second case is the Huangshan Mountain area of China. This is known for its dramatic scenery and it is a popular destination for Chinese and foreign tourists. Local residents have been eagerly involved in the opportunities associated with the growth of tourism. During the infrastructure development phase of the project, the Huangshan Mountain Scenic Development Area (HMSDA) strongly encouraged and guided community members to construct their own, supplementary tourist businesses. This has, according to Hatton, directly enhanced local incomes and prosperity. Likewise, it has helped to preserve the natural environment. Before local people routinely cut trees and killed wild animals, but since the project started they have opted to work in tourism and to promote their forests and wildlife as tourist attractions. Here, too, the local standard of living has improved manifold, with the result that even the young people are devoted to the project and participate in the benefits of tourism's growth (Hatton 1999: 26–29).

Mitigating factors

There are several mitigating factors that facilitate or limit higher levels of community and individual empowerment. Education is important in determining whether or not destination residents are able and willing to participate in decision making. Education theory argues that knowledge is power, and advocates of community empowerment contend that increased levels of community and individual awareness about tourism lead to higher levels of all four types of empowerment (Friedmann 1996). Thus, through increased awareness, community members will be better able to influence decision making and initiate development programmes on their own without a great deal of external interference (Cornwall 2003; Lyons *et al.* 2001; Timothy 2000).

Conversely, ignorance or lack of awareness keeps grassroots stakeholders under the control of more powerful elites. This in fact was a strategy commonly used by colonial powers and ruling classes. By keeping their subjects in a state of relative ignorance, rulers were (and continue to be) able to reduce the powerless to 'cultures of silence' (Botchway 2001). In subtle ways, government officials sometimes perpetuate a feeling among community members that they are not clever enough or possess adequate knowledge about tourism to be able to participate (Timothy 1999; Tosun 2000). For instance, many government officials in developing countries discount community input and suggest through official media statements that the local people do not understand the real issues involved. Likewise, many government agencies overtly involve local elites in planning and they suggest that they are the ones in possession of enough understanding to make any kind of contribution (Timothy 1999).

As noted above, gender is an important factor in determining the level of community empowerment (Aitchison 2001; Gombe 2002; Lincoln *et al.* 2002; Pilcher 2001; Slater 2001). In nearly all traditional societies, especially in the developing world, women have been among the most marginal segments because of the persistence and iniquities of the traditions of patriarchal power that exist throughout many of the world's cultures. Women are most typically excluded from policy making and their interests are often marginalized in so-called 'participatory processes' (Timothy 2001). In most cases, particularly in the developing world, although women have little voice in development decision making, they are often faced with the greatest tasks and challenges associated with the implementation of plans and projects (Cornwall 2003). In the context of forest management and ecotourism, this may take the form of patrolling the forests at night to prevent illegal cutting, or carrying a large share of the burden in building ecolodges for tourists.

Racism is a problem in most countries of the world and it is perpetuated when powerful minorities (or majorities) keep majority populations (or minorities) in a state of oppression and ignorance. One of the best examples of this in recent history is South Africa, where the white ruling minority adopted a segregationist policy and enacted many discriminatory laws until the liberation of the African population in 1994 (Murray 2000). For Lincoln *et al.* (2002: 275),

> membership of a minority may bring it an automatic powerlessness in a society led by a majority. Laws, rules and attitudes most commonly reflect those in positions of power and act against those that differ from the norm.

From a perspective of tourism development, heritage tourism in South Africa has traditionally been dominated by a Eurocentric view of history while the black past has been intentionally obscured (Goudie *et al.* 1999; Timothy and Boyd 2003). The same was true in the United States until quite recently with respect to Native American and African-American pasts (Timothy and Boyd 2003). In colonial societies, preserved and promoted heritages typically reflect primarily the interests of the colonists as indigenous pasts are relegated. Similarly, in terms of living culture, most minority groups have lacked control of how their cultures have been portrayed to tourists (Goudie *et al.* 1999: 25). Fortunately, this is beginning to change in some parts of the world as indigenous rights are being reasserted, for example among the Maoris of New Zealand, Native Americans and, in its very nascent form, in South Africa (Cloher and Johnston 1999; Goudie *et al.* 1996; Walker 1987).

A final issue is that of corruption which can also occur even with the existence of empowerment initiatives. This is especially the case when some groups or individuals become empowered while others are not. Power imbalances are thus created which are conducive to corruption as people take advantage of their (perceived) 'empowered' status to exercise their power in an informal, illegal manner over those who are less empowered. This form of control is deeply embedded in most societies of the developing world and is often manifest in tourism development. In this case, corruption becomes a major barrier to true empowerment as inequities are created in destination communities that are often not formally acknowledged or interwoven into development initiatives towards empowerment.

Drawbacks to destination community empowerment

While the benefits of empowerment from the grassroots perspective would appear to outweigh any potential pitfalls, there are notwithstanding problems associated with the empowerment process. Perhaps one of the most important of these is the potential to create disharmony among community members: although the notion of empowerment in theory should create more harmonious interpersonal and inter-organizational accord (and hence spread power to the otherwise powerless), it may sometimes backfire. Negative relationships between stakeholders may result and function to generate (new patterns of) inequality among destination residents.

In traditional societies this is especially a result of conventions of power in traditional societies, which are typically patriarchal and focus on a central figure, to which all others are subject. This creates further problems in the process of empowerment, because in some locations women, for instance, have been beaten and forced to divorce as a result of their increased authority and participation in community-based development decision making (Cornwall 2003). Thus, the very process of empowerment in its initial phases simultaneously creates disempowerment among women as conventional power brokers (men) become threatened by the apparent increase in power among women (Timothy 2001). Economic, psychological and social empowerment are by far the most common forms of empowerment among women, particularly as education and self-employment are key towards empowering women and other minorities.

Increased power transfer to the community and individual level can also divide communities if it is felt that preferential treatment is offered to some but not others. Elitism that existed at the national level may now be transferred to the local and individual level. The result may be the creation of disharmony and discontent among destination residents. According to Brown (2002: 11), intervention to empower at the grassroots level can 'exacerbate disparities and . . . marginalize some sections of society'. This may also simply be a perceptual problem in the public domain, because the people who are able and willing to devote time, effort and energy to become involved more deeply in tourism development projects may not be supported unanimously by the community. Rather, they might be seen as (new, even unwelcome) elites among a group of people struggling to survive (Laverack and Wallerstein 2001).

Similarly, complete empowerment, which hardly exists if at all anywhere in the world, could become dangerous where the community in question decides to carry out some kind of radical change to its environment without full consideration of the potential consequences and their workings over time. Examples exist of attempts to drain an ecologically valuable swamp to erect a shopping mall or, as happened in Eastern Europe after the collapse of Communism, the decision to destroy all signs of the recent ideological past from the landscape, history books and public memory (Timothy and Boyd 2003).

Conclusion

Power struggles have been, and continue to be, part and parcel of everyday life in most parts of the world. Unfortunately, as they continue, the most disadvantaged groups, such as ethnic and racial minorities, women and the poor, remain the most powerless. This is only now beginning to change in the developing world, but here as in the most advanced industrial societies, there exist many levels of community empowerment. Current trends indicate that empowerment can occur at an individual and community level and from economic, political, social and psychological perspectives. Economic empowerment exists when people are able to gain from the economic benefits of tourism. This is especially important for women in the developing world, who have now found higher levels of economic freedom by working in tourism and by producing products for tourist consumption. Signs of political empowerment include individuals and entire communities being able to initiate tourism development programmes and participate in existing ones that are in harmony with local socio-cultural and ecological conditions. To have a voice to influence preordained plans is not enough; true empowerment means control and authority to accept or reject. Harmony and equality reflect social empowerment, wherein the values and opinions of all people concerned are respected. This does not suggest, however, that a community and all its stakeholders must be of the same mind or homogenous in their views of tourism, but rather that everyone enjoys an equal opportunity to participate. Likewise, social empowerment also implies that indigenous knowledge will be utilized in understanding local socio-cultural and ecological conditions within which tourism development will occur. Finally, psychological empowerment is achieved when communities begin to take pride in their culture, history and ability to control tourism and initiate positive development programmes.

Empowerment, as through the decentralization of power to the communities and individuals most affected by development, is now seen as a critical underlying principle of tourism development efforts. Unfortunately, there are unique socio-cultural and political obstacles, in all places, to its fulfilment. Nonetheless, empowerment enables communities to exercise greater self-determination in their own futures by either initiating, modifying or accepting development programmes, or rejecting them. This is unless, of course, new tourism development efforts (and the so-called 'empowerment' that precedes them) create new forms of social division, imbalance, marginalization and the weakening of communities as certain individuals become more empowered than others. Despite its potential pitfalls, a process or state of participation in issues of power is seen as a requirement for achieving sustainability in the context of tourism, and there is little doubt that many more communities throughout the world will achieve higher levels of empowerment as they become more aware of the need to be actively involved in tourism, as gender and racial gaps begin to close, and as governments continue to realize the value of decentralizing power into the hands of grassroots stakeholders.

References

Aitchison, C. (2001) 'Theorizing other discourses of tourism, gender and culture', *Tourist Studies*, 1(2): 133–147.

Allen, J. (2003) *The Lost Geographies of Power*. Oxford: Blackwell.

Arnstein, S. (1969) 'A ladder of community participation', *American Institute of Planners Journal*, 35: 216–224.

Baez, A.L. (1996) 'Learning from experience in the Monteverde Cloud Forest, Costa Rica', in M.F. Pearce (ed.) *People and Tourism in Fragile Environments*. Chichester: Wiley.

Berger, D.J. (1996) 'The challenge of integrating Maasai tradition with tourism', in M.F. Price (ed.) *People and Tourism in Fragile Environments*. Chichester: Wiley.

Botchway, K. (2001) 'Paradox of empowerment: reflections on a case study from northern Ghana', *World Development*, 29(1): 135–153.

Brown, K. (2002) 'Innovations for conservation and development', *Geographical Journal*, 168(1): 6–17.

Campbell, H. and Marshall, R. (2000) 'Public involvement and planning: looking beyond the one to the many', *International Planning Studies*, 5(3): 321–344.

Castro, A.P. and Nielsen, E. (2001) 'Indigenous people and co-management: implications for conflict management', *Environmental Science and Policy*, 4: 229–239.

Chambers, R. (1993) *Challenging the Professions: Frontiers for Rural Development*. London: Intermediate Technology Publications.

Choguill, M.B.G. (1996) 'A ladder of community participation for underdeveloped countries', *Habitat International*, 20(3): 431–444.

Cloher, D.U. and Johnston, C. (1999) 'Maori sustainability concepts applied to tourism: a North Hokianga study', *New Zealand Geographer*, 55(1): 46–52.

Cornwall, A. (2003) 'Whose voices? Whose choices? Reflections on gender and participatory development', *World Development*, 31(8): 1325–1342.

Davidson, R. and Maitland, R. (1999) 'Planning for tourism in towns and cities', in C.H. Greed (ed.) *Social Town Planning*. London: Routledge.

Dobbs, L. and Moore, C. (2002) 'Engaging communities in area-based regeneration: the role of participatory evaluation', *Policy Studies*, 23(3/4): 157–171.

Elmhirst, R. and Silvey, R. (2003) 'Engendering social capital: women workers and rural urban networks in Indonesia', *World Development*, 31(5): 865–879.

Faust, B.B. and Smardon, R.C. (2001) 'Introduction and overview: environmental knowledge, rights, and ethics: co-managing with communities', *Environmental Science and Policy*, 4: 147–151.

Friedmann, J. (1996) 'Rethinking poverty: empowerment and citizens' rights', *International Social Science Journal*, 148: 161–172.

Gombe, C. (2002) 'Indigenous pottery as economic empowerment in Uganda', *International Journal of Art and Design Education*, 21(1): 44–47.

Goudie, S.C., Khan, F. and Kilian, D. (1996) 'Tourism beyond apartheid: Black empowerment and identity in the "New" South Africa', in P.A. Wells (ed.) *Keys to the Marketplace: Problems and Issues in Cultural and Heritage Tourism*. Enfield Lock : Hisarlik Press.

—— (1999) 'Transforming tourism: black empowerment, heritage and identity beyond apartheid', *South African Geographical Journal*, 81(1): 22–31.

Gow, D. and Vansant, J. (1983) 'Beyond the rhetoric of rural development participation: how can it be done?', *World Development*, 11(5): 427–446.

Hall, D.R. (1998) 'Rural diversification in Albania', *GeoJournal*, 46(3): 283–287.

Hatton, M.J. (1999) *Community-based Tourism in the Asia-Pacific*. Ottawa: Canadian International Development Agency.

Healy, K. and Zorn, E. (1983) 'Lake Titicaca's Campesino controlled tourism', *Grassroots Development*, 6(2): 5–10.

Ironside, R.G. (2000) 'Canadian northern settlements: top-down and bottom-up influences', *Geografiska Annaler B*, 82(2): 103–114.

Jenkins, C.L. (1980) 'Tourism policies in developing countries: a critique', *International Journal of Tourism Management*, 1(1), 36–48.

Johnston, A.M. (2003) 'Self-determination: exercising indigenous rights in tourism', in S. Singh, D.J. Timothy and R.K. Dowling (eds) *Tourism in Destination Communities*. Wallingford: CAB International.

Koczberski, G. (2002) 'Pots, plates and Tinpis: new income flows and the strengthening of women's gendered identities in Papua New Guinea', *Development*, 45(1): 88–91.

Laverack, G. (2001) 'An identification and interpretation of the organizational aspects of community empowerment', *Community Development Journal*, 36(2): 134–145.

Laverack, G. and Wallerstein, N. (2001) 'Measuring community empowerment: a fresh look at organizational domains', *Health Promotion International*, 16(2): 179–185.

Lincoln, N.D., Travers, C., Ackers, P. and Wilkinson, A. (2002) 'The meaning of empowerment: the interdisciplinary etymology of a new management concept', *International Journal of Management Reviews*, 4(3): 271–290.

Logan, B.I. and Moseley, W.G. (2002) 'The political ecology of poverty alleviation in Zimbabwe's communal areas management programme for indigenous resources (CAMPFIRE)', *Geoforum*, 33: 1–14.

Lyons, M., Smuts, C. and Stephens, A. (2001) 'Participation, empowerment and sustainability: (how) do the links work?', *Urban Studies*, 38(8): 1233–1251.

McIntyre, G., Hetherington, A. and Inskeep, E. (1993) *Sustainable Tourism Development: A Guide for Local Planners*. Madrid: World Tourism Organisation.

Mahony, K. and van Zyl, J. (2002) 'The impacts of tourism investment on rural communities: three case studies in South Africa', *Development Southern Africa*, 19(1): 83–103.

Mitchell, R.E. and Reid, D.G. (2001) 'Community integration: island tourism in Peru', *Annals of Tourism Research*, 28: 113–139.

Motteux, N., Binns, T., Nel, E. and Rowntree, K. (1999) 'Empowerment for development: taking participatory appraisal further in rural South Africa', *Development in Practice*, 9(3): 261–273.

Mowforth, M. and Munt, I. (1998) *Tourism and Sustainability. New Tourism in the Third World*. London: Routledge.

Murray, G. (2000) 'Black empowerment in South Africa: "patriotic capitalism" or a corporate black wash?', *Critical Sociology*, 26(3): 183–204.

Nel, E. and Binns, T. (2002) 'Place marketing, tourism promotion, and community-based local economic development in post-apartheid South Africa: the case of Still Bay – the "Bay of Sleeping Beauty"', *Urban Affairs Review*, 38(2): 184–208.

Ngwa, N.S.E. and Fonjong, L.N. (2002) 'Actors, options and the challenges of forest management in Anglophone Cameroon', *GeoJournal*, 57(1/2): 93–109.

Okoji, M.A. (2001) 'Depletion of forest resources in south eastern Nigeria: who loses?', *The Environmentalist*, 21: 197–203.

O'Neal, G.S. and O'Neal, R.A. (2003) 'Community development in the USA: an empowerment zone example', *Community Development Journal*, 38(2): 120–129.

Osborne, S.P. (1994) 'The language of empowerment', *International Journal of Public Sector Management*, 7(3): 56–62.

Pilcher, J.K. (2001) 'Engaging to transform: hearing black women's voices', *Qualitative Studies in Education*, 14(3): 283–303.

Pinkett, R. and O'Bryant, R. (2003) 'Building community empowerment and self-sufficiency: early results from the Camfield Estates-MIT Creating Community Connections project', *Information, Communication and Society*, 6(2): 187–210.

Scheyvens, R. (1999) 'Ecotourism and the empowerment of local communities', *Tourism Management*, 20: 245–249.

—— (2002) *Tourism for Development: Empowering Communities*. Harlow: Prentice Hall.

—— (2003) 'Local involvement in managing tourism', in S. Singh, D.J. Timothy and R.K. Dowling (eds) *Tourism in Destination Communities*, Wallingford: CAB International.

Simmons, D.G. (1994) 'Community participation in tourism planning', *Tourism Management*, 15: 98–108.

Simpson, L., Wood, L. and Daws, L. (2003) 'Community capacity building: starting with people not projects', *Community Development Journal*, 38(1): 277–286.

Slater, R.J. (2001) 'Urban agriculture, gender and empowerment: an alternative view', *Development Southern Africa*, 18(5): 635–650.

Sofield, T.H.B. (2003) *Empowerment for Sustainable Tourism Development*. Amsterdam: Pergamon.

Stiefel, M. and Wolfe, M. (1994) *A Voice for the Excluded: Popular Participation in Development*. London: Zed Books.

Telfer, D.J. (2003) 'Development issues in destination communities', in S. Singh, D.J. Timothy and R.K. Dowling (eds) *Tourism in Destination Communities*. Wallingford: CAB International.

Thomas, S. (1995) *Share and Share Alike: Equity in Campfire*. London: International Institute for Environment and Development.

Timothy, D.J. (1996) 'Tourism planning in the developing world: a case study of Yogyakarta, Indonesia', unpublishd doctoral dissertation, University of Waterloo.

—— (1999) 'Participatory planning: a view of tourism in Indonesia', *Annals of Tourism Research*, 26(2): 371–391.

—— (2000) 'Building community awareness of tourism in a developing country destination', *Tourism Recreation Research*, 25(2): 111–116.

—— (2001) 'Gender relations in tourism: revisiting patriarchy and underdevelopment', in Y. Apostolopoulos, S. Sönmez and D.J. Timothy (eds) *Women as Producers and Consumers of Tourism in Developing Regions*. Westport, CT: Praeger.

—— (2002) 'Tourism and community development issues', in R. Sharpley and D.J. Telfer (eds) *Tourism and Development: Concepts and Issues*. Clevedon: Channel View Publications.

Timothy, D.J. and Boyd, S.W. (2003) *Heritage Tourism*. Harlow: Prentice Hall.

Timothy, D.J. and Tosun, C. (2003) 'Appropriate planning for tourism in destination communities: participation, incremental growth and collaboration', in S. Singh,

D.J. Timothy and R.K. Dowling (eds) *Tourism in Destination Communities*. Wallingford: CAB International.

Timothy, D.J. and Wall, G. (1997) 'Selling to tourists: Indonesian street vendors', *Annals of Tourism Research*, 24(2): 322–340.

Timothy, D.J. and White, K. (1999) 'Community-based ecotourism development on the periphery of Belize', *Current Issues in Tourism*, 2(2/3): 226–242.

Tosun, C. (1998) 'Roots of unsustainable tourism development at the local level: the case of Urgup in Turkey', *Tourism Management*, 19: 595–610.

—— (2000) 'Limits to community participation in the tourism development process in developing countries', *Tourism Management*, 21: 613–633.

Tosun, C. and Timothy, D.J. (2003) 'Arguments for community participation in the tourism development process', *Journal of Tourism Studies*, 14(2): 2–15.

Walker, E.G. (1987) 'Indian involvement in heritage resource development: a Saskatchewan example', *Native Studies Review*, 3(2): 123–137.

Wall, G. (1995) 'People outside the plans', in W. Nuryanti (ed.) *Tourism and Culture: Global Civilization in Change*. Yogyakarta: Gadjah Mada University Press.

Wallace, G.N. and Pierce, S.M. (1996) 'An evaluation of ecotourism in Amazonia, Brazil', *Annals of Tourism Research*, 23(4): 843–873.

Watt, S., Higgins, C. and Kendrick, A. (2000) 'Community participation in the development of services: a move towards community empowerment', *Community Development Journal*, 35(2): 120–132.

Wilson, P.A. (1996) 'Empowerment: community economic development from the inside out', *Urban Studies*, 33(4/5): 617–630.

Zanetell, B.A. and Knuth, B.A. (2002) 'Knowledge partnerships: rapid rural appraisal's role in catalyzing community-based management in Venezuela', *Society and Natural Resources*, 15: 805–825.

10 Prosecuting power

Tourism, inter-cultural communications and the tactics of empowerment

Tim Coles and Nicolai Scherle

Tourism and power: towards a tactical approach to empowerment

There has been a recent trend to identify tourism as a central component in major global issues. Tourism continues to be used in the prosecution of global terrorism (09/11 Commission 2004), while the potential of travel to inspire peace has been enthusiastically if simplistically championed (WTO 2003). Climate change poses a significant risk to tourism, according to the World Tourism Organisation (WTO 2005), although tourism may not so much be the victim as the perpetrator of global environmental change (Hall and Higham 2005). Where once tourism exposed the gaps between the 'haves' from the 'global north' and the 'have nots' from the 'south', to the world's media tourism has been rehabilitated, cast in a positive light by the WTO (2004), the World Bank (Younis 2004) and the United Nations (UNEP 2002) as a means to tackle global inequalities, to alleviate poverty in even the poorest countries and regions, and to empower local peoples.

That tourism may lead to community empowerment is no great revelation (Scheyvens 1999, 2000; Sofield 2003; Timothy, Chapter 9 in this volume): discourses of empowerment stemmed from debate of *Our Common Future* in 1987 and discussion of the relevance of the principles of sustainable development for tourism (Mowforth and Munt 1998; Liu 2003). As prevalent and influential as these principles may have become (Weaver 2004), here we reflect whether they may have overshadowed other lines of enquiry, in particular on the political economy of empowerment through tourism (Britton 1991; Bianchi 2002; Williams 2004). Empowerment has been considered more generally alongside other desiderata for tourism development initiatives (Ryan 2002; Smith and Duffy 2003). Rarely, though, have there been critical accounts of the precise tactics used by stakeholders involved in political struggles over the distribution of capital, resources, information and power. As Mowforth and Munt (1998: 3) have reflected,

> much tourism analysis has played down relationships of power, which remain implicit, or are absent . . . [moreover] where power is invoked in a discussion of tourism it has tended to be in passing: references to ideology, discourse, colonialism, imperialism and so on . . . in a rather unstructured, even anecdotal fashion.

This problematic treatment is compounded in many instances because the politics of empowerment are mediated across cultural as well as political divides. Non-governmental organisations (NGOs) from European states may, for instance, encourage local people in developing countries to develop tourism (Kopp 2003). Latent in this scenario are the cultural backgrounds, values and identities brought to the politico-economic episode by both parties, as well as the culturally constructed nature of the concepts 'power' and 'empowerment' (Morriss 2002; Lukes 2005). The positionality of the actors will inevitably influence their understandings of power, their valorisations of empowerment and the actions taken in the social relations of power.

This chapter explores how specific power and empowerment strategies and tactics are devised, deployed and utilised in business-to-business relationships between small and medium-sized tourism enterprises (SMTEs) from Morocco and Germany (see also Scherle 2004, 2006). Morocco has set itself ambitious macro-economic development goals in which tourism has featured as a prominent driver (CGEM 2000). Part of the approach has been to empower local businesses by encouraging visitors, foreign direct investment and knowledge transfer from major host markets such as Spain, France and Germany (Scherle 2006). We explore what Few (2002: 29) terms the 'micro-sociological' aspects of power; that is, the 'coupling of power and agency', not just 'the structural features or implications of the social distribution of power' between partner businesses. We focus on how businesses develop 'actor strategies' for coping with their strategic partners and their implications for relative empowerment. Few (2002, after Brown and Rosendo 2000: 212) describes actor strategies as 'the way social groups use their available power resources, or their knowledge and capability to resolve their particular problems'. We also consider the specific tactics (measures, devices, technologies), or the micro-practices used as part of the actor strategies.

Allen (2003) has cautioned against detailed inspection of tactics at the micro-level because of the potential for taxonomy to frustrate critical enquiry. To a degree, we would agree: our intention is not to describe and classify, rather to cast the spotlight on how distinctive tactics manifest, feature in and contribute to unfolding power relations between tourism enterprises across cultures. Steven Lukes's (1974, 2005) readings of power are used as an initial framework (see also Coles and Church, Hall – chapters 1 and 11 – in this volume; Reed 1997; Doorne 1998). Lukesian perspectives have found notable application in organisational studies (Knights and Willmott 1999; Buchanan and Badham 1999; Buchanan and Huczynski 2004). The influence of Lukes's ideas are considered in the next section and it is argued that they fulfil two interpretative functions: as one of a number of 'deeper systems of decision-making/non-decision-making . . . that constitute modes of power such as domination' (Few 2002: 29), they help explain how particular strategies toward power relations emerge; and they serve as a conceptual platform from which to explore further the micro-practices of power in organisational politics (Buchanan and Badham 1999). This discussion raises a number of practical considerations for the study of power and organisations, not least how to connect power with studies of inter-cultural contact and communication between businesses. This is discussed in the third part of the chapter and it precedes an empirical account of the business relations between Moroccan and German SMTEs to mediate products for German inbound tourists to Morocco. Finally, the conceptual implications for the study of tourism and empowerment are discussed.

Tourism, transactions and 'turf games'

Tourism experiences are produced by inter-organisational exchanges (Watkins and Bell 2002), and these transactions increasingly take place in networks stretched over long distances, across international boundaries and between different cultures (Clancy 1998; Vorlaufer 1998; Mosedale 2005). As demonstrated elsewhere in this volume, there are several theoretical approaches to power relations and in studies of organisational behaviour this same plurality is evident (McKinlay and Starkey 1998; Clegg 1989; Knights and Willmott 1999; Buchanan and Huczynski 2004). Prominent among these contested approaches to inter- and intra-organisational power relations have been Steven Lukes's (1974) writings. Suffice to record here that Lukes's major conceptual achievement is the identification of three dimensions to power which are outlined briefly in Table 10.1 (see Table 1.3) and these are discussed in some detail in Chapter 1.

Table 10.1 Lukes (1974) in a nutshell

Dimension	*Key characteristics*
One	Power that is exercised to secure a decision in situations where there is some observable conflict or disagreement.
Two	Power that is exercised to keep issues on or off the decision-making agenda, so that potential conflicts or disagreements are precluded and therefore unobservable.
Three	Institutionalised power that is exercised to define social reality. To the extent that norms and meanings are internalised, people accept and thereby reproduce the power-invented definition of reality, even when this is against their 'real' interests.

Sources: Knights and Willmott (1999: 95) after Lukes (1974).

Lukes's ideas, like those of Foucault (McKinlay and Starkey 1998; Knights and Willmott 1999; Knights and McCabe 2002), offer a framework through which to explore the motives and behaviours of individuals and collectives and the operation of particular modes of power. A Lukesian approach has a basic appeal in its behavioural focus on outcomes. The explanatory framework it offers mimics superficially the formalised nature by which the power relationships among actors are played out between and within business enterprises. Decision and non-decision making, observable and latent conflict, ideology and the control of the agenda identified by Lukes have immediate resonances with the processes, settings and structurings in commercial behaviour over bids, tenders, contracts, delivery, evaluation and the like.

General criticisms of this approach have been rehearsed, such as the difficulties in empirical verification (Lukes 1974), as well as how to deal with multiple power relations among stakeholders and where unitary interests are lacking (Lukes 2005: 64). Knights and Willmott (1999: 97) argue that there are difficulties because Lukes's perspective assumes a priori knowledge of 'real' interests. Through Foucauldian readings of the organisation, they contend that interests are a product of the exercise of power not, as Lukes postulates, pre-existing, preordained features that dictate the course and nature of power relations. Indeed, they question the notion of 'real' interest. Whereas Lukes considers real interests to be manipulated by the exercise of the third dimension of power (Lukes 1974), Knights and Willmott (1999: 98) point to the difficulties in (legitimately) justifying such attributions.

Rather than favour one conceptual fiat over another, Buchanan and Badham (1999) adopt a more pragmatic view and so indirectly hint at the limitations of Lukesian thinking. Sensitive to several theoretical canons, they identify power variously as an individual property, a relational property, and as an embedded property. In the latter, power is understood to be function of regulations, structures, systems and the like in which the organisation or the individual within an organisation function, and it is here that they point to the contribution of Lukes, like Bachrach and Baratz (1962, 1970) before him, in setting the 'rules of engagement' for decision- and non-decision-making. Somewhat uncontroversially, they contend that such an embedded perspective is incapable alone of providing a satisfactory explanation for the way in which power functions within organisations. Their treatment of power as an individual property is effectively an alternative discussion of power as a capacity, a disposition (see Coles and Church, Chapter 1 in this volume), with the identification of a series of what they term 'structural' and 'individual' sources of power from which basic valorisations of 'power to' (potentially) achieve particular outcomes are possible.

Table 10.2 French and Raven's (1959) bases of power

Base of power	*Basis of power is . . .*
Reward power	P's perception that O has the ability to mediate rewards for P.
Coercive power	P's perception that O has the ability to mediate punishments for P.
Legitimate power	Perception by P that O has a legitimate right (authority) to prescribe behaviour for P.
Referent power	P's identification with O or the feeling of 'oneness' of P with O, or a desire for such an identity. O has attributes and abilities that are seen as charismatic by P.
Expert power	Perception that O has some special knowledge or expertise that makes P in some way perceive O as superior.
Informational power	O has information that P perceives as important.

Sources: adapted from French and Raven (1959: 155–164) and Raven (1965: 374–379).

Similarly, Buchanan and Badham's (1999) discussion of power as a relational property in part merely rehearses arguments on power as a relational construct (Latour 1986). Significantly, though, they draw on perspectives from social psychology to explore how power is perceived by actors relative to one another in order to interpret how power relations unfold. A more tactical approach ensues concerning how power is named, located, valorised and exercised at the level of the individual in their strategies to guide them through the turf games within organisations. Based originally on the work of French and Raven (1959; Raven 1965, 1974, 1992) six bases of power are identified: reward, coercive, referent, legitimate, expert (French and Raven 1959) and informational power each form a basis for the unfolding of power in social relations (Raven 1965) (Table 10.2). Bacharach and Lawler (1980: 34) refined French and Raven's work to differentiate between bases of power – 'what parties control that enables them to manipulate the behaviour of others' – and sources of power, or 'how parties come to control the bases of power'. They identified four bases and four sources of power (Table 10.3), while Benfari *et al.* (1986) identified eight bases towards the effective use of power – reward, coercion, authority, referent, expert, information, affiliation, group – and that bases may have a positive or negative function (Raven 1992). Subsequent studies have suggested that from the bases and sources of power a series of discrete influence tactics exist that are intended to help secure their practitioners' desired outcomes (Table 10.4; Kipnis *et al.* 1980, 1984; Yukl and Falbe 1990; Yukl and Tracey 1992; Raven 1992). In addition, a series of diagnostics emerge that reflect an individual's power relative to other actors' (Huczynski 1996; Buchanan and Badham 1999).

Table 10.3 Bacharach and Lawler's (1980) bases and sources of power

Bases of power	*Sources of power*
Coercive power • Ability to apply the threat of sanctions.	Office or structural position • The position or structural location might provide a party access to various bases of power.
Remunerative power • Control of material resources and rewards.	Personal characteristics • Charisma and leadership as manifestations of personal abilities and characteristics that key individuals have apart from other sources of power.
Normative power • Control of symbolic rewards.	Expertise • Specialised information that actors have come to control and apply to particular issue.
Knowledge • Control of unique information that is required to make a decision.	Opportunity • Located in informal settings, come to control information that is of importance to others.

Source: abridged from Bacharach and Lawler (1980: 34–36).

Table 10.4 Power influence tactics in organisations

Scale of influence tactics	*Definition*
Pressue tactics	Use of demands, threats or intimidation to secure compliance.
Upward appeals	Use of persuasion to persuade another that the request for compliance is approved by higher management.
Exchange tactics	Use of explicit or implicit promises that rewards or tangible benefits will result from compliance with request.
Coalition tactics	Aid of others is used to persuade an individual to comply with a request.
Ingratiating tactics	By means of getting an individual in a good mood, requests for compliance will be met more positively.
Rational persuasion	Use of logical arguments and factual evidence to persuade an individual to comply with a request.
Inspirational appeals	Emotional request of an individual to secure compliance by appealing to the values and ideals of an individual.
Consultation tactics	Use of consultation or the opportunity to input to decision-making or planning in order to secure compliance to a request.

Source: abridged from Yukl and Falbe (1990: 133).

It is not the intention to enter into a detailed investigation or discussion of these typologies. As Allen (2003: 101) notes, 'there is little to be gained by assuming that all practices reveal a kernel of power' but conversely Buchanan and Badham's work emphasises that it would be remiss to overlook the operation of power tactics altogether in four respects. First, they point to the need to consider the individual as opposed to the group. As Bruins (1999: 7) suggests, there has been a tendency in power discourse to theorise and explain at the inter-group (societal) level rather than the inter-personal level. Certainly, constituencies, their common positions and their collective power relations are important in tourism (see Gill, Church and Ravenscroft – chapters 6 and 8 – in this volume); however, individual relationships are a common medium for social relations in the production of tourism. The tourism sector is characterised by small- and medium-sized enterprises and individuals within these organisations routinely transact business with one another (Dahles and Bras 1999; Ateljevic and Doorne 2000).

Second, compromises have to be made and accepted in order to investigate the exercise of power in and between organisations. For instance, Buchanan and Badham's work could be criticised as committing what Morriss (2002: 15) terms the 'exercise fallacy' (Coles and Church, Chapter 1 in this volume). Simply put, the exercise fallacy reduces the identification and assessment of power to the observation of its exercise and it impedes the observation or measurement of power as disposition. While power may indeed be a more complex conceptual construct than some of their simplified instrumental devices might suggest, Buchanan and Badham (1999) demonstrate there are differences in the way power is theorised by scholars and how it is understood and applied in everyday lives and operations. Admittedly, power may be exercised in a quite sophisticated manner by certain individuals and their organisations; however, frequently it functions in quite straightforward, even imperfect ways as a means to particular ends. Its exercise is based on a simple calculation by individual actors which might involve the recognition and assessment of the bases and sources of power of each of the protagonists in power struggles (Raven 1992). Decisions are made on the basis of what Bacharach and Lawler (1980: 7) term a series of 'competitive tactical encounters' in which influence, authority, bargaining and compromise contribute to the negotiation of the best resolution at any given time (Buchanan and Badham 1999: 41).

A third reason for studying tactics is that 'turf games' get messy. While individuals may adopt broad strategies reminiscent of Lukes's three dimensions of power, they may deploy more complicated arrays of tactics to enhance or maintain their perceived power. Basic auditing, however conceptually crude or not, reveals the relativities of power between organisations and individuals as the protagonists understand them (Raven 1992). Furthermore, studying tactics reveals that it is not just the apparently more powerful that participate in power struggles and 'even being down-trodden, voiceless and marginalized is to possess a power source which can be exploited, if and when circumstances allow' (Buchanan and Badham 1999: 49). As Bugental and Lewis (1999) have noted, the seemingly powerless adopt control-oriented strategies when challenged: to be powerless does not necessarily imply that an individual is bereft of deliberate strategies or tactics towards power.

Finally, tactics are important because studies of inter- and intra-organisational power relations reveal the importance of organisational settings in framing and shaping behaviours (Knights and Willmott 1999). Often these studies have, though, considered organisations from the same or similar cultural backgrounds. Not only is power culturally constructed (see Morriss 2002), but so too are business practices and so-called 'power differences', the perception of which are vital to commercial transactions between businesses from different cultures. As Bourdieu (1989: 14 – authors' translation) has described:

> If a French person communicates with an Algerian, one does not only have two persons communicating with each other; it is moreover France communicating with Algeria. There are two histories communicating with each other; it is the whole colonisation, the whole history of a simultaneously economic and cultural . . . system of rule. And the same happens between an American and a French person.

In bilateral interactions across cultures with asymmetric power relations, power often remains hidden by the more powerful partner because of the desire not to highlight the inequity of the situation. Controversially, some writers also argue that the less powerful partner recognises the power difference, but also normally avoids raising it out of embarrassment at being perceived as feeling culturally inferior (cf. Moosmüller 1997). In the next section, we turn to a discussion of inter-cultural contacts between businesses, and we attempt to situate more precisely how power strategies and tactics may feature in communications between tourism businesses from different cultures.

Tourism, inter-cultural contact and communication

As Jack and Phipps (2005: 6) argue, tourism as an 'inter-cultural activity, constructed within and through language, has been largely ignored in tourism research until very recently'; simply put, what limited attention there has been has focused primarily on host–guest encounters (Reisinger and Turner 2003; Hunter 2001) rather than business-to-business relations (cf. Scherle 2004, 2006). In contrast, inter-cultural communication has become a major focus for management studies. Mainly driven by Hofstede's work (1982, 1999), studies of inter-cultural communication have courted controversy (Moosmüller 1997; Thomas 2003). There are several reasons for this, but one of the more important is that there is a propensity to reduce culture to a mere factor, practically an ingredient that has to be present in successful international business (Johnson and Turner 2003; Rugman and Hodgetts 2003; Gesteland 2005). Inter-cultural communication is, however, a far more complex affair. Partners from different cultures have to cope with cultural differences which otherwise impede communication and, in the worst case, result in a culture shock and breakdown in social relations (Casmir and Asuncion-Lande 1989; Ward *et al.* 2001; Pütz 2003).

Inter-cultural communication depends heavily on the notion of inter-cultural competence. With its roots in linguistics, inter-cultural competence is the mutual avowal of the interactants' cultural identities where both interactants engage in behaviour perceived to be appropriate and effective in advancing both cultural identities (Collier 1989). Kühlmann and Stahl (1998) identify seven characteristics to facilitate inter-cultural competence: tolerance of ambiguity, behavioural flexibility, target orientation, sociability, empathy, polycentrism and meta-communicative competence. While each of these facets operates differentially, all should be present in competent and effective inter-cultural engagements, and abstract tolerance alone is rarely effective (Moosmüller 1997).

Two schools of thought have emerged concerning the relevance of culture in international business (Table 10.5). 'Culturalists' advocate that management techniques are intricately and mutually implicated with culture, or international business is 'culture-bound'. Businesses have to be more culturally competent, sensitive to the needs and operations of their partners in other cultural settings, thereby adjusting and gearing their own operations accordingly. For instance, prior to EU enlargement, *The Economist* (2003) reported that officials from the European Commission had described Polish officials as the most difficult among the 15 new member states. A series of 'golden rules' had been devised to facilitate their induction into the EU culture of doing business.

The 'culturalist' position is challenged by the 'universalists'. For them, management techniques are universal and therefore independent of culture-specific influences; business is 'culture free', low levels of inter-cultural competence are

Table 10.5 Approaches to business management in inter-cultural situations

Position	*Universalism*	*Culturalism*
Cultural doctrine:	Society has a culture	Society is a culture
Management assumption:	'One best way'	'Several good ways'
Management paradigm:	Culture-free management	Culture-bound management
Typical business concepts:	– Planning instruments – Investment analysis – Budget planning – Production control	– Management style – Motivation – Flexible working – Closer working relationships in all levels of organisation

Source: abridged from Kutschker (2001).

required, and operations need minimal, if any, adjustment to cater for bilateral business co-operations. Closely linked to these two postions are the concepts of convergence and divergence. The latter holds that culture is a central construct of organisational and management behaviour and hence should be more deeply appreciated. Differing values and behaviours, differing stages of development and the uneven distribution of resources will ensure diversity and inequality that must be embraced in business. Convergence theory holds that technology, structure and a global orientation by many firms render inter-cultural management unnecessary (Warner and Joynt 2002).

The culture-free and -bound positions have been contested (see Osterloh 1994; Bosch 1997; Hofstede 1999; Kutschker and Schmid 2002). While it may be tempting to use broad cultural designations, cultures are constantly evolving and highly differentiated into micro-cultural sub-systems. Such fissures and cleavages hint at how tensions form; they help to explain divergence and exceptional behaviours in bilateral co-operations (Moosmüller 1997); and significantly they expose the use of stereotypes in inter-cultural management. Some writers have argued that cultural stereotypes are used to simplify heterogeneity within complex cultural groups (Lippmann 1965) although racism and discrimination are perhaps equally valid explanations. The use of stereotypes may manifest the level of inter-cultural competence, but they may also be a demonstration of, and used as, a particular tactic to impose power by ordering other cultures and establishing a degree of authority. Inter-cultural business relations rely not just on the interpretation of who is *present* in the other, foreign culture, but more who is *perceived* to be present (Roth 1996).

In the context of inter-cultural business communications in tourism, the preceding discussion raises two important points. The first concerns the paradox that power is central to the mediation of these inter-cultural business relations, but power discourse has not been wired into studies of inter-cultural business communications. This is a notable omission because, even where inter-cultural competence is high, conflicts cannot be eliminated (Bruck 1994; Stüdlein 1997; Gilbert 1998). In fact, they are relatively common occurrences and the types of 'competitive tactical encounters' involving power are routine in cross-cultural business relations. Questions arise therefore of how the individual protagonists perceive their positions over power, what power strategies and tactics they are prepared to use in conflict situations, and how their approaches to power map against their other organisational characteristics.

Conflicts frequently have a cultural background (Moran *et al.* 1993; Steinmann and Olbrich 1994) and they occur with the interaction of interdependent people who perceive incompatible goals and interference from each other in achieving those goals (Hocker and Wilmot 1995). Kopper (1992) identifies two principal, idealised approaches to conflict management (Table 10.6). Individualistic cultures deal with conflicts in a more dispassionate, matter-of-fact way. They are designed to get beyond the implications for particular individuals and instead work towards the future of the organisation. They are based – in theory at least – on a rational and objective assessment of the circumstances. When mapped against inter-cultural management approaches (Table 10.5), there are obvious overlaps with the culture-free approach. In collectivist cultures, conflict management is more culturally bound. Conflicts are viewed negatively, as more destructive forces; they are constructed and resolved in more emotive, subjective and ambiguous ways; and they are associated with individuals and groups whose status is questioned during the conflict.

Of course, individual and collectivist organisations may use any or all of the tactics outlined in Table 10.4 depending on the precise circumstances of the conflict. However, when read against tables 10.5 and 10.6, there is a suggestion that individualist, culture-free organisations may be more predisposed towards rational persuasion, exchange and rewards tactics given their devotion to the one best way. In contrast, collectivist organisations where culture and business are intricately and

Table 10.6 Inter-cultural differences in the context of conflict management

Aspects	*Individualist cultures*	*Collectivistic cultures*
Conflicts are seen as:	positive and productive	negative and destructive
Optimum approach:	conflict solution	conflict prevention
Conflicts are:	handled and solved	avoided and oppressed
Conflicts and persons are:	mostly separated	closely connected
The conflict parties orient to:	proceedings and solutions	people and 'face'
The conflict style is:	rational and factually oriented	intuitive and emotional
The procedure in the context of solving conflicts is:	open and direct	ambiguous and indirect

Source: adapted from Kopper (1992: 239).

messily bound, inspirational appeals, coalition tactics and consultation tactics may be more prevalent. This concern for tactics should not mask the importance of conflict management as a source of power. For conflicts to be resolved, especially where the protagonists originate from individualistic and collectivist cultures, there has to be a mutual recognition of each other's approach. Joint dialogue may advance creative solutions to improve future co-operations and it improves the actors' insights into, and understandings of, their counterparts' behavioural patterns (Bergemann and Volkema 1989; De Dreu 1997; Schneider and Barsoux 1999).

The second important issue in the consideration of inter-cultural business interactions in tourism is that the injection of power discourse requires us to rethink the structure and outcomes of international business relations. Cross-border activities may generate important economic outputs, but they are rarely straightforward and the economic gains for their protagonists are culturally preconditioned (Stüdlein 1997). Until now, readings of power in transnational business relations have consistently implied that there must be 'winners' and 'losers' in the process whereby one party's gain is the other's loss of power (cf. Bastakis *et al.* 2004: 153–154). Such views are rooted in Weberian notions of power, domination and authority and they suggest strongly that power is a finite capacity such that struggles over it are always set in a 'zero-sum game'. As Parsons (1963) has observed, power does not necessarily have to have a finite capacity. Empowerment is a culturally constructed and hence differentially understood term. Perceptions of whether (economic) empowerment has taken place and the relative shifts in empowerment vary depending on who makes them and their background. Significantly, this introduces the possibility that both parties may perceive themselves to be empowered through their unfolding inter-cultural commercial relations, or there is at least a perceived 'win–win', positive-sum outcome.

In the next section, we turn to consider the development of tourism and the emerging policy context in Morocco. Businesses from Morocco and Germany, with their contrasting and complex cultures which in part reflect very different colonial and migratory histories (cf. Bourdieu 1989), provide an opportunity to explore the processes and outcomes of inter-cultural interactions in a situation where the state has promoted tourism as a source of empowerment for indigenous entrepreneurs.

Key issues in recent tourism development in Morocco

According to Gray (2002), Morocco has been more successful than other North African states in harnessing the economic potentials of tourism. A systematic approach to tourism development was first evident in the 1960s with the creation of a five-year followed by a three-year plan to encourage modernisation (Kagermeier 1999). These plans extolled the advantages of international tourism. Package holidays, in particular, stimulated hard currency receipts, new employment opportunities, infrastructural improvements in local economies as well as domestic tourism (Müller-Hohenstein and Popp 1990). In the last forty years, there has been steady growth in tourism albeit the performance of the sector has fluctuated (cf. Barbier 1999; Sebbar 1999; Berriane and Popp 1999; Gray 2002; Kester 2003). In 1996, there were 2.7 million arrivals in Morocco and receipts were US$1.278 billion (Gray 2002: 398). In 2001, Morocco attracted an estimated 40 per cent of North African international tourist arrivals (second only to Tunisia) and 58.9 per cent of international tourism receipts (leading the market) (Kester 2003: 211). Its primary markets are mass package tourism based on the beach resorts and towns of Casablanca, Fez and Marrakech, independent travellers, and increasingly adventure tourists taking advantage of desert and mountain in the interior (Gray 2002: 399).

Recent tourism policy has been influenced by two critical issues. First, the Moroccan government has recognised the great contribution of tourism to the economy, and that the economy has frequently fallen victim to demand fluctuations (Gray 2002; Kester 2003). To support viable, sustained, long-term growth there is neither sufficient bedspace capacity of the quality and quantity required, nor a suitably diversified product portfolio. Second, since 2000 King Mohammed VI has directed the government to introduce a comprehensive programme of economic and social reform, with tourism as a key driver (Loverseed 2002). Early in 2001, the government in co-operation with the Central Association of Moroccan Business (CGEM) established a series of strategic guidelines and policy aims to harness more fully the economic potential of tourism.

This policy context has fostered, and will continue to demand, inter-cultural business exchanges. Tourism enterprises are encouraged to become more independent from the state to reap greater rewards from tourism and to spread the benefits through the country (Gray 2002). Not only does this desired process of empowerment compel them to seek out new markets and segments, but also it requires them to seek greater value from their transactions with tourists and their intermediaries from their existing (core) markets. This may be through increased visitor numbers, greater consumption of tours, excursions and peripheral offers, and/or the extraction of more favourable terms with their strategic partners. Rather than empower local businesses, such a policy has the potential to achieve the opposite by enhancing the power of overseas operators to act as information brokers and 'gatekeepers' to bedspaces (Ioannides 1998; Bastakis *et al.* 2004). Greater overseas visitor numbers and spend may be desired but, crucially, guests have to be 'delivered' from foreign markets. Moroccan destination marketing organisations overseas comprise one dimension of the infrastructure of supply (Loverseed 2002). 'Local' travel agents and tour operators in overseas markets function as important conduits of information about Morocco as well as retailers of Moroccan vacations especially in new niche markets such as golf tourism, meetings and incentives, desert safari tourism, skiing, mountaineering and trekking. Diversification of this type has created opportunities for, and requires the involvement of, Moroccan SMTEs. While these businesses are for the Moroccan government most welcome sector participants, they are also some of the most economically marginal; most sensitive to market fluctuations and the power of other businesses; and in need of assistance, co-operation and collaboration from Moroccan and overseas businesses and non-governmental organisations (Scherle 2006).

While Morocco retains strong links with France as the former colonial power, Moroccan tourism policy targets Germany, its businesses and NGOs because German tourists are already favourably predisposed towards North Africa as a destination, and there is a belief that visitors may be attracted away from Morocco's principal competitor, Tunisia (Scherle 2006). Thus, the success of current tourism policy and the prospects for long-term empowerment of Moroccan businesses depends in no small measure on the nature of cross-cultural communications and the way in which power relations are played out between indigenous enterprises and their strategic partners from countries like Germany.

Conflicts between German and Moroccan businesses

In order to explore the relationships between Moroccan and German businesses, a series of interviews was conducted between July 2000 and January 2002 (see Scherle 2004, 2006). A qualitative approach is especially well suited to sensitive and thorough analysis of culturally oriented research problems (Wiseman and Koester 1993; Kopp 2003). Qualitative methods are increasingly emerging as popular and powerful methods for exploring business practices in the tourism sector (Watkins and Bell 2002; Davies 2003; Bastakis *et al.* 2004). In total over 60 interviews were completed and during these interviews business attribute data were collected alongside information on their bipartite business relations with partners in Germany and Morocco respectively. In both samples, over 50 per cent were established businesses that had been operating for over ten years, over 60 per cent had less than 50 employees, and there were similarities in the levels of professional training (Table 10.7). Few had direct training in inter-cultural business management. Nevertheless, all but one German business had been co-operating for over a year, there was a relatively even spread of the durations of inter-cultural co-operations among firms in the two samples, and approximately 60 per cent of the individual German and Moroccan respondents had experience of inter-cultural business relations. Not indicated on this table, though, was that the primary mechanism for co-operation among the participating tour operators was a non-contractual arrangement, a collaboration form that allowed its actors an optimum of flexibility.

On the basis of the writings on inter-cultural management strategies and cultural conflict resolution outlined above, two broad sets of issues emerged as significant which are reported in more depth below: the sources of conflicts for the German and Moroccan interviewees in their collaborations with their Moroccan and German counterparts respectively, and the methods of conflict resolution.

Sources of conflict for German businesses

For the German respondents, the most frequently mentioned sources of conflict were adherence to time agreements for meetings and the way in which Moroccan businesses dealt with disputes (Table 10.8). For 20 per cent of German respondents, there were problems associated with how business was conducted on the Moroccan

Table 10.7 A comparison of the main attributes of the German and Moroccan respondents

Attributes	*German respondents no.*	*(%)*	*Moroccan respondents no.*	*(%)*
Average age of business (years)				
< 1	—		—	
1 – 5	2	(6.7)	5	(16.7)
5 – 10	5	(16.7%)	7	(20.0)
> 10	23	(76.7)	19	(63.3)
Size of business (employees)				
< 5	10	(33.3)	3	(10.0)
6 – 50	9	(30.0)	19	(63.3)
51–100	5	(16.7)	4	(13.3)
101–500	2	(6.7)	4	(13.3)
> 500	4	(13.3)	—	
Business involvement in inter-cultural co-operation (years)				
< 1	1	(3.3)	—	
1 – 5	10	(33.3)	9	(30.0)
5 – 10	9	(30.0)	12	(40.0)
> 10	10	(33.3)	9	(30.0)
Respondent's professional training				
apprenticeship	8	(26.7)	7	(23.3)
university	17	(56.7)	18	(60.0)
other	1	(3.3)	—	
no response	4	(13.3)	5	(16.7)
% of respondents with inter-cultural experience	19	(63.3)	18	(60.0)
% of respondents with intercultural training	3	(10.3)	1	(3.3)

Sources: Scherle (2004, 2006).

side in terms of attitudes to and style of invoicing, employee motivation and the general conduct of business transactions.

German business persons perceived conflicts in how their Moroccan partners connected their organisational behaviours to their religious and cultural practices. Ramadan, the month of fasting, and its implications for co-operation, was a particular and frequent source of contention. Moroccan working practices were comparatively rigid, a feature which becomes more significant around major religious festivals and events. As the manageress of a small incentive operator

Table 10.8 Frequencies by which typical sources of conflict were mentioned by German and Moroccan businesses

Source of conflict	*German businesses*	*Moroccan businesses*
Adherence to time agreements / punctuality	13	5
Dealing with conflicts	9	5
Decision-making	4	0
Delegation of tasks	5	1
Employee motivation	6	1
Information exchange	5	6
Investments	3	1
Invoicing	6	2
Marketing/public relations	1	2
Performance	2	5
Programme development	2	5
Style of working	6	4
Total number of businesses	30	30

Sources: Scherle (2004, 2006).

observed, Ramadan impacts not only on the potential to conduct 'everyday business', it is perceived to have a direct impact on person-to-person interactions:

> I have experienced little willingness to compromise, to adapt to Western culture; Ramadan being of course a good example. . . . This fasting has a tremendous impact. You notice an irritability, and the whole tempo is of course slowed down. . . . It affects our direct partners, who I'm talking about, who on their part are also dependent on their partners, i.e. the hoteliers, the people who rent out four-wheel-drive vehicles, the restaurants and so on, so that this slowing down practically multiplies. . . . And it is a really big problem that the German tour operators who should know about it also don't understand. They just come and say 'We need an answer tomorrow!', and when they [the Moroccans] say 'that is not possible, you know that', then they [the German tour operators] say 'that's your problem!'

Three features are revealed by this response: first, how easily and explicitly German interviewees' intercultural ignorance was exposed; second, the degree to which they were intolerant of differences, although many had inter-cultural experiences; and third, the magnitude of their concerns over the commercial difficulties presented to their businesses. These responses are clearly problematic in terms of racial stereotyping and show little attempt on the part of interviewees to respond positively or sensitively to cultural difference. Many other interviewees articulated the consequences of Ramadan for incoming tourists, such as restrictions in service which, in turn, lead to customer complaints and strained co-operations. Perhaps more concerning is that German business persons wanted a liberalisation of trade from the perceived constraints exerted by religious doctrine (Nienhaus 1996). However, they misjudged how difficult change would be. For instance, the Tunisian government's attempts to alter business practice during Ramadan, because of its far-reaching economic and social consequences, failed largely because of bitter resistance on the part of religious leaders (Dülfer 1997).

Personal relations played a central role in Morocco to a level that frequently troubled the German respondents. Friends and relations were frequently employed in the Moroccan businesses. Moroccan interviewees stressed that such interpersonal contacts are established and cultivated in Arab countries not only at business meetings but also in private settings. The intention is to create a network of trust which fulfils a crucial function giving security in countries that are traditionally associated with corruption (Heine 1996; Kuran 2004). Despite a basic awareness of these linkages, German respondents underestimated the role and outcomes of employing relatives. Within the (larger) German tour operators with pronounced horizontal and vertical business structures, such personalised business structures were judged as disconcerting. They had a propensity to induce acute conflicts in the day-to-day negotiation of business. A product manager of a leading German company summed up the frustrations of many German interviewees:

> One Moroccan firm we work with is closely linked to family. You'll hardly find a firm where people not belonging to the family occupy key positions. . . . Qualifications aren't important at all! That means, from the very beginning you have to expect great weaknesses in personnel because there are always positions, important positions, which are occupied in this firm by people who don't have the capacity, the know-how.

The product manager added that he could understand practically no personnel decision made on the Moroccan side in the last few years. While this may have been melodramatic, the consequences of 'familial' business structures for practical co-operation remained tangible such that German partners assumed the lead in their dealings with Moroccan partners:

> Over the years we have learned to take on more and more responsibility. That is, when they weren't pure agency tasks such as bus planning or day trip planning, tour leaders, local guides, then we took everything over . . . so that we relied less and less on the partner's agreement, especially in fundamental decisions, such as personnel decisions.

The cultural dimension also caused complications in the conception and implementation phases of a co-operation as the following makes clear:

> A lot of Moroccan partners present themselves at [trade] fairs in an excellent light – what they have, what they can do, and whom they know – under the motto: 'In each hotel there's a relative of mine'. As a circumstance, this is

> of course not bad at all, but in the end it's often just a promise, which is not compatible with reality. Clearly a lot of things are being described, but in the end there's nothing behind then. You really have to filter. . . . I mean, of course I can't look into anyone's head or into the enterprise if I am not familiar with how they work there. But generally it is of course always rather difficult, because everybody promises everything and then you face again that issue of 'Inshallah' [If Allah (God) wants it].

Set against this backdrop, it is hardly surprising that ever more German tour operators prefer to co-operate with an incoming agency that offers a bicultural background/management. For instance, there is evidence of an increasing number of agencies being conducted by Moroccan–German couples, where the German partner is seen as a mediator (bridging the gap) between the cultures (Scherle 2006).

The gendered social relations of tourism production were identified by several interviewees as sources of irritation. Despite all attempts at modernisation in recent years, many North African women are still bound to a religious and patriarchal value system oriented around the Koran (Moser-Weithmann 1999). In a male-oriented tourism industry such in Moroccon, the presence of a high proportion of women in senior positions among their German counterparts created friction. Overall, this issue was not as problematic as either working practices or familial structure but it resulted in several poignant, troubling experiences for the female interviewees concerned. One product manager of a medium-sized German tour operator observed that in her experience,

> Not every Muslim, Arab or Moroccan is open-minded enough to let a woman give him orders. I noticed that when I started to work in Morocco. The first three months were very hard, because they just tried to say: 'You woman, you European, you can't tell us anything.' Until one day the penny dropped, the tune changed, and after about three months I noticed, ah, they're accepting me as a woman, they're accepting what I say and they see that the work I do is good and gets results.

Sources of conflict for Moroccan businesses

In their collaborations with their German partners, Moroccan interviewees identified economic sources of conflict, although this strength of feeling is not really adequately registered in Table 10.8. The apparent profitability of German businesses, asymmetrical flows of information and capital, and the perception that the viability of Moroccan companies was irrelevant to their German partners were frequent, strongly articulated complaints. One product manager of a medium-sized incoming agency from Casablanca epitomised this view: 'The negative aspects in co-operation are reflected most especially in the fact that continually less is paid and more quality is expected. Our outgoings are too high in relation to the price paid.' Relatively frequently Moroccan interviewees bemoaned financial arrangements in bilateral co-operation. For example, the manager of a small agency from Tangiers articulated the view that German mismanagement and bankruptcy induces problems for the (seemingly) powerless Moroccans:

> In Germany there are at the moment many firms which go bankrupt. Many of these agencies have no finances! That is a risk and we have been affected twice: once by an Austrian agency with 120,000 Schillings [outstanding debt] and a second time with a German agency with 16,000 Marks. You know that that is a lot of money for us. When we lose 16,000 Marks that corresponds to the profit from 15 groups [of visitors]. . . . We cannot cope with such a loss. That is difficult for us.

Clearly, Moroccan interviewees were aggrieved at being expected to trust the good name of the German partners without the same status being afforded them in return. In fact, because of disadvantageous experiences with their German partners over finance, 'trust' had been divorced from its purely moral meaning. Instead, as the ironic comments below make clear, trust had assumed a more calculative meaning as an outcome, not the foundation of a transaction:

> There is one important thing: for us, trust is synonymous with money. Personally, if I don't know you – even if I find you congenial – you would not be allowed to send me your group before we have checked the billing. That's trust! . . . That's the reason why trust is synonymous for us with money. Pay me in time, I trust you. We could allow some days of delay, but it should not last too long.

Some Moroccan tour operators repeatedly referred to financial problems which stemmed from rationalisation in the German tourism sector through takeovers and insolvencies (Vorlaufer 1993, 1998; Freyer 2000; Bastakis *et al.* 2004). This had increased competition, especially among small Moroccan agencies, for collaborations with the diminishing number of German players in the market. The transnational activities of the larger overseas operators were viewed as threatening smaller-scale indigenous enterprises in Morocco. Whether rational or not, fear of globalisation was followed by the view that the takeover of local businesses would then put further pressure on other types of businesses elsewhere in the tourism value chains in a destination:

> Globalization is a problem for many incoming-agencies. In three or four years the huge tour-operators will take over all the small agencies. That's not good at all for small destinations like Agadir. Our firm is not so much affected, because we co-operate with solid and independent enterprises. But of course that cannot rule out the possibility that one day we also might be taken over by a huge tour-operator. That's really bad for a destination like Agadir.

Many small operators had to forge alliances with one or more of the best-known (larger) German tour operators. While this broadened their product range, they had hoped to obtain further insurance against market conditions and fluctuating demand. Moroccan interviewees felt this had served to empower the German partners because there were fewer of them, they were being chased by broadly the same number of potential Moroccan partners, and they felt they could establish better terms from their potential partners. The situation conspired to make the smaller Moroccan businesses feel more vulnerable still: many forms of co-operation were not based on contractual agreements, with the result that German tour operators could switch partners and operating conditions relatively easily. For some Moroccan interviewees, these 'partnerships' with German businesses were 'unfair' practices. One agency manager in the incentives sector noted:

> Difficulties and misunderstandings arise in the course of co-operation because the partner is looking for a new product every time although Morocco remains unchanged. We cannot always create a new product for him. Also, our partner doesn't like our way of working. We are always supposed to change our methods because of this and follow theirs. Moreover, we once went to court, because a German customer gave us a cheque that bounced. In such difficult situations we try to find a compromise. But if we are in a weak position we sometimes have to give up.

In practice, changing demand conditions induced further dilemmas. Product portfolio diversification remains a central dimension of Moroccan tourism policy, and it was welcomed by the German partners. However, the Moroccan businesses were worried about product development costs and the ephemerality of tourist demand. Their concern was that just as the new product would be introduced, demand would be diverted elsewhere. This perspective may be read from a Western perspective as representing a 'lack of initiative and entrepreneurial spirit', but it is not without some justification because Gray (2002) reports that Morocco suffers from a lack of repeat visitors. Interculturalists might argue that this relates to the so-called 'fatalism hypothesis', which has its origins in the view of Islamic theologians that human lives are largely preordained (Dülfer 1997; Kutschker and Schmid 2002). Whatever interpretation is placed on this situation the Moroccan respondents clearly felt the pressure to diversify reflected the asymmetrical power relations with their German business partners.

Conflict resolution: inter-cultural perspectives

Given the tensions stemming from cultural differences and market pressures, especially to diversify, it is no surprise that respondents highlighted conflicts that arose. When conflict occurred three principal approaches to resolution were identified as overwhelmingly popular and highly successful for both German and Moroccan interviewees: 'open discussions when needed' (>90 per cent of respondents); 'employee discussions' (>50 per cent); and 'regular team meetings' (>32 per cent). In each case less than 7 per cent of the businesses that employed these tactics reported poor success. 'Legal resolution' and 'management consultants' were used as conflict resolution by less than 5 per cent and 2 per cent of businesses respectively and each had poor success. Fewer than 17 per cent of businesses attempted to solve conflicts by 'pressure and compulsion', 'arbitration', 'inter-cultural training', and 'reputation' alone, and these were viewed by less than 7 per cent as successful (see Scherle 2006 for a fuller discussion).

Thus, the most frequently applied and successful methods were those that had a clearly personalised and discursive nature. The small and medium-sized nature of the businesses interviewed, combined with the often long-term nature of their bilateral relationships, meant that many had great volumes of social capital invested with their partners. Rather than recourse to the law or investment in more expensive solutions such as management consultants or inter-cultural training programmes, the businesses could fall back on their close personal contacts in order to fashion a solution through dialogue. Moreover, such solutions were often characterised by remarkable pragmatism on both sides. As one manager of a German tour operator wryly observed:

> I always say that there is a legal and a business solution . . . and we are always for the business solution, although I am a lawyer. This sector is much too small for legal solutions. In the 30 or 20 years I have been working, I have only twice gone for the legal solution and every other time the business solution. You keep running into people! And you cannot work in the future with someone you are at legal loggerheads with. . . . Everybody turns up again, our sector is much too small for that.

A senior employee of a Moroccan incoming agency in Agadir noted in a similarly realistic manner, albeit with a slight tone of despair, that 'we compromise so as not to lose our partner and to have work all year round'. This sentiment was repeated frequently, and it encapsulated a common anxiety among Moroccan partners: not only could they be the victims of commercial exploitation by their overseas partner, but also that they were too vulnerable to failure as a result of the increasing globalisation of tourism production and consumption. Increased concentration among German tour operators and the disappearance of previous partners provoked many small and medium-sized Moroccan operators to sense themselves as more open to exploitation because of the greater difficulties in establishing beneficial collaborations. As the managing director of an incoming agency in Agadir opined:

> It is becoming more and more difficult to find business partners, as globalisation is made for those who are already strong and who become even stronger with globalisation. When you talk about globalisation you have to automatically think of the three large groups in the Moroccan market: TUI, Neckermann [now Thomas Cook] and LTU. These are the biggest, who will benefit, and for us everything will get harder.

There were also complaints that the purchasing power of the large groups is growing, and that local tourism businesses may be further marginalised, especially in the centres of mass tourism (primarily Agadir) where the global players are established and dominant in all types of tourism. Interviewees noted variations in outlook for co-operations in different markets. Moroccan agencies specialising in narrow market segments (such as desert and trekking tourism, thalassic therapy) reported better prospects. They looked back on long years of often exclusive co-operation with small and medium-sized niche operators in Germany, whose customer numbers have remained relatively constant. Given the diversification of tourism production pursued by the Moroccan government especially in peripheral areas, incoming agencies in established co-operations, and who are able to respond flexibly to market conditions, considered they would have an advantage in deflecting some of the competitive pressures from larger operators. However, this would be reliant on a highly personalised style of co-operation, sometimes even based on genuine friendships. Both would facilitate the solution of potential conflicts in an uncomplicated manner, but ironically both were the sort of personal relationships that troubled many German businesses.

Discussion: reconciling inter-cultural positions

When the empirical material is mapped against the conceptual framework (Tables 10.1–10.6), significant juxtapositions begin to emerge. German businesses mainly subscribed to a 'culture-free' approach with their Moroccan partners (Tables 10.5–10.6). They were, by and large, above all concerned with the effective and optimal operation of commercial partnerships. Profit and efficiency were motive forces whereas cultural aspects were mainly seen as a burden not as a resource. German respondents' relationships with Moroccan partners were on the whole positive, although their behaviour was characterised by relatively low levels of inter-cultural competence and a propensity to revert to stereotypical views of Moroccan business, culture and society. Cultural barriers to effective business such as time management, conflict resolution and gender-specific experiences proved more irksome than business-related differences. Strict adherence to religious doctrine as well as the frequent use of family and relatives in the Moroccan labour force were disconcerting.

The 'culture-free' approach of the German businesses is further evident in their strategies and tactics for dealing with their Moroccan partners. Their initial strategies have strong resonances with Lukes's second and third dimensions of power. German businesses attempted to keep culture off the commercial agenda through their individual business practices. Furthermore, they attempted to establish what was in their Moroccan counterparts' 'real' interests: namely, to adopt a much more systematic approach to business operations, reminiscent of German (i.e. Western) methods and best practices in which business is for business's sake. Rather than an asset, the entwining of family and culture in operations presented a threat to the 'legitimate' and expert power plays in their culture-free approach (Table 10.2). Knowledge, remunerative power and, where necessary, coercive power were the bases of power while in their structural position, expertise and opportunity were the sources of power (Table 10.3). Tactics of rational persuasion, exchange and pressure were used as basic starting points in their commercial dealings with Moroccan partners (Table 10.4).

The German businesses' preference for non-contractual arrangements might be seen as hypocritical in this context, but it reflects their relative empowerment. Where German businesses perceived themselves in control of a commercial relationship, they felt suitably empowered to work with non-contractual arrangements. Although they were prepared to tolerate what they saw as the negative aspects of working with Moroccan partners, ultimately they perceived control because they felt their commercial lessons were being received and they held the ultimate sanction: they could terminate the relationship at any time to lessen their exposure to risk. Non-payment was also used as a punitive measure to reinforce control. German businesses felt that Moroccan entrepreneurs needed a commercial arrangement with them more than they needed one with the Moroccans. Although Moroccan partners were demanding better terms and conditions, they were chasing relationships with a declining number of German operators, businesses on whom they relied to deliver large numbers of visitors, and businesses who understood their dependency through the Moroccan policy context that promoted diversification.

Culture and business were on the whole closely integrated for Moroccan enterprises. They did not isolate obvious differences in cultural traits with their German colleagues as principal sources of conflict, nor did they revert to stereotypes of German culture as explanations of business tensions. Major complaints surrounded unequal power relations, commercial demands and flows of information which favoured German businesses, who were perceived to value their own profitability at the expense of their Moroccan partners. Importantly, economic considerations were not separated from cultural ones. To Moroccan enterprises, the unfamiliar business practices and allegedly unfair position their partners had assumed were read as synonymous with globalising Western culture, values and social practices; that is, the 'German business model' was read by the Moroccans not as culture free, not detached from culture, but as bound to Western culture.

Moroccan businesses used overt strategies more reminiscent of Lukes's first dimension of power. For them, there were certain fundamental issues that they overtly sought to get on the agenda in their dealings with their German partners. These included consideration of basic terms such as payment and cash flow, as well as (quality) standards of service and the nature of routine operations. Knowledge and remunerative power were the bases of power (Table 10.3) while expertise and opportunity typically formed the sources of power. Upward appeals, ingratiation, persuasion and inspirational appeals were the tactics initially used in their dealings with German partners (Table 10.4). Moroccan enterprises had bases of expert power and informational power (Table 10.2) given their knowledge of local culture, markets and service opportunities, but their reward and hence coercive power through their ability to frustrate service relations was latent power because of the way the German enterprises controlled the agenda. The Moroccan respondents had more modest expectations from their relationships with German partners. As lifestyle enterprises, empowerment was manifest in their being able to enter commercial arrangements with overseas partners, retain them and to grow the relationship; to extract even modestly better terms and conditions was even more desirable given the marginal nature of many of the businesses and the substantial benefits the German businesses could deliver. Moroccan businesses felt disempowered partly because government policy made them more vulnerable through the need to develop new products and markets. It was the Moroccan businesses, not the German partners, who had to carry the costs and risks of product development.

On this basis, one may have expected the German and Moroccan businesses to adopt different conflict resolution strategies. This was not the case. Both were loathe to seek redress in the law or costly commercial solutions. Instead, both often reverted to the long-term social capital they had developed with their co-operators as the basis for reconciliation. Once conflict precipitated, German businesses, like their Moroccan partners, participated in solutions which at first glance subscribed to the culturally bound approach. Personal relationships, discussions and meetings were primarily used to resolve differences and to reinforce the commercial relationships. When unpacked further, it is clear that some elements of the individualistic culture approach to conflict management were present: namely, conflicts occur and have to be solved; they are rooted in 'fact'; and they are open and direct. Similarly, several characteristics of a collective 'industry' culture were evident such as the perception of conflicts as negative and destructive, closely associated with individuals and status, and intuitively played out.

Among the German respondents, disagreements heralded a change in tactics. Prior to conflict, German businesses adopted more sophisticated, subtle power tactics based on what to exclude from the agenda and how to manipulate the Moroccans' real interests. Once disagreement emerged, they retreated back to positions of domination, authority, control of the agenda and coercion through the threat of sanctions, were they unable to find a solution. However, it was precisely in such circumstances when the apparently powerless could deploy their control strategies and where the Moroccans were further empowered. Disagreement cut through layers of pretence and dependency. There was then a more open discussion of cost-benefits to the respective parties. German businesses and their officers were independently responsible to investors and shareholders (cf. Watkins and Bell 2002: 15): for reasons of the balance sheet and corporate accountability, could they really afford either to enter a potentially costly legal case at home or overseas, or to deny themselves turnover and margin because they no longer wished to sell their erstwhile partners' products? In a game of brinkmanship, the Moroccans effectively implied that the German position also had an element of bluff about it: the German businesses and their employees, like the Moroccans, were vulnerable to market conditions, albeit in a different manner. Thus, it was often in disputes that the Moroccans were further empowered and where they were able to exert the extra leverage over their German partners to improve their situation significantly. This situation partly reflects the initial asymmetrical power relations but also reveals the contradictory nature of the operation of power; that is, during disputes, when it might be expected that those with greater resources might assert themselves, it is in fact the seemingly weaker partners who through the use of particular tactics have greater opportunity to enhance their position and power.

Conclusion

The nature of tourism production requires interaction among several businesses in order to engineer the tourism experiences consumed by visitors (Watkins and Bell 2002). These interactions, like other social relations, are political episodes in which power is inescapably present and central to their resolution. Power in commercial relationships should not be solely restricted to a simple assessment of the debits and credits evident in a transaction. Rather, the construction and framing of the transaction are crucial. In the case of international tourism, the politics of production are mediated between partners across cultural as well as political divides. As the example of German and Moroccan SMTEs highlights, power is understood differently across cultures and empowerment is a culturally constructed term. The politics of production were contested, with organisations and individuals in each country making specific valorisations of their sources and bases of power prior to formulating actor strategies and devising an array of tactics to cope with their overseas partners. German–Moroccan relations demonstrate that power relations can resolve in a 'win–win' relationship at the end of which both parties may perceive themselves to be empowered. German businesses perceived in improving service delivery for their customers and a higher standard of collaboration that the Moroccans were acknowledging their business practices, and that they had the potential at least to withdraw at any time. Domination, authority and coercion may have been elements in their conflict resolution strategies but these revealed their vulnerability through their responsibilities to deliver high quality and more diverse opportunities for their paying customers and value to their managers and shareholders. In contrast, the Moroccan businesses perceived their greater access to information, protected and nurtured the delivery of visitors, and extracted better terms and conditions in spite of greater competition for partnerships and the threats of globalisation. As the Moroccan businesses revealed, even those who apparently have little power can form potent strategies that take advantage of what appear to be limited weakness in their opponents to empower themselves.

There are three wider implications for studies of power in tourism. First, Foucauldian perspectives may be currently in vogue in the social sciences and in tourism studies (Few 2002; Hannam 2002). Aspects of Weberian thinking may be evident in some accounts of tourism empowerment (Kayat 2002; Sofield 2003). However, there are several contested conceptual approaches to power which offer exciting multiple possibilities for excavating power relations in tourism (Haugaard 2003). Theory and concept should be used with discrimination, and vocabularies of power allow power and its function to be named in a more precise and insightful manner. As a second issue, it is important to note that a power approach has distinctive methodological connotations and it raises dilemmas for the research worker. Questions emerge as to where, when and how to look for power and the mechanisms by which it operates (Few 2002). Exercise fallacies may be common faults in power analysis from a philosophical analysis (Morriss 2002), but in a practical context power is often used within and between organisations in a far more basic, instrumental manner (Buchanan and Badham 1999). Often power relations take place in relatively modest, even banal settings and, as Clegg (1989: 17–18) reminds us, power can be 'a far less massive, oppressive and prohibitive apparatus than it is often imagined to be'. As disagreeable to some critics as an inspection of tactics at the level of the individual may be, such tactics are vital to the prosecution of power relations especially in cross-cultural situations. Webs of relations exist between organisations and individuals in the real life of tourism production rather than singular dyadic connections of the textbook. Within complex, messy arrangements it is important to identify the micro-practices, the actor strategies, the tactics and 'tricks of the trade' used in 'turf games'. Finally, this chapter has concentrated on empowerment through inter-cultural business relations. However, notions of empowerment loom large once again in the major contemporary global debates on the role of tourism in poverty alleviation, social exclusion, environmental change, and safety, peace and harmony. Empowerment may be an aspiration of politicians, policy-makers and the powerless, but it is seldom a simple process to engineer as contextual power relations will shape complex and even unexpected outcomes.

References

09/11 Commission (2004) *The 09/11 Commission Report. Final Report of the National Commission on Terrorist Attacks Upon the United States*. Authorized edition. New York: W.W. Norton.

Allen, J. (2003) *The Lost Geographies of Power*. Oxford: Blackwell.

Ateljevic, I. and Doorne, S. (2000) '"Staying within the fence": lifestyle entrepreneurship in tourism', *Journal of Sustainable Tourism*, 8(5): 378–392.

Bachrach, P. and Baratz, M.S. (1962) 'Two faces of Power', *American Political Science Review*, 56: 941–952.

Bachrach, P. and Baratz, M.S. (1970) *Power and Poverty: Theory and Practice*. New York: Oxford University Press.

Bacharach, S.B. and Lawler, E.J. (1980) *Power and Politics in Organizations. The Social Psychology of Conflict, Coalitions and Bargaining*. San Francisco, CA: Jossey-Bass.

Barbier, J. (1999) 'Tourisme et développement régional dans la stratégie touristique du Maroc', in M. Berriane and H. Popp (eds) *Le Tourisme au Maghreb: diversification du produit et développement local et régional*. Rabat: Publications de la Faculté des Lettres et des Sciences Humaines.

Bastakis, C., Buhalis, D. and Butler, R. (2004) 'The perception of small- and medium-sized tourism accommodation providers on the impacts of the tour operators' power in Eastern Mediterranean', *Tourism Management*, 25: 151–170.

Benfari, R.C., Wilkinson, H.E. and Orth, C.D. (1986) 'The effective use of power', *Business Horizons*, 29 May–June: 12–16.

Bergemann, T.J. and Volkema, R.J. (1989) 'Understanding and managing interpersonal conflict at work: its issues, interactive processes, and consequences', in M.A. Rahim (ed.) *Managing Conflict: An Interdisciplinary Approach*. Westport, CT: Praeger.

Berriane, M. and Popp, H. (eds) (1999) *Le Tourisme au Maghreb: diversification du produit et développement local et régional*. Rabat: Publications de la Faculté des Lettres et des Sciences Humaines.

Bianchi, R. (2002) 'Towards a new political economy of global tourism', in R. Sharpley and D. Telfer (eds) *Tourism and Development: Concepts and Issues*. Clevedon: Channel View.

Bosch, B. (1997) 'Interkulturelles management', in H. Reimann (ed.) *Weltkultur und Weltgesellschaft: Aspekte globalen Wandels; zum Gedenken an Horst Reimann (1929–1994)*. Opladen: Westdeutscher Verlag.

Bourdieu, P. (1989) *Satz und Gegensatz: Über die Verantwortung des Intellektuellen*. Berlin: Wagenbach.

Britton, S. (1991) 'Tourism, capital, and place: towards a critical geography of tourism', *Environment and Planning D: Society and Space*, 9: 451–478.

Brown, K. and Rosendo, S. (2000) 'Environmentalists, rubber tappers and empowerment: the politics and economics of extractive reserves', *Development and Change*, 31: 201–227.

Bruck, P.A. (1994) 'Interkulturelle Entwicklung und Konfliktlösung. Begründung und Kontextualisierung eines Schwerpunktthemas für universitäre Forschung', in K. Luger and R. Renger (eds) *Dialog der Kulturen. Die multikulturelle Gesellschaft und die Medien*. Wien: Österreichischer Kunst- und Kulturverlag.

Bruins, J. (1999) 'Social power and influence tactics: a theoretical introduction', *Journal of Social Issues*, 55(1): 7–14.

Buchanan, D. and Badham, R. (1999) *Power, Politics and Organizational Change. Winning the Turf Game*. London: Sage.

Buchanan, D. And Huczynski, A. (2004) *Organizational Behaviour. An Introductory Text*, fifth edition. Harlow: FT Prentice Hall.

Bugental, D.B. and Lewis, J.B. (1999) 'The paradoxical misuse of power by those who see themselves as powerless: how does it happen?', *Journal of Social Issues*, 55(1): 51–54.

Casmir, F.L. and Asuncion-Lande, N. (1989) 'Intercultural communication revisited: conceptualization, paradigm building, and methodological approaches', in J. A. Anderson (ed.) *Communication Yearbook*, twelfth edition. Newbury Park, CA: Sage.

CGEM (2000) *Contrat Programme 2000–2010. La relance de la croissance du Royaume à travers un développement accéléré de son tourisme*. No place: CGEM.

Clancy, M. (1998) 'Commodity chains, services and development: theory and preliminary evidence from the tourism industry', *Review of International Political Economy*, 5(1): 122–148.

Clegg, S. (1989) *Frameworks of Power*. London: Sage.

Collier, M.J. (1989) 'Cultural and intercultural communication competence: current approaches and directions for future research', *International Journal of Intercultural Relations*, 13(3): 287–302.

Dahles, H. and Bras, K. (eds) (1999) *Tourism and Small Entrepreneurs. Development, National Policy, and Entrepreneurial Culture: Indonesian Cases*. New York: Cognizant Communication Corporation.

Davies, B. (2003) 'The role of quantitative and qualitative research in industrial studies of tourism', *International Journal of Tourism Research*, 5(2): 97–112.

De Dreu, C.K.W. (1997) 'Productive conflict: the importance of conflict management and conflict issue', in C.K.W. De Dreu and E. Van de Vliert (eds) *Using Conflict in Organizations*. London: Sage.

Doorne, S. (1998) 'Power, participation and perception: an insider's perspective on the politics of the Wellington Waterfront redevelopment', *Current Issues in Tourism*, 1(2): 129–166.

Dülfer, E. (1997) *Internationales Management in unterschiedlichen Kulturbereichen*, fifth edition. Munich: Oldenbourg.

Economist, The (2003) 'Could and should do better: Poland should be smart as well as tough', *The Economist* 22 November (A Survey of EU Enlargement): 10.

Few, R. (2002) 'Researching actor power: analyzing mechanisms of interaction in negotiations over space', *Area*, 34(1): 29–38.

French, J.R.P and Raven, B.H. (1959) 'The bases of social power', in D. Cartwright (ed.) *Studies in Social Power*. Ann Arbor, MI: Institute for Social Research.

Freyer, W. (2000) 'Globalisierung in der Tourismuswirtschaft', in S. Landgrebe (ed.) *Internationaler Tourismus*. Munich: Oldenbourg.

Gesteland, R.R. (2005) *Cross-cultural Business Behaviour: Negotiating, Selling, Sourcing and Managing across Cultures*, fourth edition. Copenhagen: Copenhagen Business School Press.

Gilbert, D.-U. (1998) *Konfliktmanagement in international tätigen Unternehmen. Ein diskursethischer Ansatz zur Regelung von Konflikten im interkulturellen Management.* Sternenfels: Verlag Wissenschaft und Praxis.

Gray, M. (2002) 'The political economy of tourism in North Africa: comparative perspectives', *Thunderbird International Business Review*, 42(4): 393–408.

Hall, C.M. and Higham, J. (2005) 'Introduction: tourism, recreation and climate change', in C.M. Hall and J. Higham (eds) *Tourism, Recreation and Climate Change*. Clevedon: Channel View.

Hannam, K. (2002) 'Tourism and development I: globalization and power', *Progress in Development Studies*, 2(3): 227–234.

Haugaard, M. (2003) 'Reflections on seven ways of creating power', *European Journal of Social Theory*, 6(1): 87–113.

Heine, P. (1996) *Kulturknigge für Nichtmuslime: Ein Ratgeber für alle Bereiche des Alltags*, second edition. Freiburg: Herder.

Hocker, J.L. and Wilmot, W.W. (1995) *Interpersonal Conflict*, fourth edition. Madison, WI: Brown and Benchmark.

Hofstede, G. (1982) *Culture's Consequences. International Differences in Work-related Values*. Newbury Park: Sage.

——(1999) 'The universal and the specific in 21st-century global management', *Organizational Dynamics*, 28(1): 34–43.

Huczynski, A.A. (1996) *Influencing within Organizations: Getting in, Rising up and Moving on*. Hemel Hempstead: Prentice Hall.

Hunter, W.C. (2001) 'Trust between culture: the tourist', *Current Issues in Tourism*, 4(1): 42–67.

Ioannides, D. (1998) 'Tour operators: the gatekeepers of tourism', in D. Ioannides and K.G. Debbage (eds) *The Economic Geography of the Tourist Industry: A Supply-side Analysis*. London: Routledge.

Jack, G. and Phipps, A. (2005) *Tourism and Intercultural Exchange. Why Tourism Matters*. Clevedon: Channel View.

Johnson, D. and Turner, C. (2003) *International Business. Themes and Issues in the Modern Global Economy*. London: Routledge.

Kagermeier, A. (1999) 'Neue staatlich geförderte Tourismusprojekte in Marokko und Tunesien und ihre Rolle für die Entwicklung peripherer Räume', in H. Popp (ed.) *Lokale Akteure im Tourismus der Maghrebländer. Resultate der Forschungen im Bayerischen Forschungsverbund FORAREA 1996–1998*. Passau: LIS Verlag.

Kayat, K. (2002) 'Power, social exchanges and tourism in Langkawi: rethinking resident perceptions', *International Journal of Tourism Research*, 4: 171–191.

Kester, J.G.C. (2003) 'Databank: international tourism in Africa', *Tourism Economics*, 9(2): 203–221.

Kipnis, D., Schmidt, S.M. and Wilkinson, I. (1980) 'Intraorganizational influence tactics: explorations in getting one's own way', *Journal of Applied Psychology*, 65: 440–452.

Kipnis, D., Schmidt, S.M., Swaffin-Smith, C. and Wilkinson, I. (1984) 'Patterns of managerial influence: shotgun managers, tacticians and bystanders', *Organizational Dynamics*, Winter: 58–67.

Knights, D. and McCabe, D. (2002) 'A road less travelled. Beyond managerialist, critical and processual approaches to total quality management', *Journal of Organizational Change*, 15(3): 235–234.

Knights, D. and Willmott, H. (1999) *Management Lives. Power and Identity in Work Organizations*. London: Sage.

Kopp, H. (ed.) (2003) *Area Studies, Business and Culture: Results of the Bavarian Research Network forarea*. Münster: LIT.

Kopper, E. (1992) 'Multicultural workgroups and project teams', in N. Bergemann and A. Sourisseaux (ed.) *Interkulturelles Management*. Heidelberg: Physica.

Kühlmann, T.M. and Stahl, G.K. (1998) 'Diagnose interkultureller Kompetenz. Entwicklung und Evaluierung eines Assessment Centers', in C.I. Barmeyer and J. Bolten (eds) *Interkulturelle Personalorganisation*. Sternenfels: Verlag Wissenschaft und Praxis.

Kuran, T. (2004) *Islam and Mammon: The Economic Predicaments of Islamism*. Princeton, NJ: Princeton University Press.

Kutschker, M. (2001) 'Internationale Kooperationen kleiner und mittelständischer Betriebe', paper presented at the FORAREA Conference Kultur im Internationalen Management, in Berlin, December.

Kutschker, M. and Schmid, S. (2002) *Internationales Management*. Munich: Oldenbourg.

Latour, B. (1986) 'The powers of association', in J. Law (ed.) *Power, Action and Belief: A New Sociology of Knowledge*. London: Routledge.

Lippmann, W. (1965) *Public Opinion*. New York: The Free Press.

Liu, Z. (2003) 'Sustainable tourism development: a critique', *Journal of Sustainable Tourism*, 11(6): 459–475.

Loverseed, H. (2002) *Travel and Tourism in Morocco*. London: Mintel International Group Ltd.

Lukes, S. (1974) *Power: A Radical View*. Basingstoke: Macmillan.

—— (2005) *Power: A Radical View*, second edition. Basingstoke: Palgrave.

McKinlay, A. and Starkey, K. (eds) (1998) *Foucault, Management and Organization Theory*. London: Sage.

Moosmüller, A. (1997) *Kulturen in Interaktion. Deutsche und US-amerikanische Firmenentsandte in Japan*. Münster: Waxmann.

Moran, R.T., Harris, P.R. and Stripp, W.G. (1993) *Developing the Global Organization: Strategies for Human Resource Professionals*. Houston, TX: Gulf Publishing Company.

Morriss, P. (2002) *Power. A Philosophical Analysis*. Manchester: Manchester University Press.

Mosedale, J.T. (2005) 'Capital mobility in the tourism sector: an analysis of integrated corporations', unpublished paper presented at 'The End of Tourism? Mobility and Local–Global Connections', 23–24 June, Centre for Tourism Policy Studies, University of Brighton, Eastbourne, UK.

Moser-Weithmann, B. (1999) 'Kulturüberprägung durch Tourismus – Sozio-kulturelle Konfliktsituationen tunesischer Frauen im touristisch bedingten Akkulturationsprozess', in H. Popp (ed.) *Lokale Akteure im Tourismus der Maghrebländer. Resultate der*

Forschungen im Bayerischen Forschungsverbund FORAREA 1996–1998. Passau: LIS Verlag.

Mowforth, M. and Munt, I. (1998) *Tourism and Sustainability. Development and New Tourism in the Third World*. London: Routledge.

Müller-Hohenstein, K. and Popp, H. (1990) *Marokko: Ein Islamisches Entwicklungsland mit kolonialer Vergangenheit*. Stuttgart: Klett.

Nienhaus, V. (1996) 'Wirtschaftsordnungen im Islam', *Geographische Rundschau*, 48(6): 366–371.

Osterloh, M. (1994) 'Kulturalismus versus Universalismus. Reflektionen zu einem Grundlagenproblem des interkulturellen Managements', in B. Schiemenz and H.-J. Wurl (eds) *Internationales Management: Beiträge zur Zusammenarbeit*. Wiesbaden: Gabler.

Parsons, T. (1963) 'On the concept of political power', *Proceedings of the American Philosophical Society*, 107(3): 232–262.

Popp, H. (ed.) (1999) *Lokale Akteure im Tourismus der Maghrebländer. Resultate der Forschungen im Bayerischen Forschungsverbund FORAREA 1996–1998*. Passau: LIS Verlag.

Pütz, R. (2003) 'Kultur und unternehmerisches Handeln –Perspektiven der Transkulturalität als Praxis', *Petermanns Geographische Mitteilungen*, 147(2): 76–83.

Raven, B.H. (1965) 'Social influences and power', in I.D. Steiner and M. Fishbein (eds) *Current Studies in Social Psychology*. New York: Wiley.

—— (1974) 'The comparative analysis of power and power preference', in J.T. Tedeschi (ed.) *Perspectives on Social Power*. Chicago, IL: Aldine.

—— (1992) 'A power/interaction model of interpersonal influence: French and Raven thirty years later', *Journal of Social Behaviour and Personality*, 7: 217–244.

Reed, M.G. (1997) 'Power relations and community-based tourism planning', *Annals of Tourism Research*, 24(3): 566–591.

Reisinger, Y. and Turner, L.W. (2003) *Cross-cultural Behaviour in Tourism: Concepts and Analysis*. Oxford: Butterworth-Heinemann.

Roth, K. (1996) 'Europäische Ethnologie und interkulturelle Kompetenz', in K. Roth (ed.) *Mit der Differenz leben: Europäische Ethnologie und Interkulturelle Kommunikation*. Münster: Waxmann.

Rugman, A.M. and Hodgetts, R.M. (2003) *International Business*, third edition. Harlow: FT Prentice Hall.

Ryan, C. (2002) 'Equity, management, power sharing and sustainability – the issues of the "new" tourism', *Tourism Management*, 23: 17–26.

Scherle, N. (2004) 'International bilateral business in the tourism industry: perspectives from German–Moroccan co-operations', *Tourism Geographies*, 6(2): 229–256.

—— (2006) *Bilaterale Unternehmenskooperationen im Tourismussektor vor dem Hintergrund ausgewählter Erfolgsfaktoren*. Wiesbaden: Gabler (Management International Review Edition).

Scheyvens, R. (1999) 'Ecotourism and the empowerment of local communities', *Tourism Management*, 20: 245–249.

—— (2000) 'Promoting women's empowerment through involvement in ecotourism: experiences from the Third World', *Journal of Sustainable Tourism*, 8(3): 232–249.

Schneider, S.C. and Barsoux, J.-L. (1999) *Managing across Cultures*. London: Prentice Hall.

Sebbar, H. (1999) 'Evolution du tourisme au Maroc', in M. Berriane and H. Popp (ed.) *Le Tourisme au Maghreb: diversification du produit et développement local et régional*. Rabat: Publications de la Faculté des Lettres et des Sciences Humaines.

Smith, M. and Duffy, R. (2003) *The Ethics of Tourism Development*. London: Routledge.

Sofield, T.H.B. (2003) *Empowerment for Sustainable Tourism Development*. Oxford: Pergamon Press.

Steinmann, H. and Olbrich, T. (1994) 'Unternehmensethik und internationales Management. Implementationsprobleme einer Unternehmensethik der internationalen Unternehmung', in B. Schiemenz and H.-J. Wurl (eds) *Internationales Management: Beiträge zur Zusammenarbeit*. Wiesbaden: Gabler.

Stüdlein, Y. (1997) *Management von Kulturunterschieden. Phasenkonzept für internationale strategische Allianzen*. Wiesbaden: Deutscher Universitäts-Verlag.

Thomas, A.B. (2003) *Controversies in Management. Issues, Debates, Answers*, second edition. London: Routledge.

United Nations Environment Programme (UNEP) (2002) How tourism can contribute to socio-cultural conservation. Online document available from: www.uneptie.org/pc/tourism/sust-tourism/soc-global.htm [last accessed 14 November 2005].

Vorlaufer, K. (1993) 'Transnationale Reisekonzerne und die Globalisierung der Fremdenverkehrswirtschaft: Konzentrationsprozesse, Struktur- und Raummuster', *Erdkunde*, 47(4): 267–281.

—— (1998) 'Die Globalisierung der Tourismuswirtschaft', in H. Gebhardt, G. Heinritz and R. Wiesner (eds) *Europa im Globalisierungsprozeß von Wirtschaft und Gesellschaft*. Stuttgart: Steiner.

Ward, C., Bochner, S. and Furnham, A. (2001) *The Psychology of Culture Shock*, second edition. London: Routledge.

Warner, M. and Joynt, P. (2002) *Managing Across Cultures: Issues and Perspectives*, second edition. London: Thomson.

Watkins, M. and Bell, B. (2002) 'The experience of forming business relationships in tourism', *International Journal of Tourism Research*, 4(1): 15–28.

Weaver, D. (2004) 'Tourism and the elusive paradigm of sustainable development', in A.A. Lew, C.M. Hall and A.M. Williams (eds) *A Companion to Tourism*. Oxford: Blackwell.

Williams, A.M. (2004) 'Towards a political economy of tourism', in A.A. Lew, C.M. Hall and A.M. Williams (eds) *A Companion to Tourism*. Oxford: Blackwell.

Wiseman, R. L. and Koester, J. (eds) (1993) *Intercultural Communication Competence*. Newbury Park: Sage.

World Tourism Organisation (WTO) (2003) *Tourism, Peace and Sustainable Development for Africa: Luanda, Angola 29–30 May 2003*. Madrid: World Tourism Organisation Seminar Proceedings.

—— (2004) Ministers and senior officials converge in Cambodia and Vietnam to combat poverty through tourism. Online document (press release) available from: www.world-tourism.org/newsroom/Releases/2004/june/41MeetAP.htm [last accessed 14 November 2005].

—— (2005) Climate change poses risk to tourism WTO warns. Online document (press release) available from: www.world-tourism.org/newsroom/Releases/2005/november/climate.htm [last accessed 14 November 2005].

Younis, E. (2004) Sustainable tourism and poverty alleviation. Online document. Available from: wbln0018.worldbank.org/eurrp/web.nsf/Pages/ Paper+by+Yunis/ $File/EUGENIO+YUNIS.PDF [last accessed 14 November 2005].

Yukl, G. and Falbe, C.M. (1990) 'Influence tactics and objectives in upward, downward and lateral influence attempts', *Journal of Applied Psychology*, 75(2): 132–140.

Yukl, G. and Tracey, J.B. (1992) 'Consequences of influence tactics used with subordinates, peers and the boss', *Journal of Applied Psychology*, 77(4): 525–535.

11 Tourism, governance and the (mis-)location of power

C. Michael Hall

Power, tourism and the social sciences

Power is one of the core concepts of the social sciences. Elster (1976: 249) describes it as 'the most important single idea in political theory, comparable perhaps to utility in economics'. Unfortunately, issues of power have not been at the centre of understanding tourism. Arguably, this situation is now beginning to change. Issues of power have started to rear their head with respect to such matters as tourism promotion and planning as well as tourism development in the less developed countries (eg. Reed 1997; Morgan and Pritchard 1999; Cheong and Miller 2000). More recently, there has been some limited interest in political security issues in tourism and the extent to which personal mobility is regulated as a result of new state approaches to the management of the terrorist threat (O'Byrne 2001; Hall 2002; Hall *et al*. 2004). Beyond such apparent enthusiasm, a more acute comment on tourism studies is that they are often not only cast as atheoretical but they may also be read as effectively apolitical.

This chapter relates the concept of power to the study of tourism governance which has become increasingly multi-scalar in character (Hall 2004) (Figure 11.1). Under conditions of contemporary globalisation the strict territorial basis of state authority, power and legitimacy, which has been the basis for sovereign governance for most of the past 150 years, has become challenged. Such transformative processes are uneven in time, space and scale. Nevertheless, key economic, financial, cultural, environmental and political issues are presently being dealt with in an emerging framework of post-sovereign governance in which such issues are increasingly handled by the transfer of goal-specific authority from states to supranational organisations and to local or subnational polities (Morales-Moreno 2004; O'Neil and Argent 2005). Under this particular set of conditions the governance of a number of issue areas will be maintained not just by territorial state-bounded authorities (as in much of the past) 'but rather by a network of flows of information, power and resources from the local to the regional and multilateral levels and the other way around' (Morales-Moreno 2004: 108).

Tourism is directly and indirectly implicated in such processes. Direct illustration of the multi-scale role of tourism governance can be seen in the activities of the World Tourism Organisation (WTO) and the World Travel and Tourism Council (WTTC), both of which have been influential in encouraging national and local

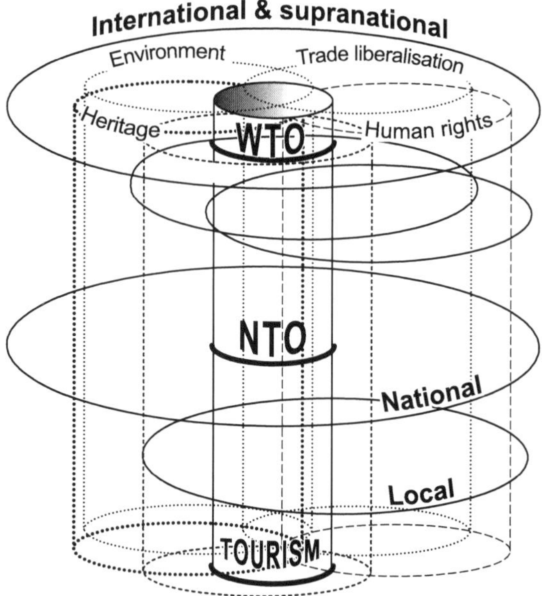

Figure 11.1 Multi-scaled tourism analysis

Source: author.

states to adopt certain regulatory approaches towards managing tourism as well as, of course, attaching value to tourism in the first place. At the level of the local state, tourism has long been endorsed as a mechanism for economic development by growth coalitions. However, the activities of the local state itself in tourism are now being undertaken at a broader scale than ever before; that is, local states are, in fact, functioning as international political actors themselves. It is in terms of the development of post-sovereign policy arenas, such as the environment, trade and investment regimes, heritage and culture, and human rights, where tourism is likely to be most impacted and implicated. Yet the interactions, overlaps and outcomes of such multi-scale (global, supranational, national, regional and local) policy fields has been little charted in tourism discourse, particularly with reference to the neo-liberal dimensions of much of this discourse (Peck and Tickell 2002), nor has the means by which actors operating at one level influence the behaviour of actors at other levels. In order to undertake such an analysis, it is vital to understand the manner in which power is distributed through the policy arenas of emerging post-sovereign governance.

Governance, politics and power

The study of governance and politics is inexorably the study of power. However, why do we need to study power? According to Morriss (1987), it is because of the practical, moral and evaluative contexts: first, we are interested in power because we want to know how things are brought about; second, because moral responsibility for the use of power can be attributed through the assessment of power;

and third, people are not just interested in the judgement of individuals but in the evaluation of society (see also Church and Coles, Chapter 12 in this volume).

At a superficial level, Morriss's response to the need to study power is similar to Lasswell's (1936) comment about politics: namely, politics is about power, who gets what, where, how and why. From this perspective, decisions affecting the regulation of the movement of capital and people, the nature of state involvement in tourism, the institutional structures of tourism development, management, marketing and promotion, and the selection of state-sanctioned place images all emerge from a political process. This process involves the values of actors (individuals, interest groups and public and private organisations) in a struggle for power. As Lindblom (1959: 82) noted, 'one chooses among values and among policies at one and the same time'. Similarly, Simmons *et al.* (1974: 457) noted that 'it is value choice, implicit and explicit, which orders the priorities of government and determines the commitment of resources within the public jurisdiction'.

Governance is essentially about power. The study of power arrangements is vital in the analysis of the political dimensions of tourism because power governs 'the interplay of individuals, organisations, and agencies influencing, or trying to influence the direction of policy' (Lyden *et al.* 1969: 6). For example, within the processes of tourism development and management certain issues may be suppressed, relationships between parties and stakeholders altered, or there may be deliberate inaction. Critical to this may be the design and structure of institutional or organisational arrangements for tourism (Hall and Jenkins 1995), such as the relationships between institutions at different scales of regulation. As Schattsneider (1960: 71) reminds us,

> All forms of political organisation have a bias in favour of the exploitation of some kinds of conflict, and the suppression of others, because organisation is the mobilisation of bias. Some issues are organised into politics while some others are organised out.

Those who benefit from tourism may well be placed in a preferred position to defend and promote their interests through the structures and institutions by which communities are managed. Significantly, the influential models of community tourism promoted by Murphy (1985) clearly fail to address issues of the distribution of power and representation in a community-based approach. Indeed, there is a wider tendency in tourism studies to romanticise the collective capacity of local communities to undertake participative decision-making, particularly when exclusion of some stakeholders is a necessary component of practical consensus (Connelly and Richardson 2004). As Millar and Aiken (1995: 629) observed,

> Communities are not the embodiment of innocence; on the contrary, they are complex and self-serving entities, as much driven by grievances, prejudices, inequalities, and struggles for power as they are united by kinship, reciprocity, and interdependence. Decision-making at the local level can be extraordinarily vicious, personal, and not always bound by legal constraints.

Power

That any definition of power can be 'neutral' or 'descriptive' (Nagel 1975) has been hotly contested. Some scholars have argued that 'power' is an 'essentially contested' concept (Gallie 1955–56; see Coles and Church, Chapter 1 in this volume). Guzzini (2002), for example, notes that any neutral definition of 'power', such as that proposed by Oppenheim (1981), seems elusive, exactly because power is used as an explanatory variable and there is no neutral concept of power for the dependence of theory, empirical and conceptual analyses on meta-theoretical commitments. Similarly, Gray (1983: 94) compared individualist (voluntarist) and structuralist (determinist) positions and concluded that,

> since judgements about power and structure are theory-dependent operations, actionists and structuralists will approach their common subject-matter – what goes on in society – using divergent paradigms in such a fashion that incompatible explanations (and descriptions) will be produced.

Indeed, for Miller (1987), in reference to the Weberian and Foucauldian notions of power (see Chapter 1), the basic individualist-intersubjectivist divide is so great that it is merely sufficient to register that a 'considerable distance separates a notion of power understood as the exercise by A of power over B, contrary to B's preferences, and a notion of power as a multiplicity of practices for the promotion and regulation of subjectivity' (Miller 1987: 10). The value-laden nature of defining and interpreting power was also identified by Lukes (1974: 26) who observed that 'its very definition and any given use of it, once defined, are inextricably tied to a given set of (probably unacknowledged) value-assumptions which predetermine the range of the empirical application':

> Thus, any given way of conceiving of power (that is, any given way of defining the concept of power) in relation to the understanding of social life presupposes a criterion of significance, that is, an answer to the question what makes A's affecting B significant? . . . but also . . . any way of interpreting a given concept of power is likely to involve further particular and contestable judgements.
>
> (Lukes 1977: 4–5, 6)

However, such a position – which is also adopted by Connolly (1974) – was judged by some as a form of value-dependent relativism that failed to establish a particular category of social science concepts (e.g. Giddens 1979; Oppenheim 1981). Nevertheless, the connection between ideas of power and of responsibility gave sway to the importance of understanding the morality of power and the political space of what we study (Lukes 2005):

> When we see the conceptual connection between the idea of power and the idea of responsibility we can see more clearly why those who exercise power are not eager to acknowledge the fact, while those who take a critical perspective of existing social relationships are eager to attribute power to those

> in privileged positions. For to acknowledge power over others is to implicate oneself in responsibility for certain events and to put oneself in a position where justification for the limits placed on others is expected. To attribute power to another, then, is not simply to describe his role in some perfectly neutral sense, but is more like accusing him of something, which is then to be denied or justified.
>
> (Connolly 1974: 97)

For Lukes and Connolly the notion of power therefore implies counterfactuals; that is, it could be done otherwise (see Morriss 1987 on the details of counterfactuals). Lukes (1974, 2005) indicated that Bachrach and Baratz's (1962, 1970) conceptualisation of power with respect to the importance of non-decision-making served to redefine what counts as a political issue in the sense that what is not done is as important as what is done, and often more so (see Chapter 1). To be 'political' therefore means to be potentially changeable (Hoffmann 1988; Guzzini 2002). Following Guzzini (2000, 2001), this provides a constructivist dimension to the analysis of power because concept formation is part of the social construction of knowledge; thus, the defining and assigning of power is a 'political' or power exercise in itself and hence part of the social construction of reality.

Therefore the study of power, in and of itself, runs counter to those who seek to 'depoliticise' policy fields such as through the notion of 'securitisation' (Wæver 1995), thereby reducing the visibility of the intersection of knowledge, interests, values and power. Such a critique may also be applied to discussion in much of tourism studies about the inherent value of partnership and networks which appear to exclude explicit or significant consideration of the power dimensions of such social relationships (e.g. Bramwell and Lane 2000; Murphy and Murphy 2004). Nevertheless, as Guzzini (2002: 16) highlights, present international affairs that imply 'new forms of collective management of international politics', such as through new regimes, have led to an 'increased role of politics and power, more widely conceived'. 'It is this context of both an expansion of "politics" as a potential field of action, and a perceived contraction of "politics" as real room of manoeuvre' that has triggered the need to examine both 'the new direct and indirect ways to control agendas and regimes and on the increasingly perceived impersonal rule of the international scene' (Guzzini 2002: 17), such as those witnessed with respect to contemporary tourism governance.

In seeking to understand both the concept of the power and its application, Lukes (1974, 1979, 2005) uses a distinction between a concept and its different conceptions:

> What I propose to do instead is to offer a formal and abstract account of the concepts of power and authority respectively which inhere within the many conceptions of power and authority that have been used by particular thinkers within specific contexts, in development from and in reaction to one another. Any given conception of power and of authority (and of the relation between them) can be seen as an interpretation and application of its concept.
>
> (Lukes 1979: 634)

In Lukes's original formulation, power may be conceptualised as 'all forms of successful control by A over B – that is, of A securing B's compliance' (Lukes 1974: 17). Power was therefore conceived as a relational phenomenon with Lukes constructing a 'cluster' (Debnam 1984) or typology of power and related concepts in an effort to clarify their meaning and relationship (see Table 1.2). However, as Lukes (2005) later acknowledged, this initial binary conceptualisation failed to adequately address issues of power among multiple actors. Moreover, it focuses on the exercise of power and the power of some over others for the purposes of the securing of compliance. Although this form is accurate in a number of situations with a narrow range of interests it does not adequately deal with more complex situations of multiple interests. Lukes (2005) argues that the conceptual map of power also needs to be understood in terms of the issue-scope, the contextual range, i.e. where it is operative, the intention, and its exercise. Interestingly, such observations are reminiscent of policy flow analysis (Simmons *et al.* 1974) that greatly influenced Hall and Jenkins's (1995) formulation of the tourism public policy process. Indeed, Lukes (2005: 110) suggests that the real question he is asking in *Power: A Radical View* is 'how do the powerful secure the compliance of those they dominate?', which is, as he notes, addressing a question that 'is not without interest'.

Despite Lukes's (2005) reconsideration of the application of power he continues to argue for a three-dimensional approach to power. Lukes (1974, 2005) identified the following three different approaches in the analysis of power, each focusing on different aspects of the decision-making process (Table 1.3):

- one-dimensional view emphasising observable, overt behaviour, conflict, pluralism and decision-making;
- two-dimensional view which recognises decisions and non-decisions, observable (overt or covert) conflict, and which represents a qualified critique of the behavioural stance of the one-dimensional view, in particular with respect to recognition of the values and institutional practices within a political system that favour the interests of some, relative to others; and a
- three-dimensional view which focuses on decision-making and control over the political agenda (not necessarily through decisions), and which recognises observable (overt or covert) and latent conflict and hegemony.

Each arises out of, and operates within, a particular political perspective as the concept of power is 'ineradicably value-dependent' (Lukes 1974: 26). For example, a pluralist conception of the tourism policy-making process, such as that which underlies the notion of community-based tourism planning (e.g. Murphy 1985), will focus on different aspects of the decision-making process rather than structuralist conceptions of politics which highlight social relations within the framed consumption of tourist services (e.g. Urry 1990; Britton 1991). These distinctions are significant for our understanding of tourism. As Britton (1991: 475) recognised,

> we need a theorisation that explicitly recognises, and unveils, tourism as a predominantly capitalistically organised activity driven by the inherent and defining social dynamics of that system, with its attendant production, social, and ideological relations. An analysis of how the tourism production system markets and packages people is a lesson in the political economy of the social construction of 'reality' and social construction of place, whether from the point of view of visitors and host communities, tourism capital (and the 'culture industry'), or the state – with its diverse involvement in the system.

However, given the need to understand the dominant groups and ideologies operating within the political and administrative system which surrounds tourism, it seems reasonable to assume that the use of wide conceptions of power (for instance, capable of identifying decisions, non-decisions and community political structure) will provide the most benefit in the analysis of the political dimensions of tourism (Hall 1994; Hall and Jenkins 1995). It is to this proposition to which we turn in the next sections.

One-dimensional views of community decision-making

A somewhat romantic and naive view of tourism holds that everyone has (or should have) equal access to power and representation (see also Timothy, Chapter 9 in this volume). To an extent this has been one of the driving elements behind utilising the community as an appropriate framework for planning for tourism development as there appears to be an inherent assumption that it is somehow 'closer to the people'. However, public participation in tourism planning has long been recognised as imperfect (cf. Jenkins 1993). Nevertheless, a one-dimensional view of power in communities suggests that even though imperfect, the community decision-making process is at least observable as it operates through the overt action of pluralist interests (Dahl 1961; Debnam 1984). In addition, political issues are regarded as coming into existence when they command 'the attention of a significant segment of the political stratum' (Dahl 1961: 92). This concept of power has the advantage that it can be relatively easily operationalised. As Lukes (1986: 2) observes, when B seeks to resist the power of A it is 'relevant in the sense that, if it is actualized, it provides the test by which one can measure relative power, where parties conflict over an issue'.

The latter observation is significant in that power relativities shift according to the issue under discussion (Lukes 1986: 8). Moreover, power is not evenly distributed within a community. Some groups and individuals have the ability to exert greater influence over the tourism planning process than others through access to financial resources, expertise, public relations, media, knowledge and time to put into contested situations and the nature of what is discussed, or not (Hall and Jenkins 1995; see also Gill, Church and Ravenscroft, Coles and Scherle – chapters 6, 8 and 10 – in this volume). Therefore, in some circumstances, the level of wider public involvement in tourism planning may be more accurately described as a form of tokenism in which decisions, or the direction of decisions, has already been prescribed by local government by virtue of other policies and decisions (see also Timothy, Chapter 9 in this volume). Communities rarely have the opportunity to say 'no' in the longer term, nor can decisions be easily undone.

Two-dimensional views of community decision-making: the two faces of power

Bachrach and Baratz (1970) identified two major weaknesses in the pluralist approach to power: first, it overlooks the fact that power may be exercised by confining the scope to political decision-making; and second, the pluralist model provides no criteria for determining what are the significant issues.

Two-dimensional views of community decision-making focus on decision-making *and* non-decision-making as well as observable (overt and covert) conflict (Bachrach and Baratz 1962, 1970). Bachrach and Baratz (1970: 44) defined a non-decision as 'a decision that results in suppression or thwarting of a latent or manifest challenge to the values or interests of the decision-maker'. A non-decision is a means by which demands for change in the existing allocation of benefits and privileges in the community may be suffocated before they are even voiced; alternatively, they may be kept covert or killed-off before they gain access to the relevant decision-making arena; or, failing all these things, maimed or destroyed in the implementation stage of the policy process (Lukes 1974). Non-decision-making exists 'to the extent that a person or group – consciously or unconsciously – creates or reinforces barriers to the public airing of political conflicts, that person or group has power' (Bachrach and Baratz 1970: 8). The role of non-decision-making is now widely acknowledged in the political literature given that political actors, 'organizations and collectivities can leave selected topics undiscussed for what they consider their own advantage' (Holmes 1988: 22).

Bachrach and Baratz (1970: 11) also stress the importance of the 'mobilization of bias' which is 'the dominant values and the political myths, rituals, and institutional practices which tend to favor the vested interests of one or more groups, relative to others'. Non-decision-making is also the 'primary method for sustaining a given mobilization of bias' (Bachrach and Baratz 1970: 43–44). Instances of two-dimensional views of power abound. A classic example of non-decision-making is when electors are given a number of options with respect to development proposals. However, certain potential options, such as to reject a development altogether, may not be made available. A variation of non-decision-making is the concept of non-implementation; that is, although policy is developed or regulation enacted, it is not actually enforced (Mokken and Stokman 1976). Bachrach and Baratz's (1970) method for empirical application of the concept of non-decision-making consists of three stages: first, the study of the actual decision-making process within the political arena and the resultant outcomes; second, the determination of the remaining overt and covert grievances of the apparently disfavoured group; and finally, the determination of 'why and by what means some or all of the potential demands for change have been denied an airing' (Bachrach and Baratz 1970: 49).

Public–private partnerships have generated considerable interest by commentators on tourism in relation to non-decision-making. Harvey (1989: 7) recognised that,

> the new entrepreneurialism of the smaller state has, as its centrepiece, the notion of a 'public–private partnership' in which a traditional local boosterism is integrated with the use of local government powers to try [to] attract external sources of funding, new direct investments, or new employment sources.

In a classic restatement of non-decision-making, Goodwin (1993: 161) argued that, in order to ensure that urban leisure and tourism development projects were carried out,

> local authorities have had planning and development powers removed and handed to an unelected institution. Effectively, an appointed agency is, in each case, replacing the powers of local government in order to carry out a market-led regeneration of each inner city.

In this case, partnership does not include all members of a community. Those who do not have enough money, or are not of the right lifestyle, or simply do not have sufficient power, are ignored. The institutional arrangements for supposed community development were therefore able to ignore the demands of some members of that community. Similarly, Sadler (1993: 190) argued,

> The kind of policy which had been adopted – and which was proving increasingly ineffective even in terms of its own stated objectives . . . rested not so much on a basis of rational choice, but rather was a simple reflection of the narrow political and intellectual scope for alternatives. This restricted area did not come about purely or simply by chance, but had been deliberately encouraged and fostered.

The two-dimensional perspective of power underlines the importance of identifying '*potential issues* which non-decision-making prevents from being actual' (Lukes 1974: 19, original emphasis). Political issues have an organisational aspect. Therefore, research on tourism's political dimensions needs to connect the substance of policy with the process of policy-making including the relationship between power, structure and ideology (Hall and Jenkins 1995). Clearly, there are 'politically imposed limitations upon the scope of decision-making', such that 'decision-making activities are channelled and directed by the process of non-decision-making' (Crenson 1971: 178) often by dominant business interests. As Crenson (1971: 181) recognised, pluralism is 'no guarantee of political openness or popular sovereignty', and 'neither the study of [overt] decision-making' nor the existence of 'visible diversity' will tell us anything about 'those groups and issues which may have been shut out of a town's political life'. Studies of the political aspects of tourism should therefore attempt to understand not only the politically imposed limitations upon the scope of decision-making, but also the political framework within which the research process itself takes place (Hall 1994), or what Schattsneider (1960) described as 'the rules of the game'.

Within many current discussions in the tourism literature of the establishment of collaborative networks and partnership arrangements (see Hall 2000), including those in place marketing (e.g. Kotler *et al.* 1993), tourism policy networks are typically portayed as interdependent, co-equal, patterned relationships (Klijn 1996). However, different political actors occupy different positions and can carry different weight within such networks. Some sit in positions with extensive opportunity contexts, filling 'structural holes' (Nohria 1992: 10); others may be reluctant participants or may not even be able to participate at all. Organisations and actors also differ with respect to resource dependencies (Rhodes 1981), leading to differences in their relative power to influence decision-making processes. As Clegg and Hardy (1996: 678) remind us, 'we cannot ignore that power can be hidden behind the facade of "trust" and the rhetoric of "collaboration", and used to promote vested interest through the manipulation of and capitulation by weaker partners'.

Three-dimensional views of power

The three-dimensional view of power (Lukes 1974, 2005) incorporates Dahl's (1961) observable power in decision-making settings and Bachrach and Baratz's (1962, 1970) power through non-decision-making, but adds to these the third dimension of institutional bias, hegemony and the manipulation of preferences. The three-dimensional view of power 'allows for consideration of the many ways in which potential issues are kept out of politics, whether through the operation of social forces and institutional practises or through individuals' decisions' (Lukes 1974: 24).

The third dimension arises out of Lukes's (1974) critique of non-decision-making with respect to non-intentional use of power as well as the role of power in shaping preferences. Lukes (1974: 22) argues that Bachrach and Baratz (1970) do not recognise that the phenomenon of collective action is not necessarily 'attributable to particular individual decisions or behaviour, nor that the mobilisation of bias results from the form of *organisation*, due to "systemic" or organisational effects'. Lukes (1974: 23) then emphasises the role that power has in shaping human preferences because, for him, 'to assume that the absence of grievances equals genuine consensus is simply to rule out the possibility of false or manipulated consensus by definitional fiat' (Lukes 1974: 24). Moreover, 'A may exercise power over B . . . by influencing, shaping or determining his very wants' (Lukes 1974: 23). To Lukes (1974: 23), such an approach is 'the most effective and insidious use of power'.

At first glance, a three-dimensional view of power in tourism decision-making may appear to be quite problematic. After all, 'how can one study, let alone explain, what does not happen?' (Lukes 1974: 38). Nevertheless, as Crenson (1971: vii) recognised, the way 'things do not happen' is as important as what does: 'the proper object of investigation is not political activity but political inactivity'. Similarly,

Lukes (2005: 134) notes that a key concern in the examination of power, 'is the shaping of agents' desires and beliefs by factors external to those agents'. Indeed, Lukes (1974) argued that third-dimensional power may be revealed when it is not in accordance with an individual or group's 'real interests' (Lukes 1974: 24–25). This is a concept similar to Marxian ideas of false consciousness (Hyland 1995) and Gramsci's concept of hegemony (Lukes 2005).

Lukes (1974) used the work of Crenson (1971) on environmental pollution in Gary – a steel town in Indiana, USA – to illustrate the operation of the third dimension of power. Although the extent of environmental pollution was clearly against the real interests of the populace, opposition to such pollution did not precipitate from within Gary because of its nature as a 'company town' and the apparently associated view that industrial operations were an inherent good. For Lukes (1974: 24–25, original emphasis), this situation illustrates that in the resulting conflict of interests one must recognise a '*latent conflict*, which consists in a contradiction between the interests of those exercising power and the *real interests* of those they exclude'.

Non-decisions and latent conflicts provide evidence for the existence of the third dimension of power. However, such evidence may be persuasive but not conclusive in character given the invisibility of power. This contrasts sharply with the first dimension where ready-made positive evidence in the form of explicit preferences and outcomes is readily available (Barnes 1993). The third dimension of power is clearly related to the analysis of structural dominance in the restriction of human agency. However, critics such as Giddens (1979) and Hyland (1995) argue that such structural domination is beyond the scope of any focus on intentionally exerted power of individual actors, although it is also apparent that the analysis of power cannot be separated from the analysis of structural dominance. Perhaps more significantly, Lukes's third dimension also intersects with Foucault's (1972, 1980) power/knowledge framework which also acknowledged the relational nature of power: 'in reality power means relations, a more-or-less organised, hierarchical, co-ordinated cluster of relations' (Foucault 1980: 198). To Foucault, knowledge and power are inseparable (see also Winter, Coles and Church – chapters 5 and 1 – in this volume). Power can be assessed through knowledge because knowledge itself has a function of power: 'Once knowledge can be analysed in terms of region, domain, implantation, displacement, transposition, one is able to capture the process by which knowledge functions as a form of power and disseminates the effects of power' (Foucault 1980: 69). In this power–knowledge relationship, power in turn impacts the formation of knowledge. According to Foucault (1980: 51, 59), 'the exercise of power itself creates and causes to emerge new objects of knowledge and accumulates new bodies of information . . . Far from preventing knowledge, power produces it'.

In seeking to operationalise Foucauldian notions of power, we arrive at the importance of locating issues of power within particular issue and locational contexts, even though we must also acknowledge that such loci of power relations will be connected to the myriad of other issues and sets of interests (Foucault 1980: 188). Indeed, the value of a Lukesian approach to power is highlighted in the multilayering of observations of power occurring in the three dimensions. As Lukes (2005: 98) observes, 'Foucault's first way of interpreting the key idea central to his view of power – that power is "productive" through the social construction of subjects, rendering the governed governable – made no sense'. Lukes's (1974, 2005) three dimensions therefore provide an empirical strength often missing in Foucauldian analyses. While the latter acknowledge the role of structural dominance, often they fail to record the actions of individual actors in relation to specific issues and interests. For example in tourism studies, Cheong and Miller (2000) in their Foucauldian reading present a single-dimensional account of power with little empirical grounding. Indeed, arguably there is a substantial amount of such writing in tourism where authors have exhorted the notion of a tourist gaze without interrogation of the concepts of power and knowledge on which it is grounded (see also comments by Crouch, Cater, Winter – chapters 2, 3 and 5 – in this volume). Similarly, they have given little thought to the role that individual actors play with respect to power relations from a decision- and non-decision-making perspective. Nevertheless, such criticisms aside, tourism provides a number of good examples that illustrate three-dimensional notions of power with respect to knowledge and interests.

Heritage tourism and the third dimension

Heritage provides a useful setting in which to investigate the third dimension of power. Institutional representations and reconstructions of heritage are often not fully inclusive. Particular ideologies are represented to the tourist through museums, historic houses, historic monuments and markers, guided tours, public spaces, heritage precincts and landscapes in a manner that may act to legitimate current social and political structures. As Norkunas (1993: 5) recognised,

> 'The public would accept as 'true' history that is written, exhibited, or otherwise publicly sanctioned. What is often less obvious to the public is that the writing or the exhibition itself is reflective of a particular ideology.

Monterey, California provides a valuable example of the political nature of tourism at the local state level in relation to heritage. Monterey has a substantial heritage tourism industry that is based on the historic significance of the region in terms of United States' expansionism, and a literary and industrial heritage in the form of Cannery Row, made famous by Steinbeck, and Fisherman's Wharf. Different tourist landscapes, such as those in Monterey, can be read as distinct cultural texts, a kind of outdoor museum which display the artefacts of a community and society. Each of these tourist cultural texts reveals certain ideological assumptions and power relationships which underly the tourist environment as a form of cultural production. Indeed, Norkunas (1993: 10) notes that the 'ideology of the powerful is systematically embedded in the institutions and public texts of tourism and history'.

The rich and complex ethnic history of Monterey is almost completely absent in the 'official' historic tours and the residences available for public viewing. In Monterey, as in many other parts of the world, heritage is presented in the form of the houses of the aristocracy or elite. History is 'flattened' or conflicting histories suppressed. In a restatement of Lukes's third dimension of power, Norkunas (1993: 36) notes,

> This synopsis of the past into a digestible touristic presentation eliminates any discussion of conflict; it concentrates instead on a sense of resolution. Opposed events and ideologies are collapsed into statements about the forward movement and rightness of history.

Such a situation occurs with very little or no overt conflict over heritage representation despite the clear presence of potentially alternative perspectives on heritage in the form of ethnic and minority groups which have a long-standing history and relationship to Monterey. Heritage is not contested in the public sphere. Despite the presence of 'democratic' institutions and channels for the representation of diversity in the community, conflicts and issues are ignored in the public history of Monterey. Therefore, what gets reproduced

> is the image of elite Americans of European descent who control, and have always controlled the destiny of the city . . . public history texts as well as tourist texts are involved in a form of dominance, a hegemonic discourse about the past that legitimates the ideology and power of present groups.
>
> (Norkunas 1993: 26)

The recent industrial past has also been de-emphasised in the heritage product of Monterey. As in many Western urban industrial communities, economic restructuring within the new global economy has led to the demise of many industrial operations, such as canning. The industrial waterfront has now become a leisure space that combines shopping and entertainment with residential and tourist development. Industrial heritage is typically a central component of waterfront redevelopment. Heritage precincts are established which tell the reader the economic significance of the area, not of the lives of those who contributed to wealth generation. Narratives of labour, class and ethnicity are typically replaced by romance and nostalgia. Previous overt conflict in the community, whether between ethnic groups, classes or, more particularly, in terms of industrial and labour disputes are either ignored or glossed over in 'official' tourist histories. The overt conflict of the past has been reinterpreted by local elites to create a new history in which heritage takes a linear, conflict-free form. In the case of Monterey, the past is reinterpreted through the physical transformation of the canneries. As Norkunas (1993: 50–51) laments,

> Reinterpreting the past has allowed the city to effectively erase from the record the industrial era and the working class culture it engendered. Commentary

on the industrial era remains only in the form of touristic interpretations of the literature of John Steinbeck.

Power and tourism: multi-scaled governance

The different examples of the operation of the three dimensions of power used so far are primarily from the level of the local state. Undoubtedly in tourism terms, this is the level dealt with by most accounts of politics. However, as noted above, tourism governance is multi-scaled, with roles increasingly being played by supranational actors, such as the WTO. Moreover, the nature of tourism governance is complicated by the situation that tourism is not an issue that tends to register on the political agenda of most national governments. Although economically significant, tourism is not a decisive vote winner nor will the nature of tourism policy be likely to be a key factor in a national election. At the local or regional level where there is either a very high proportion of the population employed in the sector or a high visitor-to-permanent population ratio then tourism may be an electoral issue. However, in general tourism is not a political issue. Yet, arguably, it is in this situation that the analysis of tourism power possibly becomes even more significant because it is an area that is not open to public scrutiny or influence. Indeed, given the economic, social and environmental effects of tourism and the lack of public debate over tourism, it is remarkable just how little attention is given to the way in which tourism is governed and directed.

Much of the lack of debate at the nation and local state level may be partly explained through the development of sub-governments (Cigler 1991); that is, closely linked sets of administrative and private sector interests in which the interests of institutions, such as a national or regional tourism organisation are seen to be synonymous with the interests of key players in the tourism industry, including industry associations. If a stable relationship exists between the members of the triangle then it means that the sub-government is relatively impervious to outside influences on policy formulation and implementation. Indeed, in most parts of the developed world the institutional arrangements for tourism only serve to reinforce this role with regulatory structures often being altered to encourage such private–public sector partnerships which usually have little input from the general public.

For the most part, this has been a function of overt, one-dimensional decision-making that has been able to occur without debate primarily because issues have never become part of the unfolding political agenda; they are practically just agreed as they are. In other words, the decisions have not been hidden but neither have they been subject to systematic public policy debate. In fact, it is remarkable that in countries such as Australia and New Zealand, which have enacted substantial neo-liberal trade and promotion regimes for many industry sectors that now have to pay for their own promotion and research, tourism has been exempt from such changes and it continues to receive substantial assistance in the form of state funding for destination promotion and marketing (Hall 2000). Perhaps it is a reflection of a sentiment that tourism is somehow a ‘benign’, ‘harmless’ or ‘friendly’ industry that contributes to the wider public good? A more likely explanation is that the actual proportion of total direct budget expenditure on tourism by national governments is tiny, and that its allocation, as well as the wider determination of tourism policy per se, remains in reality in the sub-government realm unthreatened by any outside interest.

Nevertheless, there are a number of significant occasions where tourism-related policy issues are clearly examples of non-decision-making. One of the clearest examples in terms of an intermestic (the collision between *inter*national and do*mestic*) tourism policy arena is bidding to host the Olympic Games. Given the financial and other long-term costs of hosting the Olympic Games, a case can be made that it is in the interests of inhabitants of a bid city that they be given the opportunity to debate the merits of hosting the games and even being able to say 'no'. Yet in almost every case bid city authorities – or to be more precise a bidding organisation comprised of local and national state officials, and members of local growth coalitions – seek to have the merits of an Olympic bid undebated for fear that it may harm bid prospects. In a classic Lukesian-style scenario, candidate cities are charged with demonstrating that the bid is supported by, and hence in the interests of, the population. To help achieve this opinion polls are usually conducted that 'demonstrate' public support for the event, although in the non-democratic tradition of the Olympic political legacy direct political participation through a ballot is generally not undertaken. Hitherto latent conflict may then become overt as sections of the public feel it is not in their best or real interests to have the games. Rather, investment may be better allocated to other social and economic outcomes.

Examples of changing the rules of the game and the creation of a mobilisation of bias in the case of the Sydney 2000 Summer Olympic Games included:

- Passing legislation to assist in the development and regeneration of projects associated with the games by means of an amendment on page 163 of New South Wales (NSW) threatened species legislation, 'not a bill where you would find such a change, and you have to wonder why they put it there' (Totaro 1995: 1).
- Legislation passed under the New South Wales government's *Olympic Co-ordination Authority Act* allowed (somewhat ironically given the games' green image) all projects linked with the games to be suspended from the usual environmental impact statements requirements that would have had a public input provision (Totaro 1995).
- The creation of a 'community of interest' (Greiner 1994: 13) between the Olympic bid interests and the Sydney and national media.

As Nick Greiner (1994: 13), NSW Premier at the time of the bid commented,

> Early in 1991, I invited senior media representatives to the premier's office, told them frankly that a bid could not succeed if the media played their normal 'knocking role' and that I was not prepared to commit the taxpayers' money unless I had their support. Both News Ltd and Fairfax subsequently went out

> of their way to ensure the bid received fair, perhaps even favourable, treatment. The electronic media also joined in the sense of community purpose.

All of which clearly begs the question of whose community purpose?

The Olympic Games example also highlights the multi-scalar dimension of tourism policy issues. For the event to be run, local and national interests not only have to be involved, but also the supranational interests of the Olympic Movement have to be satisfied. The Olympic Movement consists of unelected members who seek to use the games to promote Olympian virtues. Indeed, while bids include technical information on the event infrastructure and the environmental impacts of the event, they fail to include a social impact statement. In the case of London, opposition to the bid was kept off the agenda. News media management strategies instead focused on articulating that the games were in the real interest of the UK population both within and outside London (see Table 11.1). This was despite the fact that polls indicated significant 'minority' groups opposed to the games, although the public opinion polls were never subject to scrutiny. To some extent, the network of institutions that utilise overt and covert decision-making in the bidding and hosting process for an Olympics is similar to the web of institutions that operate in much of tourism policy.

Table 11.1 'Real Interests' and the London 2012 Olympic bid

Public support is regarded as a critical element of an Olympics Games bid. Without it it would be extremely difficult politically to justify the huge amounts of public expenditure on non-community sports infrastructure and other facilities as well as the high degree of political and market regulation that surround the games. At the time that Olympic officials visited London in 2005 to review the city's bid to host the 2012 Summer Games, Kevan Gosper, the Australian former vice-president of the International Olympic Committee (IOC), stated that public support would be crucial in deciding the 2012 host city. 'Public support is very high on the list of priorities – it is, for example, the first thing I look at in a bid. . . . The government, community and public must be right behind London's bid. If I was running a candidature perceived to be a front-runner, I'd make sure everyone identified with it, felt confident about it and wanted to win' (quoted in Kelso 2005a).

Just prior to the bid visit it was reported that 74% of those polled supported bringing the games to the capital (Kelso 2005a). However, at the same time that the London bid committee was seeking to convince the IOC as to the virtues of London's bid, a loose coalition of community and environmental interests, NoLondon2012, was seeking to publicly oppose it. This led the UK culture secretary, Tessa Jowell, to plead with the group to keep quiet for the duration of the evaluation commission's visit, and not to undertake their protests and marches so as not to damage London's chances. The reasons for the opposition of NoLondon2012 and the government response are a classic example of the ability of issues to emerge in the public arena as a result of shifts in the conflict of interests from the covert to the overt. According to Kelso (2005b), 'NoLondon2012 is concerned about the environmental implications of a London games, the loss of amenities and housing during construction, and the diversion of funds from grassroots sport. It claims that the bid company has misled local people and that consensus of support for the bid in east London has led to the suppression of debate'.

Sources: author, adapted from Kelso (2005a, 2005b).

The WTO was originally founded in 1925 as the International Union of Official Tourist Publicity Organisations based at The Hague. After the Second World War, it was renamed the International Union for Official Tourism Organisations (IUOTO) and moved to Geneva. In 1975 it was renamed the World Tourism Organisation, with its first general assembly in Madrid where the Secretariat was based at the invitation of the Spanish government.

The WTO is now a formal member of the United Nations network. The WTO provides an example of an international actor that serves to both act as a contributor to the regulatory regime of tourism – for example with respect to the collection and preferred harmonisation of statistics – as well as a promoter of tourism in its own right. For instance, it is able to influence individual governments' and (world) region's intermestic tourism policies and tourism development strategies. Furthermore, it exercises influence in international regulatory developments in environment, trade and human mobility. This potentially conflicting role of both promoter and regulatory institution is increased by the funding arrangements of the organisation. It is primarily funded by members' subventions, but it is increasingly generating external funding streams to finance activities and industry partnerships. The increased focus on industry partnerships and trade liberalisation within the policy advice that the WTO generates, provides a relatively narrow neo-liberal framework within which tourism and leisure mobility is examined. As Burns (1994, in Davidson and Maitland 1997: 119) observed,

> it is clear . . . that WTO is actively promoting the expansion of tourism at a global level. WTO survives not so much through its membership fees (governments and affiliates) but through spin-off activities such as consulting and project management. It therefore actually needs more tourism!

In such a situation the potential for a 'no' or 'minimal' tourism option to be provided in its policy advice is unlikely. Policy solutions to issues of regulating tourism are typically seen in terms of self-regulation rather than government regulation. Indeed, the transnational industry counterpart and often partner to the WTO, the WTTC, a corporate-based organisation by invitation only, is an even greater advocate of a neo-liberal agenda.

> WTTC's mission is to raise awareness of the full economic impact of the world's largest generator of wealth and jobs – Travel & Tourism. Governments are encouraged to unlock the industry's potential by adopting the Council's policy framework for sustainable tourism development.
>
> (WTTC in Hall 2005: 132)

Unfortunately, the policy framework established by the WTTC in its 2003 *Blueprint for New Tourism* did not even acknowledge global environmental change as a factor that might influence tourism in the future even though it had a supposed focus on sustainability. This is, therefore, an instance, arguably, of the exercise of the second or third dimensions of power through the absence of information and avoidance of what is otherwise a significant issue for interests which are otherwise excluded from WTTC agenda formulation.

The neo-liberal stances of the WTO and the WTTC are an important part of the manner in which issues or non-issues at the supranational level percolate through to influencing the terrain of thinking at the national and even local level. The WTO and WTTC are significant knowledge generators in tourism; they shape definitions of a 'tourist' such that they distinguish tourism from other forms of mobility; and they direct the collection of statistics, estimates of its tourism impacts, and the mechanisms for tourism planning and promotion. The generation of such knowledge assumes great significance within the third dimension of power not least because, ideologically, it helps structure both the 'rules of the game' in tourism development and also how those rules are developed. For instance, they invite or select some individuals and universities to participate in their proceedings and deliberations, and not others. In effect, they provide a de facto 'truth regime' in which expert knowledge and disciplinary discourses are endorsed by them to project apparent 'truths' regarding tourism that supply systematic procedures for the generation, regulation and production of international tourism. Therefore, in looking at the second and third dimensions of power with respect to the WTO and the WTTC, it is perhaps more important to recognise the potential options that are never considered rather than the paths that have been taken.

Concluding observations: power, presence and its visibilty

In examining the various dimensions of power and its relationship to tourism discussed above several key observations can be made. First, power lies at the heart of the interplay of values, interests and tourism policy. Power exists in both observable (overt and covert) cases but in many instances as a latent feature. Second, the relative absence of these issues from the radar of tourism analysis only serves to lessen dramatically the potential understanding of tourism as a phenomenon in contemporary society and the pattern of its development. Third, there is the potential for certain institutions and coalitions – or what we have termed 'subgovernments' – to control many aspects of tourism policy and development. This is despite the rhetoric, at least in academic circles, of community-based tourism. Fourth, the issue of power – who controls – in tourism is multi-scalar in character, with power relations operating both vertically and horizontally in regulatory structures. Finally, power is real and its operation can be readily studied. Arguably, this is where the explanatory power of the Lukesian multidimensional approach to political power has much to merit it above (or indeed alongside) competing approaches, such as Foucault's.

The creation and representation of tourism and knowledge of tourism is a political process. Some issues are organised into tourism policy, some issues are organised out. The same applies to tourism research and scholarship. As Morriss (1987) argued, there is a moral element in the analysis of power. In general, that moral element is not located in the analysis of tourism, which is often presented in an 'amoral' and 'objective' fashion. Perhaps that is also a reflection that there is also little consideration of power in tourism. Yet, as one starts to investigate the decision-making processes in tourism-related issues it soon becomes apparent that this is not the case. Just as importantly, as noted above, it also becomes vital to identify what is not included in the tourism agenda that might otherwise reasonably be included. The failure of the WTO and the WTTC to deal with issues of global environmental change is a stark example. If power is conceived broadly and realistically in its complexity in the ways suggested above, it can be hypothesised to be at work where narrower views of power will see none. Yet the power reality is that, whether it be growth coalitions, supranational organisations or even parts of the academy, to start looking at decision-making and therefore power there may not only be disincentives to undertake such studies but possibly even outright opposition.

References

Bachrach, P. and Baratz, M.S. (1962) ‘Two faces of power’, *American Political Science Review*, 56: 947–952.

Bachrach, P. and Baratz, M.S. (1970) *Power and Poverty: Theory and Practice*. New York: Oxford University Press.

Barnes, B. (1993) ‘Power’, in R. Bellamy (ed.) *Theories and Concepts of Politics: An Introduction*. Manchester: Manchester University Press.

Bramwell, B. and Lane, B. (eds) (2000) *Tourism Collaboration and Partnerships: Politics, Practice and Sustainability*. Clevedon: Channel View.

Britton, S.G. (1991) ‘Tourism, capital and place: towards a critical geography of tourism’, *Environment and Planning D: Society and Space*, 9(4): 451–478.

Cheong, S. and Miller, M. (2000) ‘Power and tourism: a Foucauldian observation’, *Annals of Tourism Research*, 27(2): 371–390.

Cigler, A.J. (1991) ‘Interest groups: a subfield in search of an identity’, in W. Crotty (ed.) *Political Science: Looking to the Future (Volume 4, American Institutions)*. Evanston, IL: Northwestern University Press.

Clegg, S.R. and Hardy, C. (1996) ‘Conclusion: representations’, in S.R. Clegg, C. Hardy and W.R. Nord (eds) *Handbook of Organization Studies*. London: Sage.

Connelly, S. and Richardson, T. (2004) ‘Exclusion: the necessary difference between ideal and practical consensus’, *Journal of Environmental Planning and Management*, 47(1): 3–17.

Connolly, W.E. (1974) *The Terms of Political Discourse*. Oxford: Martin Robertson.

Crenson, M.A. (1971) *The Un Politics of Air Pollution: A Study of Non-decisionmaking in the Cities*. Baltimore, MD and London: Johns Hopkins Press.

Dahl, R.A. (1961) *Who Governs? Democracy and Power in an American City*. New Haven, CT: Yale University Press.

Davidson, R. and Maitland, R. (1997) *Tourism Destinations*. London: Hodder and Stoughton.

Debnam, G. (1984) *The Analysis of Power: A Realist Approach*. London: Macmillan.

Elster, J. (1976) ‘Some conceptual problems in political theory’, in B. Barry (ed.) *Power and Political Theory: Some European Perspectives*. London: John Wiley.

Foucault, M. (1972) *The Archeology of Knowledge* (trans. A.M. Sheridan Smith). New York: Pantheon.

—— (1980) *Power/Knowledge: Selected Interviews and Other Writings 1972–1977*. New York: Pantheon.

Gallie, W.B. (1955–56) 'Essentially contested concepts', *Proceedings of the Aristotelian Society*, 56: 167–198.
Giddens, A. (1979) *Central Problems in Social Theory: Action, Structure and Contradiction in Social Analysis*. London: Macmillan.
Goodwin, M. (1993) 'The city as commodity: the contested spaces of urban development', in G. Kearns and C. Philo (eds) *Selling Places: The City as Cultural Capital, Past and Present*. Oxford: Pergamon Press.
Gray, J. (1983) 'Political power, social theory and essential contestability', in D. Miller and L. Siedentop (eds) *The Nature of Political Theory*. Oxford: Clarendon Press.
Greiner, N. (1994) 'Inside running on Olympic bid', *The Australian*, 19 September 13.
Guzzini, S. (2000) 'A reconstruction of constructivism in international relations', *European Journal of International Relations*, 6(2): 147–182.
—— (2001) 'The significance and roles of teaching theory in international relations', *Journal of International Relations and Development*, 4(2): 98–117.
—— (2002) '"Power" in international relations: concept formation between conceptual analysis and conceptual history', unpublished paper prepared for the Forty-third Annual Convention of the International Studies Association, 24–27 March, New Orleans.
Hall, C.M. (1994) *Tourism and Politics: Power, Policy and Place*. Chichester: John Wiley.
—— (2000) 'Rethinking collaboration and partnership: a public policy perspective', in B. Bramwell and B. Lane (eds) *Tourism Collaboration and Partnerships: Politics, Practice and Sustainability*. Clevedon: Channel View.
—— (2002) 'Travel safety, terrorism and the media: the significance of the issue-attention cycle', *Current Issues in Tourism*, 5(5): 458–466.
—— (2004) 'Multi-scaled tourism governance: implications for the analysis of tourism policy', unpublished paper presented at the Thirteenth Nordic Symposium in Tourism and Hospitality Research, Aalborg, Denmark, November.
—— (2005) *Tourism: Rethinking the Social Science of Mobility*. Harlow: Prentice Hall.
Hall, C.M. and Jenkins, J. (1995) *Tourism and Public Policy*. London and New York: Routledge.
Hall, C.M., Duval, D. and Timothy, D. (eds) (2004) *Safety and Security in Tourism: Relationships, Management and Marketing*. New York: Haworth Press.
Harvey, D. (1989) 'From managerialism to entrepreneurialism: the transformation in urban governance in late capitalism', *Geografiska Annaler*, 71B: 3–17.
Hoffmann, J. (1988) *State, Power and Democracy: Contentious Concepts in Practical Political Theory*. New York: St Martin's Press.
Holmes, S. (1988) 'Gag rules or the politics of omission', in J. Elster and R. Slagstad (eds) *Constitutionalism and Democracy*. Cambridge: Cambridge University Press.
Hyland, J.L. (1995) *Democratic Theory: The Philosophical Foundations*. Manchester: Manchester University Press.
Jenkins, J. (1993) 'Tourism policy in rural New South Wales – policy and research priorities', *GeoJournal*, 29(3): 281–290.
Kelso, P. (2005a) 'As Olympic officials arrive in London, poll shows public support for bid – but little confidence', *The Guardian*, 15 February.
—— (2005b) 'Protest stirs in troubled east', *The Guardian*, 15 February.
Klijn, E. (1996) 'Analyzing and managing policy processes in complex networks', *Administration and Society*, 28(1): 90–119.
Kotler, P., Haider, D.H. and Rein, I. (1993) *Marketing Places: Attracting Investment, Industry, and Tourism to Cities, States, and Nations*. New York: Free Press.
Lasswell, H.D. (1936) *Politics: Who Gets, What, When, How?* New York: McGraw-Hill.

Lindblom, C.E. (1959) 'The science of muddling through', *Public Administration Review*, 19: 79–88.

Lukes, S. (1974) *Power: A Radical View*. London: Macmillan.

—— (1977) 'Power and structure', in S. Lukes (ed.) *Essays in Social Theory*. New York: Columbia University Press.

—— (1979) 'Power and authority', in T. Bottomore and R. Nisbet (eds) *History of Sociological Analysis*. London: Heinemann.

—— (1986) 'Introduction', in S. Lukes (ed.) *Power*. Oxford: Basil Blackwell.

—— (2005) *Power: A Radical View*, second edition. Basingstoke: Palgrave Macmillan.

Lyden, F.J., Shipman, G.A. and Kroll, M. (eds) (1969) *Policies, Decisions and Organisations*. New York: Appleton-Century-Crofts.

Millar, C. and Aiken, D. (1995) 'Conflict resolution in aquaculture: a matter of trust', in A. Boghen (ed.) *Coldwater Aquaculture in Atlantic Canada*, second edition. Moncton: Canadian Institute for Research on Regional Development.

Miller, P. (1987) *Domination and Power*. London: Routledge & Kegan Paul.

Mokken, R.J. and Stokman, F.N. (1976) 'Power and influence as political phenomena', in B. Barry (ed.) *Power and Political Theory: Some European Perspectives*. London: John Wiley.

Morales-Moreno, I. (2004) 'Postsovereign governance in a globalizing and fragmenting world: the case of Mexico', *Review of Policy Research*, 21(1): 107–117.

Morgan, N. and Pritchard, A. (1999) *Power and Politics at the Seaside*. Exeter: University of Exeter Press.

Morriss, P. (1987) *Power: A Philosophical Analysis*. Manchester: Manchester University Press.

Murphy, P. (1985) *Tourism: A Community Approach*. New York and London: Methuen.

Murphy, P.E. and Murphy, A. (2004) *Strategic Management for Tourism Communities: Bridging the Gaps*. Clevedon: Channel View.

Nagel, J.H. (1975) *The Descriptive Analysis of Power*. New Haven, CT: Yale University Press.

Nohria, N. (1992) 'Is a network perspective a useful way of studying organizations?', in N. Nohria and R. Eccles (eds) *Networks and Organizations: Structure, Form and Action*. Boston, MA: Harvard Business School Press.

Norkunas, M.K. (1993) *The Politics of Memory: Tourism, History, and Ethnicity in Monterey, California*. Albany, NY: State University of New York Press.

O'Byrne, D.J. (2001) 'On passports and border controls', *Annals of Tourism Research*, 28: 399–416.

O'Neil, P. and Argent, N. (2005) 'Neoliberalism in antipodean spaces and times: an introduction to the special theme issue', *Geographical Research*, 43(1): 2–8.

Oppenheim, F.E. (1981) *Political Concepts: A Reconstruction*. Oxford: Basil Blackwell.

Peck, J. and Tickell, A. (2002) 'Neoliberalizing space', *Antipode*, 34: 380–403.

Reed, M. (1997) 'Power relations and community-based tourism planning', *Annals of Tourism Research*, 24: 566–591.

Rhodes, R.A.W. (1981) *Control and Power in Central–Local Relations*. Aldershot: Gower.

Sadler, D. (1993) 'Place-marketing, competitive places and the construction of hegemony in Britain in the 1980s', in G. Kearns and C. Philo (eds) *Selling Places: The City as Cultural Capital, Past and Present*. Oxford: Pergamon Press.

Schattsneider, E. (1960) *Semi-sovereign People: A Realist's View of Democracy in America*. New York: Holt, Rinehart and Winston.

Simmons, R., Davis, B.W., Chapman, R.J.K. and Sager, D.D. (1974) 'Policy flow analysis:

a conceptual model for comparative public policy research', *Western Political Quarterly*, 27(3): 457–468.

Totaro, P. (1995) 'Olympic opponents denied sporting chance', *Sydney Morning Herald*, 16 December: 1.

Urry, J. (1990) *The Tourist Gaze: Leisure and Travel in Contemporary Societies*. London: Sage.

Wæver, O. (1995) 'Securitization and desecuritization', in R. Lipschutz (ed.) *On Security*. New York: Columbia University Press.

World Travel and Tourism Council (2003) *Blueprint for New Tourism*. London: WTTC.

12 Tourism and the many faces of power

Andrew Church and Tim Coles

Tourism and the power imperative

The introduction to this book in Chapter 1 started by outlining how media commentators have made connections between tourism, terrorism and those who appear to wield power in global geopolitics. Recently, media and human rights reports have also illustrated the apparent powerlessness in the lives of some individuals involved in tourism. One of the effects of the 2004 tsunami in southern and Southeast Asia was to destroy the 'economic' spaces on and near to beaches used by locally owned businesses and independent operators. A report compiled by ActionAid International, the People's Movement for Human Rights Education and Habitat International Coalition was presented to the United Nations in February 2006 and reveals that despite the official emphasis on rebuilding, many individuals now find they are denied access to these spaces or the funding support to re-establish their tourism enterprises, while others with financial resources, political influence and claims to the land will determine their future use (ActionAid International 2006; Weaver 2006).

Natural disasters not only destroy livelihoods but they also produce a situation in which power relations can be suddenly reworked in the turbulent aftermath of the event. An understanding of the tourism outcomes of post tsunami rebuilding in Southeast Asia will not simply be provided by an assessment of visitor patterns, what is rebuilt or who gains and loses. Uncovering the resources mobilised, the tactics utilised, the modalities that emerge and the power relations involved will indicate the broader processes shaping tourism outcomes in post tsunami areas.

This simple example highlights some of the benefits of a focus on power in tourism studies as power theory provides a terminology and series of conceptual frameworks for organising analyses of the interactions between tourism and wider social processes. As this book has indicated, theories and the language of power are seldom straightforward. The diverse chapters presented here illustrate how a plurality of perspectives may be utilised to explore topics ranging from the embodied negotiations of tourism to the actions of large corporations and supra-national regulatory bodies. A multiplicity of topics and approaches was one of the aspirations for this collection because it serves to illustrate the various strengths of power-centred approaches to bring together researchers with very different theoretical perspectives to advance our understanding of a common theme, tourism. Indeed, in Chapter 2 Crouch highlights the need for contrasting views when he argues for a focus on how the tourist engages with power through encounters and spaces while acknowledging that other readings of institutions and power are also important. The introduction made clear however that, while the consideration of power generally conceived has been advancing recently in the tourism literature, the more particular connections between tourism and power theory have remained unarticulated in comparative terms.

It is far from clear why so much tourism research, including that concerned with social inequality and environmental conservation, has not addressed the detail of power theory. Undoubtedly, as this book illustrates, the theoretical terrains of power are contested and they contain many unresolved debates. These may conspire to deter researchers keen to highlight the role of tourism in shaping a fast-changing world. The limited focus on power may in part reflect the natures of dominant concerns of recent tourism research. For instance, the focus on sustainability and tourism has been an important developing area but has generated much descriptive even rhetorical writing aiming to capture new developments in tourism but which still all too often sidesteps complex critical issues such as power (Shaw and Williams 2004: Wheeller 2004).

The disciplinary origins of tourism studies may have frustrated the emergence of power as a focus of study. Tribe (1997) identified the emergence of tourism scholarship from two sources: tourism business studies, and a second, wider coalition of intellectual interests in 'non-business-related [studies of] tourism' (Tribe 2004: 50). Researchers from tourism business studies are increasingly central to the production of tourism knowledge (Tribe 2004). Their status has been accompanied by the accusation that the development of theory and concept in tourism studies more widely has been stifled by the propensity for more applied studies (Franklin and Crang 2001; Coles *et al.* 2005). This is de rigueur criticism of tourism business studies but it is not without some enduring justification. Power has been considered in mainstream critical management studies of organisational behaviour, human resources management, marketing, business ethics and corporate social responsibility (Buchanan and Huczynski 2004; Knights and Willmott 1999), from where tourism business studies takes many of its cues; however, research workers drawing on such sub-disciplinary foundations have chosen not to consider power. Chapter 10 by Coles and Scherle partly aims to show how insights from writings on business organisations and management can be used to understand power relations in the tourism industry. A similar criticism of partiality and selectivity may be levelled at some of those tourism researchers in the non-business-related studies. Given the tourism academy contains many researchers with backgrounds in anthropology, geography and sociology it is still curious that power, a core concept in social sciences generally and more importantly very recently, has not become a more prominent issue in tourism research. Nevertheless, it is here that we find most signs of progress as outlined in the introductory chapter. More confident treatments of power are evident in work on empowerment, performance and tourism (Coleman and Crang 2002; Crouch 2004), a range of writers influenced by Foucault and other post-structural theorists (Cheong and Miller 2000;

Edensor 2001; Franklin and Crang 2001; Urry 2002; Franklin 2004; Tribe 2003, 2004), feminist writings (Aitchison 2003), research drawing on postcolonial theory (Puar 2002), public policy and tourism (Hall 2000; Reed 1997; Doorne 1998; Tribe 2003, 2004) and studies of sexuality and tourism (Pritchard *et al.* 2000; Johnston 2005; Browne 2007). Collectively they have played a role in arguing that mobility should no longer be considered in social science simply as a product of other social and economic processes (Hannam *et al.* 2006). These writings, often from different theoretical perspectives, serve to highlight how tourism and identities are linked to power and how tourism and mobility are clearly implicated in power relations based on age, class, disability, gender, race and sexuality. It is not sufficient, however, when analysing power in tourism simply to make the links between tourism activities and power relations linked to salient social difference, although clearly this is a central endeavour. Instead, it is necessary to identify the specific roles of mobility and tourism in the actual operation and organisation of power.

Keen observers of power theory will identify certain theoretical perspectives already developed in the tourism power discourse that, outside of the introduction, have not been the subject of attention here. For instance, there is no chapter specifically concerned with feminist theorisations of power. This exclusion is not deliberate but a simple function of the production process. Feminist writing will undoubtedly continue to shape discussions of power and tourism in the future. Furthermore, many of the other theoretical perspectives already used in tourism power discourse have considerable additional potential. For example, writings in tourism that draw direct inspiration from Foucault have been mainly concerned with particular aspects of tourism such as the gaze, identities, semiotics and discourses. The full richness of Foucault's writings, however, have still to be fully utilised to understand power relations in the context of tourism (cf. Cheong and Miller 2000). Towards the end of his life, Foucault considered how legal policy and rights could not only be critiqued but also advanced (Foucault 1982). Gordon (2002) identifies three key conceptualisations that emerged from Foucault's later writing on politics, namely: a concern with developing relational rights as opposed to individual rights; advancing subjective capacities while reducing domination; and satisfying individual autonomy while preserving advances in social security. Such 'tensions' within power relations are undoubtedly central to pro-poor and community empowerment tourism initiatives but their implications are yet to be fully appreciated. They indicate that among pro-poor tourism researchers a focus on power and issues such as relational rights will be essential to avoid the limitations of some past writings on sustainable tourism which recommend how sustainable tourism activities might develop without revealing, or in some cases even considering, the power relations that underpin currently environmentally problematic tourism practices (Shaw and Williams 2004). The mobile and transient dimensions of tourism present significant challenges for understanding the role of rights. Shaw and Williams (2004: 25) note that 'the opaqueness of property rights circumscribes the capacity to extract income from tourism practices' so that some tourism activities such as walking around cities do not readily lend themselves to commodification. Nevertheless, as Chapter 8 by Church and Ravenscroft reveals, property rights and other discourses around rights can be central to conflicts associated with tourism and the significance of rights in relation to tourism is likely to increase in future.

In some tourism studies where the issue of power is raised, it has been treated to either a very literal conceptualisation or a generalised, even implied discussion. Attempts to valorise or consider power in a systematic manner are rare. This does not mean that power has to be measured in some quantitative sense. Rather, the implications of the diverse array of theoretical writings on power summarised in the introduction are that tourism analysis requires a more direct consideration of power, and moreover that this should take place within meaningful and theoretically informed frameworks of analysis. In the most recent discussions of tourism and power more consideration has been given to utilising concepts such as discourse, the gaze, practice, performance and ordering that are drawn from Foucauldian, post-modern and post-Foucauldian theory. Less attention has been paid to the writings of power theorists such as Lukes, Mann and Giddens whose work is more influenced by the social theory of Weber or Marx and political theorists such as Dahl. The potential of this body of power theory for developing insights into tourism and power was explored in chapters 6, 8, 10 and 11 by Gill, Church and Ravenscroft, Coles and Scherle, and Hall respectively.

The naming of power

By bringing together writings from a range of theoretical perspectives, this collection not only sets out the potential of different theoretical positions, it highlights certain lessons and lays down markers for future research into power and tourism. Collectively the chapters highlight the terminological and language issues that need to be embraced in future treatments of tourism and power. There is clearly a need in tourism studies to start to 'name' power more openly, certainly and precisely. A lack of precision in the use of terminology is a theoretical and methodological issue that has major analytical connotations.

Indeed, the different language used by authors in this collection does not mean that they are talking at cross-purposes; rather, they are considering similar concepts but with different yet complementary sets of nomenclature. For example, chapters 2, 3, 4, 6, 8 and 10 by Crouch, Cater, Shaw, Gill, Church and Ravenscroft, Coles and Scherle all consider issues of personal and group empowerment but from different theoretical and conceptual starting points each marked by their distinctive use of terminology. Timothy's (Chapter 9) review reveals that empowerment has been the subject of a great deal of interdisciplinary attention (cf. Lincoln *et al.* 2002), with ideas brought into and further developed by studies of tourism. Empowerment is a process of enabling, ascribing or authorising the relatively powerless with greater power, and it has been a central part of the sustainable development agenda that talks about the improvement of the plight of local people, especially in developing countries. As Timothy points out, it is possible to identify four stages along a pathway towards community empowerment. Consistent with the orthodoxies of the literature, Timothy identifies the first stage as 'imposed development' which he describes as an imposed, top-down approach to development that is practised by central administrators in the best interests of the destination community. There are obvious – although as yet unidentified – resonances with Lukes's third dimension of power (Lukes 1974; Table 1.3) concerned with subjective and real interests. These connections, if explored further, may offer additional insights into the operation and politics of tourism empowerment. Moreover, other similarities exist with both domination and coercion as modalities of power, both of which involve individuals complying unwillingly with the wishes of others due to perceived or actual threats (see Table 1.2). Imposed development appears, therefore, to be domination by another means which begs the question: why is there a difference in nomenclature?

It could be that scholars of tourism empowerment are simply unaware of power discourses but this innocent explanation seems unlikely as Sofield's (2003) monograph implies. In certain contexts, such as discussions of tourism in developing countries, we would contend that the terminology of empowerment appears to reflect an implicit or even unrecognised desire among some researchers to highlight the potentially positive, enabling aspects of some tourism developments rather than the repressive features of power that may be associated with such developments. Power is often presented practically as having simplistic directional outcomes either working in favour or against the interests of those to be empowered. This tends to gloss over the complex, situated and contested nature of struggles over power. Debates over empowerment are frequently set up as antagonistic struggles between 'them' and 'us' where the victory of the disempowered has to be at the expense of the previously empowered. This does not have to be the case as a more nuanced reading of power would suggest. As far back as 1963, Parsons observed that power is not necessarily the subject of a zero-sum game and, although it is not an infinite capacity, it can be produced and enhanced to the benefit of a number of parties. Empowerment is always a possibility. The organisation and operationalistion of power is a relational process and, as a number of chapters in this collection illustrate, one group's empowerment will be dependent on the responses of others. Shaw (Chapter 4) highlights how the agency of disabled people interacts with state disability legislation in the United Kingdom to produce problematic tourism outcomes. In the context of Morocco, Coles and Scherle in Chapter 10 outline how the empowerment of locally owned businesses can also be accompanied as a result of complex power relations by the expansion of the power of transnational businesses also involved in Moroccan tourism. In many writings, however, a positive view of empowerment may inadvertently contribute to a situation whereby tourism studies do not adequately name the modalities of power that contextualise the process of empowerment. The empowerment agenda concentrates on the development of power often from a 'low base' of powerlessness. A focus on power modalities is required to identify how low the benchmark is set at the start. Depending on the context, elements of Allen's (2003) eight modalities – domination, authority, coercion, inducement, manipulation, persuasion, negotiation and seduction – will usually shape the direction, pace and limits of empowerment.

Of course, in the general development literature many writers have, for good reasons, been more concerned to promote agendas for change and empowerment rather than theorise power in situations where injustice and inequality are clearly evident. However, an emphasis on the visible obscures a fuller inspection of those with power who often seek to hide and disguise their influence. In the tourism industry the relatively limited presence of transnational corporations, fragmented forms of local ownership, complex investment vehicles and often weakly developed state institutions make identifying how power is organised a complex empirical exercise. The colourful mosaic of coalitions and alliances at work in Whistler, British Columbia appears to involve modalities of power characterised by inducement, persuasion and negotiation. Yet Gill in Chapter 6 indicates how, even in a resort town like Whistler, where the key corporations and state actors are relatively easy to identify, the mechanisms by which influence is wielded can only be identified by in-depth long-term analysis. An elusive element in the development process was the involvement in already complex local political networks of senior executives who worked for the corporation that owned much of the tourism infrastructure but who were also 'local' residents due to personal property ownership. Scenarios played out in Whistler are rehearsed in many other destinations. As Gill's work makes clear, an engagement with power theory and concepts provides an analytical framework that offers insights beyond the possibilities afforded by just identifying the stakeholders. In general, arguments for local involvement and empowerment in the tourism development process have represented an important contribution from tourism studies but, as in this collection, in the future such writings on empowerment need to be accompanied by attempts to uncover the power relations that shape tourism developments.

Power and empirical studies

The different chapters in this book indicate that, even when drawing on the insights of different power theories, revealing power is a far from straightforward empirical task partly because power is so geographically and temporally uneven in its manifestation (Allen 2003). Beyond the purer, philosophical dimensions of the power debate, more practical, empirical issues have to be addressed of understanding the cultural constructs of power in relation to tourism (chapters 5, 7, 8 and 10 by Winter, Lew, Church and Ravenscroft, Coles and Scherle). Not surprisingly, the cultural meanings and symbols of power vary geographically and what may be 'named' as power in one part of the world will be perceived very differently elsewhere. Much academic research into power and tourism draws on concepts and frameworks developed in the anglophone literature (Morriss 2002). Lew's chapter 7 reveals, for instance, that power and obligation are part of the *guanxi* system in China. Obligation becomes a form of domination or authority but the definitions of these terms in Western theory (see Chapter 1 and Table 1.2) do not fully capture the deep cultural roots to this form of obligation (Lew and Wong 2004). A variety of complex social networks in which state officials are often deeply embedded maintain modalities of domination and authority based on obligation. Similar issues are evident Coles and Scherle's discussion in Chapter 10 of how inter-cultural differences in the understandings of commerce and power are central to the resolution of business relations between Moroccan and German enterprises.

Further empirical challenges revealed by the different chapters relate to Morriss's (2002) critique of studies of power for committing the 'mistakes' of the exercise fallacy or vehicle fallacy, at the expense of revealing and evaluating the real operation of power. In effect, Morriss (2002) suggests empirical studies often collect a substantial evidence base without ever really making a full pronouncement on the nature of power. Instead, they commit the exercise fallacy by concentrating on how power operates and the vehicle fallacy involves a concern only with the means of power (e.g. intermediaries and technologies).

In many respects, some contributions to this volume could be deemed guilty as charged but equally they suggest that the vehicle and exercise fallacy are far from being intellectual failings which render empirical work of little value. Studying the vehicles and exercise of power may have an important role to play, especially in a field like tourism studies where analyses of power are relatively limited. We would concur with Morriss (2002) on the existence of the exercise and vehicle fallacy but we would query: where is the distinction between the collection of evidence and the evaluation of power? The boundaries of interpretation are perhaps more fluid and, in a field where engagement with notions of power has been limited, the identification of the exercise or vehicles of power may act as an initial pronouncement of power. This is not to argue that tourism studies should employ partial or convenient approaches to understanding power; rather, those seeking to reveal or evaluate power may often need to be pragmatic in the application of concepts and the recognition of our limitations as researchers (Tribe 2004; Hall 2004). Foucault made the somewhat nebulous pronouncement that power is all around us (Westwood 2002), but this does not mean that it is always readily exposed due to its partial, relational and geographically varied manifestations. Consequently, to reveal the exercises and vehicles of power could be considered as a vital operation as part of the process of naming power and it is perhaps not neatly separated, as some lesser component, from the evaluation of power. This is particularly the case when searching for the more personal aspects of power involved with tourism. Chapters 2 and 3 in this book by Crouch and Cater draw on phenomenological and post-structural theory to highlight that the 'vehicle' of personal power in the form of the body is not readily separated from the operation of power. The body, and in Cater's case the body's association with the environment, through its performance involves tourism in the relational and 'powerful' processes by which individuals negotiate and self-define identities.

By drawing together a diverse range of power perspectives this book illustrates that, when analysing the 'messy' negotiations, tactics, resources, sites and modalities of power associated with a social phenomenon such as tourism, some of the clear-cut claims of power theory become less sustainable when empirical research is seeking both to gather evidence and to evaluate power in a relatively new field for power discourse.

Studies of tourism and power, therefore, can contribute to the wider social discourses over power by injecting into abstract theoretical debates the implications of the untidiness of mobility in contemporary society (Urry and Sheller 2004; Coles *et al.* 2005; Hall 2005). Indeed, much power theory has concentrated on the modalities and sites of power (Westwood 2002). For example, Hirst (2005) stresses the importance of the spaces of the state, the city and military buildings. Allen (2003), however, emphasises the role space plays in enabling the organisation of power in fluid ways both from a distance as well as through proximity and presence. Furthermore, Hannam *et al.* (2006) argue that mobility studies should seek to unsettle both static views of place and notions of deterritorialisation that see mobility as a given component of late modernity. An empirical focus in tourism studies that looks beyond the more 'fixed' sites and modalities will ensure that future research fully appreciates how tourism and other mobilities underpin the often elusive and hidden nature of power in contemporary society by allowing those seeking power to be both simultaneously present and absent. Large tourism conglomerates are often not visibly present in the material sites of tourism but their use of the management systems, business networks, signs, symbols and infrastructures of tourism allows them to have a major influence on the use of 'distant' sites. Urry (2003) has argued for the importance of identifying the role of mobilities and networks in the emergence of nodes and moorings where power coalesces. In addition, we would stress that understanding power also requires an empirical appreciation of how resources, tactics and strategies are organised through tourism and mobilities so that power 'reaches out' from nodes and moorings to shape increasingly fluid distant places.

Tourism mobilities, power and theory

The writings of Morriss (2002) set out a number of conceptual challenges for the examination of power and tourism that we begin to address in this collection (see Chapter 11 by Hall). As mentioned in the introduction, Morriss (2002) confronts studies of power with the need to consider the reasons why concepts of power are used, and he argues that they fulfil three main functions: practical, moral and evaluative. The different chapters here, as well as many other writings, contribute to the practical context by identifying the different ways tourism enables agents and their competitors to organise different forms of power. The moral context has perhaps received less direct attention in tourism studies although increasingly writers are seeking to explore issues of rights, responsibilities and citizenship as evident in touristed practices (Urry 2000; Butcher 2003; Smith and Duffy 2003; Carter 2004; Hall 2005; Coles 2005). The focus on community empowerment in tourism research has an implicit moral concern that empowerment will be beneficial but does not always involve identifying how the powerful influence events (Ryan 2002; Smith and Duffy 2003). All the chapters here seek in different ways to emphasise how the responsibility for power outcomes can be attributed to an extensive range of social phenomena, including: personal embodied performances (chapters 2, 3 and 4 by Crouch, Cater and Shaw); tourism businesses (chapters 3,

6 and 10 by Cater, Gill, Coles and Scherle); local political and corporate actors and coalitions (chapters 6 and 11 by Gill, Hall); voluntary organisations (Chapter 8 by Church and Ravenscroft); local community groupings (Chapter 9 by Timothy); private landowners (chapters 6 and 8 by Gill, Church and Ravenscroft); transnational enterprises (chapters 6 and 10 by Gill, Coles and Scherle); state agencies (chapters 4, 5, 7 and 10 by Shaw, Winter, Lew, Coles and Scherle) and supranational tourism organisations (chapter 11 by Hall).

Responsibility for power cannot be neatly read off from the characteristics of actors or institutions but will be shaped by spatial and temporal contexts, the actors present, and the discourses, resources and technologies that are mobilised and resonate. This diversity raises an ongoing challenge for studies of tourism and power. It may be possible to identify the moral context to power and how particular performances, practices, actors, social groupings, organisations and institutions shape power relations and outcomes. The issue remains of how to identify the commonalities in these different situations, and to show that in the tourism context there are regularised ways that power is organised, such as through state institutions, which shape power outcomes and can be understood as part of wider social processes and systems (Hall 2000, 2005). In the terminology of Morriss (2002), this involves addressing the evaluative context of power in which power theory is used to assess social systems in terms of the distribution of power and the ability of citizens to satisfy their goals.

For studies of tourism and power, the evaluative context is usually approached in a slightly narrower manner through an assessment of the role played by tourism in the distribution and operation of power in wider social systems. For example, Urry's (2002) tourist gaze seeks to link tourism to the emergence of disciplining norms and othering in capitalist society. The chapters in this book approach the evaluative context in a variety of ways. For example, in considering power, tourism and identity Crouch in Chapter 2 argues that individuals are not defined as tourists, rather broader social identities are constituted with reference to tourism. From a contrasting perspective, Church and Ravenscroft (Chapter 8) seek to reveal how modalities of power emerge by examining how resource mobilisation processes interact with wider structures and institutional frameworks that maintain the system of property rights and which allow some tourism and leisure participants to exclude others from particular locations. Hall's chapter 11 stresses that a range of multi-scalar tourism organisations contribute to wider neo-liberal governance structures that support covert decision-making and agenda manipulation.

These different ways of examining the evaluative context and how power relates to wider social processes reflect the goal of this collection to explore the potential validity of alternative conceptualisations. This resonates with some of the developments in power theory where recently there has been a noticeable willingness to contemplate the potentialities from diverse thinking and make links between different theoretical traditions (cf. Haugaard 2003). Clegg (1989) drew on both modern and post-modern social theory in his analysis of how power functions as a series of circuits. The dispositional and facilitative circuits establish the norms which act as the context to the episodic circuit involving actors and agency. Each chapter in this collection examines elements of the episodic circuit of power while a number seek to uncover what Clegg (1989) termed the 'rules of the game' developed in the other two circuits. The book as a whole, however, emphasises how an examination of the interconnections between power and tourism is an immensely wide-ranging endeavour and there are clearly areas where considerably more theoretical discussion and empirical research are needed. Two directions emerge as particularly worthy of further treatment as studies of power and tourism develop in the future. The first concerns tourism and state power, and the second relates to an understanding of mobilities, power and structuration.

Tourism, power and the state

The state has been discussed in this book in a number of chapters (5, 7, 8, 10 and 11 by Winter, Lew, Church and Ravenscroft, Coles and Scherle, Hall) but there has been no broad synthesis of the changing power of the state in relation to tourism. The state has been a key component in discourses of power from the time of Hobbes and Locke. Recent theoretical discussions of the state have noted significant transformations linked to globalisation, neo-liberal economic agendas and the rise of meta governance. Jessop (2002, 2005) in a European context suggests these changes are typified by the emergence of post-national statehood based on meta- and multi-level governance, networks, partnerships and decentralised guidance. Hannam's (2005) study of tourism in India begins to consider the significance of these complex changes for state tourism policies. Hall and Jenkins (2004: 532) still note that 'the study of tourism, politics and public policy lacks a coherent thread and a broader comparative perspective'.

Despite discussions of notions of the supposed 'hollowing out' of the state (Bramwell and Lane 2000) and considerable attention to new forms of partnerships (Laslo 2003; Judd and Simpson 2003; Bramwell 2004), tourism studies still lack a full appreciation of the state's current role in relation to tourism and hence its power. Shaw and Williams (2004: 7) argue that, although 'globalisation has modified the location of power and the nature of tourism dependency', nation-states still have an important role in regulating tourism spaces. Hall's chapter 11 highlights the changing role of that state in his discussion of multi-scalar governance and the internationalisation of the (local) state, its 'intermesticity' (Hall 2005). The local state has its hegemony over local conditions challenged by the influence of external actors such as transnational corporations or supra-national institutions. In the opposite sense, however, local governments increasingly project themselves on the world tourism stage often through a variety of collaborative international actions.

Whether the complex changes in multi-level governance lead to the emergence of what Jessop (2002) terms the 'Schumpeterian workfare post-national regime' remains to be seen. Nevertheless, their implications for the changing power of the state in relation to tourism clearly need more detailed analysis than they have currently received.

Tourism, power and mobilities

A second important future direction for power research is in the emerging area of tourism and mobilities. Unlike conceptualisations of tourism as the escape from the everyday to the exotic, the mobilities perspective sees tourism as just one of several forms of intricately connected human mobility (Urry and Sheller 2004; Coles *et al*. 2005). As with other forms of movement, tourism is embedded in people's spatio-temporal life paths; it is an integral component in a person's life, their being, identity and their expressive performances (Urry 2002; Hall 2005). The mobilities approach recognises tourism as one of the 'multiple everydays'. The rethinking of tourism as mobility is being approached from a number of theoretical directions. Hall (2005) argues for a revitalisation of the concepts of life paths, time geographies and space–time prisms stimulated initially through the work of Hägerstrand which became relatively marginalised in geography until recently (Coles *et al*. 2005).

One of the valuable features of these concepts is that they provide devices for deepening the understanding of tourism using Giddens's concept of structuration (Coles *et al*; 2005; Hall 2005). As Chapter 1 notes, Giddens has been an influential power theorist and Chapter 8 by Church and Ravenscroft explored how arguments developed by Giddens (1984) could be used to reveal interconnections between institutional arrangements, structural principles and the agency of tourism and leisure participants. The critiques of Giddens and the elusiveness of some of his theorising have been noted by a number of writers (Crespi 1992); however, the conceptual components of structuration theory have a strong focus on the implications of time and space for social life which, in turn, suggest they have further potential to reveal tourism's connections to power more generally in social systems.

The mobilities perspective on tourism and power is also being advanced from other theoretical directions particularly through the writings of Crouch (2004), Franklin (2004), and Urry (2002) (see chapters 1 and 2 for more details). Urry's (2003) use of complexity theory highlights how mobilities contribute to power being concentrated in certain 'attractors' that are the key sites and organisations in the global economy. Franklin (2004) promotes post-Foucauldian notions of ordering as providing a potential new ontological direction for tourism studies and also suggests that by taking a historical perspective it is possible to identify how high culture and key individuals, such as Thomas Cook, played a role in making tourism happen. Writings from these perspectives have tended to stress the need to understand how orderings emerge from the complexities of discourse, the gaze, practice, performance and embodiments. Hannam *et al*. (2006: 4) stress the need to track the 'power and politics of discourses and practices of mobility in creating both movement and stasis'. The aim of this book has been to suggest that understanding discourse and practice is only part of a wider endeavour needed to reveal the connections between power and tourism. Indeed, Urry (2002: 151) appears to acknowledge this to some degree when, in discussing the consequences of being the subject of the gaze, he notes that the outcomes will depend 'upon various determinants such as the relations of power within the "host" community, the time–space characteristics of visitors and the kinds of gaze involved'. The broad sweep of theoretical positions and topics covered in this book have been designed to examine these other 'determinants' especially those emerging from power relations. Certain chapters in this book have confirmed the importance of analysing practice, the body, text, discourse and performance to uncover the workings of power (see chapters 2, 3 and 5 by Crouch, Cater, and Winter). Other chapters, however, have adopted a different conceptual language and have illustrated how the modalities of power that coalesce around tourism and mobility also require an appreciation of the roles played in the organisation of power by tactics, strategies, resources, institutions and structures.

The differing theoretical perspectives highlight the roles mobilities can play in the organisation of power and power relations in wider society. Nevertheless, more remains to be done and Hannam *et al.* (2006: 15) in the inaugural editorial for the journal *Mobilities* argue that power is one of four key agendas for future research especially the 'unequal power relations which unevenly distribute motility, the potential for mobility'. Chapter 4 by Shaw highlights the complex power networks involving the state and other actors which shape the motility of people with disabilities. An optimistic conclusion would be that currently tourism studies appears to be moving closer to pinning down how mobilities in general, and tourism in particular, contribute to the organisation of power in contemporary society. Less optimistically, analysing the connections between tourism and power can be an activity fraught with potential compromises and complications. Hall's chapter 11 ends by stressing the potential opposition to research concerned with revealing the operation of power. Few (2002), along with Church and Ravenscroft (Chapter 8), has noted how those researching power in particular tourism spaces can become drawn into the politics of those locations (cf. Zahra and Ryan 2005; Flyvbjerg 2002). Urry's (2003) discussions of corruption and scandal have highlighted the complexity and messiness raised by researching such topics. Tourism research has generally been quite hesitant in considering corruption (Church 2004), but identifying corruption raises questions regarding the presentation of the judgements and the exposures produced by tourism researchers. Several studies have identified how corruption permeates the power relations between investors, developers, government officials and professional consultants (Bachvarov 1997; Pholpoke 1998; Williams and Balaz 2000). The exposure of scandal and corruption has complex consequences. For Urry (2003), the mobility and unbounded nature of informational and mediated power in capitalist societies results in the continual exposure of scandal and undermines attempts at ordering by those in power. Yet it also creates new cultural mechanisms that mainstream the acceptance of scandal and, as Urry (2003: 118) reminds us,

> so what began as the democratic attempt to reveal the many transgressions of the powerful unpredictably turns out to produce a culture that can drive out almost all forms of media reporting that do not display and enhance the virulent culture of scandal.

This is a stark and useful final reminder of the potential difficulties, tensions and contradictions associated with studying power. Future considerations of the connections between power and tourism will need to be highly reflexive to ensure that they do not unwillingly and unwittingly contribute to uneven power relations or a simple acceptance of powerlessness as an inevitable outcome of capitalist tourism development. As these challenges make ever more transparent, exposing power and powerlessness will remain a key agenda if we wish to conduct relevant and meaningful studies of tourism in shaping contemporary social life.

References

ActionAid International (2006) Tsunami response. A human rights assessment. ActionAid International, the People's Movement for Human Rights Education and Habitat International Coalition. Available at: http://www.actionaid.org/wps/content/documents/ [last accessed 20 February 2006].

Aitchison, C. (2003) *Gender and Leisure. Social and Cultural Perspectives*. London: Routledge.

Allen, J. (2003) *The Lost Geographies of Power*. Oxford: Blackwell.

Bachvarov, M. (1997) 'End of the model? Tourism in post-communist Bulgaria', *Tourism Management*, 18(1): 43–50.

Bramwell, B. (2004) 'Partnership, participation and social science research in tourism planning', in A.A. Lew, C.M. Hall and A.M. Williams (eds) *A Companion to Tourism*. Malden, MA: Blackwell.

Bramwell, B. and Lane, B. (2000) 'Collaboration and partnerships in tourism planning', in B. Bramwell and B. Lane (eds) *Tourism Collaboration and Partnerships. Politics, Practice and Sustainability*. Clevedon: Channel View.

Browne, K. (2007 forthcoming) 'A party with politics?: (Re)making LGBTQ Pride spaces in Dublin and Brighton', *Social and Cultural Geographies*.

Buchanan, D. And Huczynski, A. (2004) *Organizational Behaviour. An Introductory Text*, fifth edition. Harlow: FT Prentice Hall.

Butcher, J. (2003) *The Moralisation of Tourism. Sun, Sand . . . and Saving the World?* London: Routledge.

Carter, S. (2004) 'Mobilising *Hrvatsko*: tourism and politics in the Croatian diaspora', in T.E. Coles and D.J. Timothy (eds) *Tourism, Diaspora and Space*. London: Routledge.

Cheong, S-M. and Miller, M.L. (2000) 'Power and tourism. A Foucauldian observation', *Annals of Tourism Research*, 27(2): 371–390.

Church, A. (2004) 'Local and regional tourism policy and power', in A.A. Lew, C.M. Hall and A.M. Williams (eds) *A Companion to Tourism*. Malden, MA: Blackwell.

Clegg, S. (1989) *Frameworks of Power*. London: Sage.

Coleman, S. and Crang, M. (eds) (2002) *Tourism: Between Place and Performance*. Oxford: Berghahn.

Coles, T.E. (2005) 'Telling tales of tourism: mobility, media and citizenship in the 2004 EU enlargement', unpublished paper presented at The End of Tourism? Mobility and Global–local Connections, University of Brighton, Eastbourne, June.

Coles, T.E., Hall, C.M. and Duval, D.T. (2005) 'Mobilizing tourism: a post-disciplinary critique', *Tourism Recreation Research*, 30(2): 31–41.

Crespi, F. (1992) *Social Action and Power*. Oxford: Blackwell.

Crouch, D. (2004) 'Tourist practices and performances', in A.A. Lew, C.M. Hall and A.M. Williams (eds) *A Companion to Tourism*. Malden, MA: Blackwell.

Doorne, S. (1998) 'Power, participation and perception: an insider's perspective on the politics of the Wellington Waterfront redevelopment', *Current Issues in Tourism*, 1(2): 129–166.

Edensor, T. (2001) 'Performing tourism, staging tourism: (re)producing tourist space and practice', *Tourist Studies*, 1: 59–82.

Few, R. (2002) 'Researching actor power: analyzing mechanisms of interaction in negotiations over space', *Area*, 34(1): 29–38.

Flyvbjerg, B. (2002) 'Bringing power to planning – one researcher's praxis story', *Journal of Planning Education Research*, 21(4): 333–366.

Foucault, M. (1982) 'The subject and power', in L.D. Hubert and P. Rabinow (eds) *Michel Foucault: Beyond Structuralism and Hermeneutics*. London: Harvester Wheatsheaf.

Franklin, A. (2004) 'Tourism as an ordering. Towards a new ontology of tourism', *Tourist Studies*, 4(3): 277–301.

Franklin, A. and Crang, M. (2001) 'The trouble with tourism and travel theory?', *Tourist Studies*, 1(1): 5–22.

Giddens, A. (1984) *The Constitution of Society: Outline of the Theory of Structuration*. Berkeley, CA: University of California Press.

Gordon, C. (2002) 'Introduction', in J.D. Faubion (ed.) *Michel Foucault. Power. Essential Works of Foucault 1954–1984*. London: Penguin Books.

Hall, C.M. (2000) *Tourism Planning: Policies, Processes and Relationships*. Harlow: Prentice Hall.

—— (2004) 'Reflexivity and tourism research: situating myself and/with others', in J. Phillimore and L. Goodson (eds) *Qualitative Research in Tourism. Ontologies, Epistemologies and Methodologies*. London: Routledge.

—— (2005) *Tourism: Rethinking the Social Science of Mobility*. Harlow: Pearson.

Hall, C.M. and Jenkins, J. (2004) 'Tourism and public policy', in A.A. Lew, C.M. Hall and A.M. Williams (eds) *A Companion to Tourism*. Malden, MA: Blackwell.

Hannam, K. (2005) 'Tourism management issues in India's national parks', *Current Issues in Tourism*, 8(2/3): 165–180.

Hannam, K., Sheller, M. and Urry, J. (2006) 'Editorial: mobilities, immobilities and moorings', *Mobilities*, 1(1): 1–22.

Haugaard, M. (2003) 'Reflections on seven ways of creating power', *European Journal of Social Theory*, 6(1): 87–113.

Hirst, P. (2005) *Space and Power: Politics, War and Architecture*. Cambridge: Polity Press.

Jessop, B. (2002) *The Future of the Capitalist State*. Cambridge: Polity.

—— (2005) 'The European Union and recent transformations in statehood', in S.P. Riekmann (ed.) *Transformations of Statehood from a European Perspective*. Cambridge: Cambridge University Press.

Johnston, L. (2005) *Queering Tourism: Paradoxical Performances at Gay Pride Parades*. Abingdon: Routledge.

Judd, D.R. and Simpson, D. (2003) 'Reconstructing the local state. The role of external constituencies in building urban tourism', *American Behavioral Scientist*, 46(8): 1056–1069.

Knights, D. and Willmott, H. (1999) *Management Lives. Power and Identity in Work Organizations*. London: Sage.

Laslo, D. (2003) 'Policy communities and infrastructure of urban tourism', *American Behavioral Scientist*, 46(8): 1070–1083.

Lew, A.A. and Wong, A. (2004) 'Sojourners, *guanxi* and clan associations: social capital and overseas Chinese tourism to China', in T.E. Coles and D.J. Timothy (eds) *Tourism, Diaspora and Space*. London: Routledge.

Lincoln, N.D., Travers, C., Ackers, P. and Wilkinson, A. (2002) 'The meaning of empowerment: the interdisciplinary etymology of a new management concept', *International Journal of Management Reviews*, 4(3): 271–290.

Lukes, S. (1974) *Power: A Radical View*. London: Macmillan.

Morriss, P. (2002) *Power. A Philosophical Analysis*. Manchester: Manchester University Press.

Parsons, T. (1963) 'On the concept of political power', *Proceedings of the American Philosophical Society*, 107(3): 232–262.

Pholpoke, C. (1998) 'The Chiang Mai cable-car project: local controversy over cultural and eco-tourism', in P. Hirsch and C. Warren (eds) *The Politics of Environment in Southeast Asia Resources and Resistance*. London: Routledge.

Pritchard, A., Morgan, N.J., Sedgely, D., Khan, E. and Jenkins, A. (2000) 'Sexuality and holiday choices: conversations with gay and lesbian tourists', *Leisure Studies*, 19: 267–283.

Puar, J.K. (2002) 'A transnational feminist critique of queer tourism', *Antipode*, 34(4): 935–946.

Reed, M.G. (1997) 'Power relations and community-based tourism planning', *Annals of Tourism Research*, 24(3): 566–591.

Ryan, C. (2002) 'Equity, management, power sharing and sustainability – the issues of the "new" tourism', *Tourism Management*, 23: 17–26.

Shaw, G. and Williams, A.M. (2004) *Tourism and Tourism Spaces*. London: Sage.

Smith, M. and Duffy, R. (2003) *The Ethics of Tourism Development*. London: Routledge.

Sofield, T.H.B. (2003) *Empowerment for Sustainable Tourism Development*. Oxford: Pergamon Press.

Tribe, J. (1997) 'The indiscipline of tourism', *Annals of Tourism Research*, 24(3): 638–657.

—— (2003) 'The RAE-ification of tourism research in the UK', *International Journal of Tourism Research*, 5: 225–234.

—— (2004) 'Knowing about tourism: epistemological issues', in J. Phillimore and L. Goodson (eds) *Qualitative Research in Tourism. Ontologies, Epistemologies and Methodologies*. London: Routledge.

Urry, J. (2000) *Sociology Beyond Societies*. London: Routledge.

—— (2002) *The Tourist Gaze*. London: Sage.

—— (2003) *Global Complexity*. Cambridge: Polity.

Urry, J. and Sheller, M. (eds) (2004) *Tourism Mobilities: Places to Stay, Places in Play*. London: Routledge.

Weaver, M. (2006) 'Tsunami victims evicted by developers', *The Guardian*, 1 February: 12.

Westwood, S. (2002) *Power and the Social*. London: Routledge.

Wheeller, B. (2004) 'The truth? The whole truth. Everything but the truth. Tourism and knowledge: a septic sceptic's perspective', *Current Issues in Tourism*, 7(6): 467–477.

Williams, A. and Balaz, V. (2000) *Tourism in Transition: Economic Change in Central Europe*. London: I.B. Tauris.

Zahra, A. and Ryan, C. (2005) 'Reflections on the research process: the researcher as actor and audience in the world of Regional Tourist Organisations', *Current Issues in Tourism*, 8(1): 1–21.

Index

Note: page numbers in italics denote figures or tables